Springer Proceedings in Mathematics & Statistics

Volume 240

Springer Proceedings in Mathematics & Statistics

This book series features volumes composed of selected contributions from workshops and conferences in all areas of current research in mathematics and statistics, including operation research and optimization. In addition to an overall evaluation of the interest, scientific quality, and timeliness of each proposal at the hands of the publisher, individual contributions are all refereed to the high quality standards of leading journals in the field. Thus, this series provides the research community with well-edited, authoritative reports on developments in the most exciting areas of mathematical and statistical research today.

More information about this series at http://www.springer.com/series/10533

Matthew Ballard · Charles Doran
David Favero · Eric Sharpe
Editors

Superschool on Derived Categories and D-branes

Edmonton, Canada, July 17–23, 2016

Pacific Institute for the Mathematical Sciences

Springer

Editors
Matthew Ballard
Department of Mathematics
University of South Carolina
Columbia, SC, USA

Charles Doran
Department of Mathematics and Statistical Sciences
University of Alberta
Edmonton, AB, Canada

David Favero
Department of Mathematics and Statistical Sciences
University of Alberta
Edmonton, AB, Canada

Eric Sharpe
Department of Physics
Virginia Tech
Blacksburg, VA, USA

ISSN 2194-1009 ISSN 2194-1017 (electronic)
Springer Proceedings in Mathematics & Statistics
ISBN 978-3-030-06270-5 ISBN 978-3-319-91626-2 (eBook)
https://doi.org/10.1007/978-3-319-91626-2

Mathematics Subject Classification (2010): 14J33, 53D37, 14F05

This Springer imprint is published by the registered company Springer Nature Switzerland AG
The registered company address is: Gewerbestrasse 11, 6330 Cham, Switzerland

Preface

String Theory revolutionized not just how we view the physical world but also how we view Mathematics. Conversely, through String Theory, many physicists first became acquainted with beautiful fields of Mathematics, like Algebraic Geometry. The cross-pollination of insights and motivations between String Theory and Mathematics led to remarkable insights in both fields.

One such deep instance is that of Mirror Symmetry, a duality in String Theory that provides a powerful computational tool—allowing one to exchange difficult computations for simpler ones. The full range of consequences of Mirror Symmetry in Mathematics may never be understood. On the other hand, Mirror Symmetry has already provided spectacular insight in enumerative geometry [1] leading to a revolution in the field [2–6]. Two related mathematical proposals for Mirror Symmetry arose afterward. The Strominger–Yau–Zaslow or SYZ conjecture [7] posits that mirror manifolds arise from the process of T-dualization; each space admits torus fibrations over a common base, and the exchange between the two amounts to dualization of the torus fibers. The Homological Mirror Symmetry of Kontsevich [8] states that an equivalence of categories underlies all phenomena of Mirror Symmetry. It provides a deep and hitherto-unknown connection between the fields of Algebraic Geometry and Symplectic Geometry and has become a robust field of Mathematics itself in a short time.

This book consists of a series of introductory lectures on Mirror Symmetry and its surrounding topics. These lectures were provided by participants in the PIMS Superschool School for Derived Categories and D-Branes in July 2016. Together, they form a comprehensive introduction to the field which integrates perspectives from mathematicians and physicists alike.

The intent is to provide a pleasant and broad introduction into modern research topics surrounding String Theory and Mirror Symmetry which is approachable to readers who are new to the subject. Mathematical readers should expect to come away with a broader perspective on this field and a bit of physical intuition. Physicists will gain an introductory overview of the developing mathematical realization of physical predictions. Topics include constructions of various mirror pairs, approaches to Mirror Symmetry, connections to homological algebra, and physical motivations.

Of particular interest is the connection between GLSMs, D-branes, birational geometry, and derived categories. This is one of the broader themes of the text and is explained from a physical and mathematical perspective. The introductory lectures provided herein highlight many features of this emerging field and give concrete connections between the physics and the math.

Columbia, USA — Matthew Ballard
Edmonton, Canada — Charles Doran
Edmonton, Canada — David Favero
Blacksburg, USA — Eric Sharpe

References

1. Candelas, Philip; de la Ossa, Xenia C.; Green, Paul S.; Parkes, Linda. *A pair of Calabi-Yau manifolds as an exactly soluble superconformal theory*. Nuclear Phys. B 359 (1991), no. 1, 21–74.
2. Givental, Alexander. *A mirror theorem for toric complete intersections*. Topological field theory, primitive forms and related topics (Kyoto, 1996), 141–175, Progr. Math., 160, Birkhäuser Boston, Boston, MA, 1998.
3. Lian, Bong H.; Liu, Kefeng; Yau, Shing-Tung. *Mirror principle. I.* Asian J. Math. 1 (1997), no. 4, 729–763.
4. Lian, Bong H.; Liu, Kefeng; Yau, Shing-Tung. *Mirror principle. II.* Surveys in differential geometry: differential geometry inspired by string theory, 455–509, Surv. Differ. Geom., 5, Int. Press, Boston, MA, 1999.
5. Lian, Bong H.; Liu, Kefeng; Yau, Shing-Tung. *Mirror principle. III.* Asian J. Math. 3 (1999), no. 4, 771–800.
6. Lian, Bong H.; Liu, Kefeng; Yau, Shing-Tung. *Mirror principle. IV.* Surveys in differential geometry, 475–496, Surv. Differ. Geom., VII, Int. Press, Somerville, MA, 2000.
7. Strominger, Andrew; Yau, Shing-Tung; Zaslow, Eric. *Mirror symmetry is T-duality*. Nuclear Phys. B 479 (1996), no. 1–2, 243–259.
8. Kontsevich, Maxim. *Homological algebra of mirror symmetry*. Proceedings of the International Congress of Mathematicians, Vol. 1, 2 (Zürich, 1994), 120–139, Birkhäuser, Basel, 1995.

Contents

Symbols for "Abelian and Triangulated Categories"

Chantelle Hanratty[1]

1. $\mathcal{C}$ or $\mathcal{D}$: A specific category
2. Ob($\mathcal{C}$): The objects in the category $\mathcal{C}$
3. $\mathrm{Hom}_{\mathcal{C}}(A, B)$ or $\mathrm{Hom}(A, B)$: Morphisms (in the category $\mathcal{C}$) between the objects A and B
4. $\cong$: Isomorphic
5. F^{-1}: The inverse functor to a functor F
6. $A[n]$: The object A shifted n times in a triangulated category; $T^n\,(A)$
7. $f\,[n]$: The map $T^n\,(f)$: $A[n] \to B[n]$, where $f : A \to B$.
8. f_*, f^*: If f: $A \to B$, then f_* and f^* are the induced maps between morphism groups $\mathrm{Hom}(X, A) \to \mathrm{Hom}(X, B)$ and $\mathrm{Hom}(B, X) \to \mathrm{Hom}(A, X)$ respectively.

[1]University of Alberta, Edmonton, AB, Canada, e-mail: hanratty@ualberta.ca.

Part I
Derived Categories and Related Topics in Algebraic Geometry

Abelian and Triangulated Categories

Chantelle Hanratty

1 Preface

The following material is standard and based largely on the discussion of triangulated categories in the first chapter of [7]. Background information on categories, functors, and additive and abelian categories, as presented in [4], is also included throughout. The proposition at the end of the notes is paraphrased from Proposition 1.34 of [7]; however, the lemmas and corollaries to this proposition are independently observed in order to give a more detailed proof and motivation for the proposition. Any additional sources are cited inline throughout.

2 Introduction

Triangulated categories were developed independently by Jean-Louis Verdier and Dieter Puppe in the 1960s. Puppe's work was originally published as [2] in 1961 and Verdier's work was originally part of his unpublished PhD thesis, which was reprinted in 1996 as [10]. Today there are numerous applications to triangulated categories, such as derived categories of coherent sheaves, the theory of motives, stable homotopy theory, Fukaya categories, and stable module categories [6, P. 1] [9, P. 70].

The purpose of these notes is to provide enough background information to define triangulated categories. We begin the notes by defining additive categories. We continue by providing background information on monomorphisms, epimorphisms, kernels, and cokernels, which we use to define abelian categories. Finally we give a full

C. Hanratty (✉)
University of Alberta, 116 St & 85 Ave, Edmonton, AB T6G 2R3, Canada
e-mail: hanratty@ualberta.ca

M. Ballard et al. (eds.), *Superschool on Derived Categories and D-branes*,
Springer Proceedings in Mathematics & Statistics 240,
https://doi.org/10.1007/978-3-319-91626-2_1

definition of triangulated categories and conclude by describing some basic results on triangulated categories, leading to a long exact sequence of morphism groups.

3 Additive Categories

3.1 Definition of Additive Categories

Definition 1 An *additive category* $\mathscr{C}$ is a category satisfying the following axioms:

1. For every two objects $A, B \in \mathrm{Ob}(\mathscr{C})$, $\mathrm{Hom}_{\mathscr{C}}(A, B)$ is an abelian group.
2. Function composition is bilinear. That is for every $f, f_1, f_2 \in \mathrm{Hom}_{\mathscr{C}}(A, B)$, and $g, g_1, g_2 \in \mathrm{Hom}_{\mathscr{C}}(B, C)$ we have $g \circ (f_1 + f_2) = (g \circ f_1) + (g \circ f_2)$ and $(g_1 + g_2) \circ f = (g_1 \circ f) + (g_2 \circ f)$.
3. $\mathscr{C}$ has a zero object. That is there is an object 0 such that $\mathrm{Hom}(A, 0) \cong \mathrm{Hom}(0, A) \cong 0$ is the trivial group for any object $A \in \mathrm{Ob}(\mathscr{C})$.
4. For every two objects A_1, A_2 in $\mathscr{C}$, there exists an object A that is both a product and a coproduct (sum) of A_1 and A_2.

Note: Two equivalent definitions of a zero object are 0 such that $\mathrm{Hom}(0, 0) = 0$ or 0 that is both an initial and a final object in the category.

3.2 k-Linear Categories

Definition 2 A *k-linear category* is a special type of additive category in which all homomorphism groups are vector spaces over a field k and function composition is bilinear over k.

3.3 Additive Functors

For any functor, $F : \mathscr{C} \to \mathscr{D}$, between additive categories, and every two objects $A, B \in \mathrm{Ob}(\mathscr{C})$, we get an induced map $F : \mathrm{Hom}(A, B) \to \mathrm{Hom}(F(A), F(B))$.

Definition 3 A functor F is called an *additive functor* if the induced map is a group homomorphism. That is $F(f + g) = F(f) + F(g)$ for any $f, g : A \to B$.

Definition 4 A functor between two k-linear categories is *k-linear* if the induced map is k-bilinear. That is $F(k_1 f + k_2 g) = k_1 F(f) + k_2 F(g)$ for any $f, g : A \to B$ and $k_1, k_2 \in k$.

4 Abelian Categories

Roughly speaking an abelian category is an additive category with kernels and cokernels. This is powerful because it makes it possible to define exact sequences, homology, and cohomology. Before giving the proper definition, we will delve into the necessary background in category theory, etc., to make the definition complete.

4.1 Some Category Theory

4.1.1 Monomorphisms and Epimorphisms

In many familiar categories, such as sets and groups, monomorphisms and epimorphisms are the typical notions of injective and surjective maps respectively. In a general category theoretic sense, we get the following.

Definition 5 A map $f : A \to B$ is called a *monomorphism* if for any pair of morphisms $x : C \to A$ and $y : C \to A$ such that $f \circ x = f \circ y$, we also have $x = y$.

Definition 6 A map $f : A \to B$ is called an *epimorphism* if for any pair of morphisms $x : B \to C$ and $y : B \to C$ such that $x \circ f = y \circ f$, we also have $x = y$.

Although there are many categories in which monomorphisms are injections and epimorphims are surjections, it is important to note that these definitions do not coincide in every category.

Example 1 Let $\mathscr{C}$ be the category where the objects are Hausdorff topological spaces and the morphisms are continuous functions. Then the inclusion $\mathbb{Q} \hookrightarrow \mathbb{R}$ is an epimorphism in this category even though it is not surjective.

4.1.2 Difference Kernels and Cokernels

In order to define a category theoretic kernel and cokernel, we first need to define the notion of difference kernels and difference cokernels.

Definition 7 Let $x, y : A \to B$ be two morphisms, then a *difference kernel* of x and y is a morphism $k : K \to A$ satisfying the following two conditions.

- $x \circ k = y \circ k$
- Let $j : J \to A$ be another map satisfying the first condition. Then there exists a unique map $l : J \to K$ such that the following diagram commutes.

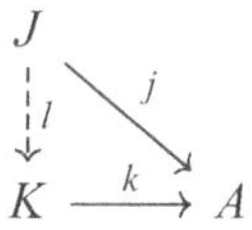

We can think of the first property as requiring x and y to be equal on K, and the second property as choosing the "largest" object with this property.

Proposition 1 *Any difference kernel is a monomorphism.*

Proof See [4, Proposition 1.61] for a proof of this statement. □

Definition 8 Let $x, y : A \to B$ be two morphisms, then a *difference cokernel* of x and y is a morphism $c : B \to C$ satisfying the following two conditions.

- $c \circ x = c \circ y$.
- Let $j : B \to J$ be another map satisfying the first condition. Then there exists a unique map $l : C \to J$ such that the following diagram commutes.

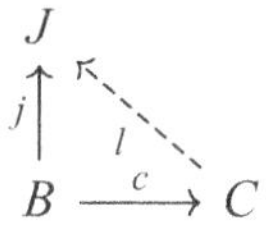

We can think of the first property as requiring x and y to be equal on C, and the second property as choosing the "largest" object with this property.

Proposition 2 *Any difference cokernel is an epimorphism.*

Proof Apply the previous proposition to the opposite category. □

4.1.3 Kernels and Cokernels

Kernels and cokernels are only well defined in a category with a zero object. Recall a zero object is an object such that for any other object A, we have $\mathrm{Hom}(A, 0)$ and $\mathrm{Hom}(0, A)$ each have exactly one element.

In such a category $0 : A \to B$ is defined as the composition of the unique maps $A \to 0 \to B$.

Definition 9 Let $f : A \to B$ be a morphism. Then we define the *kernel* of f to be the difference kernel of f and 0, and the *cokernel* to be the difference cokernel of f and 0.

Therefore a kernel is a map $k : K \to A$ such that $f \circ k = 0$, and for any other $j : J \to A$ such that $f \circ j = 0$, there exists a map $l : J \to K$ making the following diagram commute.

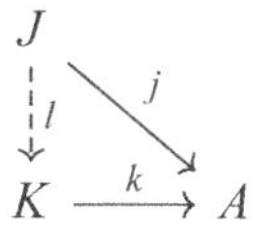

Definition 10 Similarly a cokernel is a map $c : B \to C$ such that $c \circ f = 0$. Furthermore for any other $j : B \to J$ satisfying the first condition there exists a unique $l : C \to J$ making the following diagram commute.

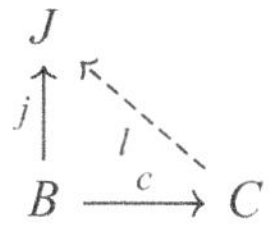

4.2 *Definition of Abelian Categories*

Definition 11 An abelian category is an additive category such that:

- If $f : A \to B$ is a morphism in the category, f has both a kernel and a cokernel.
- If $f : A \to B$ is a monomorphism in the category, then f is a kernel of some map.
- If $f : A \to B$ is an epimorphism in the category, then f is a cokernel of some map.

Note: There is an equivalent characterization of the second and third bullets that requires a certain natural map between the image of f and the coimage of f to be an isomorphism; however, we will not describe this in detail.

4.3 *Examples*

Example 2 (*Some abelian categories*)

- The category of modules over a commutative ring R is abelian.
- The category of sheaves (of abelian groups) over a topological space X is an abelian.
- For a given scheme X, the categories of coherent and quasi-coherent sheaves over X are abelian.

Example 3 (*A category that is additive but not abelian*) Let R be any non-noetherian commutative ring, and let $\mathscr{C}$ be the category of finitely generated modules over this ring. Then $\mathscr{C}$ is additive, but not abelian.

Proof (*that $\mathscr{C}$ is not abelian*) One of the equivalent definitions of a noetherian commutative ring is that every ideal is finitely generated. Since by assumption R is not abelian, there exists an ideal that is not finitely generated, call it I.

By proposition 2.3 in [1], a module over R is finitely generated iff it is isomorphic to a quotient of R^n. Therefore R/I is a finitely generated module.

R and R/I are both objects in $\mathscr{C}$, so in order for $\mathscr{C}$ to be abelian, the projection $P : R \to R/I$ must have a kernel. However, the kernel of this map is I, which is not an object in this category. Thus $\mathscr{C}$ is not abelian.

We thank the Mathematics Stack Exchange community for help with this proof.

Note: We have proved that the kernel of P, in the category of modules, is not finitely generated. However, the category of finitely generated modules could have a kernel that is not a kernel in the category of modules. Therefore for this proof to be completely correct, we also need to show that any kernel in the category of finitely generated modules is also a kernel in the category of modules. □

Example 4 (*A category that is not additive*) Let $\mathscr{C}$ be the category of sets. Intuitively it makes sense that $\mathscr{C}$ is not additive because there is no natural way to define addition of functions whose codomain is not an abelian group. More concretely, the category of sets does not have a zero object so it cannot be abelian.

5 Triangulated Categories

5.1 Some More Category Theory

Let $F : \mathscr{C} \to \mathscr{D}$ be a functor and A, B any two objects in $\mathrm{Ob}(\mathscr{C})$.

There is an induced map

$$\begin{aligned} F : \mathrm{Hom}_{\mathscr{C}}(A, B) &\to \mathrm{Hom}_{\mathscr{D}}(F(A), F(B)) \\ f &\mapsto F(f). \end{aligned}$$

Definition 12 F is *full* if the above map is surjective for each $A, B \in \mathrm{Ob}(\mathscr{C})$. That is all of the maps $F(A) \to F(B)$ come from maps $A \to B$.

Definition 13 F is *faithful* if the above map is injective for each $A, B \in \mathrm{Ob}(\mathscr{C})$.

Definition 14 ([8, P. 53]) F is *essentially surjective* if every object $B \in \mathscr{D}$ is isomorphic to an object $F(A)$ for some $A \in \mathscr{C}$. Note: We don't require that $B = F(A)$, only that they are isomorphic.

5.1.1 Equivalence of Categories

Definition 15 Two categories $\mathscr{C}$, $\mathscr{D}$ are called *equivalent* if there exists a full, faithful, essentially surjective functor $F : \mathscr{C} \to \mathscr{D}$. This equivalence is called additive if the functor is additive.

Note: There is an equivalent definition of equivalence. The categories $\mathscr{C}$ and $\mathscr{D}$ are equivalent if there are two functors $F : \mathscr{C} \to \mathscr{D}$, and $G : \mathscr{D} \to \mathscr{C}$ such that $G \circ F$ and $F \circ G$ are naturally isomorphic to the identity functor. We write this as $G = F^{-1}$.

5.2 *Definition of Triangulated Categories*

A triangulated category $\mathscr{D}$ is a special type of additive category. Such a category must have a "shift functor," a "set of distinguished triangles," and follow a set of special axioms called the TR axioms.

Definition 16 The *shift functor* is an additive equivalence $T : \mathscr{D} \to \mathscr{D}$.
Therefore

- T is a full, faithful, essentially surjective functor from $\mathscr{D}$ to itself.
- For $A, B \in \mathrm{Ob}(\mathscr{D})$ and $f, g : A \to B$ we have $T(f + g) = T(f) + T(g)$.

Definition 17 Given a fixed shift functor, a *triangle* consists of

- Three objects $A, B, C \in \mathrm{Ob}(\mathscr{D})$
- Morphisms $a \in \mathrm{Hom}(A, B)$, $b \in \mathrm{Hom}(B, C)$, $c \in \mathrm{Hom}(C, T(A))$

That is we get a diagram

$$A \xrightarrow{a} B \xrightarrow{b} C \xrightarrow{c} T(A).$$

Definition 18 In a triangulated category, we choose a subset of triangles, which we call *distinguished triangles*. This subset must follow the TR axioms which we will describe below.

5.2.1 Notation

Let $A, B \in \mathrm{Ob}(\mathscr{D})$, $f : A \to B$ and $n \in \mathbb{Z}$.
Then $A[n] = T^n(A) = T \circ \cdots \circ T(A)$ and $f[n] = T^n(f) = T \circ \cdots \circ T(f)$.
Note that since $T(f) : T(A) \to T(B)$, we get

$$f[n] : A[n] \to B[n].$$

Note: Since T is an additive equivalence, it has an inverse. Therefore letting n be negative makes sense. We use the notation $A[-n]$ to denote $T^{-1} \circ \cdots \circ T^{-1} A$.

5.2.2 Morphisms of Triangles

Let

$$A \to B \to C \to A[1]$$

and

$$A' \to B' \to C' \to A'[1]$$

be two triangles.

Definition 19 A *morphism* of triangles consists of maps $f : A \to A'$, $g : B \to B'$ and $h : C \to C'$ such that the following diagram commutes.

$$\begin{array}{ccccccc} A & \longrightarrow & B & \longrightarrow & C & \longrightarrow & A[1] \\ \downarrow f & & \downarrow g & & \downarrow h & & \downarrow f[1] \\ A' & \longrightarrow & B' & \longrightarrow & C' & \longrightarrow & A'[1] \end{array}$$

If each map f, g, h is an isomorphism, then we call the diagram an *isomorphism* of triangles.

5.3 *Axioms*

We call the axioms for triangulated categories **TR1, TR2, TR3** and **TR4**.

5.3.1 TR1

The first axiom has three parts. Each part guarantees the existence of some type of distinguished triangle.

(1) For every object A there exists a triangle

$$A \xrightarrow{id_A} A \xrightarrow{0} 0 \xrightarrow{0} A[1].$$

These triangles must always be distinguished.

(2) Distinguishedness is preserved under isomorphism. That is if two triangles are isomorphic they must both be distinguished or not distinguished.

(3)Let $f : A \to B$ be a morphism. Then there exists a distinguished triangle

$$A \xrightarrow{f} B \to C \to A[1]$$

that "completes" f. The object C is called the *mapping cone* of f.

5.3.2 TR2

Let

$$T_1 = A \xrightarrow{a} B \xrightarrow{b} C \xrightarrow{c} A[1]$$

be a distinguished triangle.

Then the triangle

$$T_2 = B \xrightarrow{b} C \xrightarrow{c} A[1] \xrightarrow{-a[1]} B[1]$$

is also distinguished.

Conversely if T_2 is distinguished, T_1 must also be distinguished.

5.3.3 TR3

Consider a commutative diagram of distinguished triangles as below.

$$\begin{array}{ccccccc} A & \longrightarrow & B & \longrightarrow & C & \longrightarrow & A[1] \\ \downarrow f & & \downarrow g & & & & \downarrow f[1] \\ A' & \longrightarrow & B' & \longrightarrow & C' & \longrightarrow & A'[1] \end{array}$$

Any such diagram can be completed to a (not necessarily unique) morphism of triangles,

$$\begin{array}{ccccccc} A & \longrightarrow & B & \longrightarrow & C & \longrightarrow & A[1] \\ \downarrow f & & \downarrow g & & \downarrow h & & \downarrow f[1] \\ A' & \longrightarrow & B' & \longrightarrow & C' & \longrightarrow & A'[1] \end{array}.$$

Note: It is straightforward to show, using **TR2**, that as long as you have any two out of three of the morphisms, f, g, h, you can prove the existence of the third.

5.3.4 TR4 The Octahedral Axiom

The **TR4** axiom is the most complicated axiom for triangulated categories. The following description is adapted from the description found in [5, PP. 21–22].

Let $u : X \to Y$ and $v : Y \to Z$ be any two maps.
Then there is a composition map $w := v \circ u : X \to Z$.

By **TR1** we can complete all three of these maps into distinguished triangles.
Let U, V, W be the mapping cones of u, v, w respectively so we get distinguished triangles

$$X \xrightarrow{u} Y \xrightarrow{u'} U \xrightarrow{u''} X[1]$$

$$Y \xrightarrow{v} Z \xrightarrow{v'} V \xrightarrow{v''} Y[1]$$

$$X \xrightarrow{w} Z \xrightarrow{w'} W \xrightarrow{w''} X[1]$$

TR4 states that there exists a distinguished triangle

$$U \xrightarrow{f} W \xrightarrow{g} V \xrightarrow{h} U[1]$$

such that the following diagram commutes. This diagram forms the vertices and edges of an octahedron, giving the axiom its name.

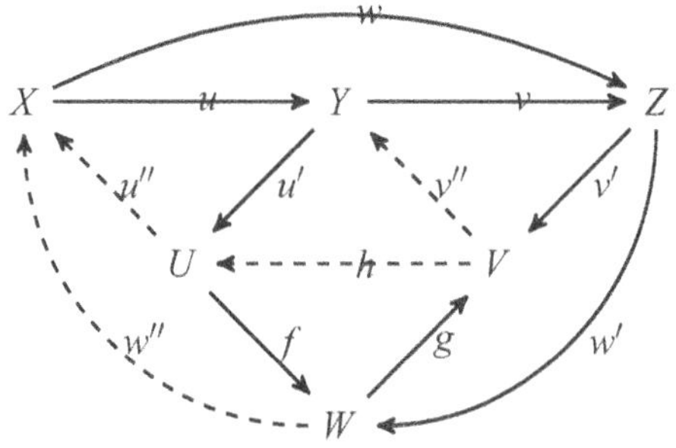

Note: The dashed lines represent maps going into $X[1]$, $Y[1]$, $U[1]$ as opposed to X, Y, U. The code for making the above commutative diagram was found at [3].

5.4 Results about Triangulated Categories

Lemma 1 *Let A be any object in a triangulated category. Then the triangle*

$$0 \xrightarrow{0} A \xrightarrow{id} A \xrightarrow{0} 0$$

is distinguished.

Proof By **TR1** we know that

$$A \xrightarrow{id} A \xrightarrow{0} 0 \xrightarrow{0} A[1]$$

is distinguished .

Therefore by **TR2** we have that

$$A \xrightarrow{0} 0 \xrightarrow{0} A[1] \xrightarrow{-\mathrm{id}} A[1]$$

is distinguished.

Applying **TR2** again we get that

$$0 \xrightarrow{0} A[1] \xrightarrow{-\mathrm{id}} A[1] \xrightarrow{-0} 0[1]$$

is also distinguished.

Since A was arbitrary, replacing A with $A[-1]$ implies that

$$0 \xrightarrow{0} A \xrightarrow{-\mathrm{id}} A \xrightarrow{-0} 0[1]$$

is distinguished.

Consider the following isomorphism of triangles.

$$\begin{array}{ccccccc} 0 & \xrightarrow{0} & A & \xrightarrow{\mathrm{id}} & A & \xrightarrow{0} & 0 \\ \downarrow{\scriptstyle 0} & & \downarrow{\scriptstyle -\mathrm{id}} & & \downarrow{\scriptstyle \mathrm{id}} & & \downarrow{\scriptstyle 0} \\ 0 & \xrightarrow{0} & A & \xrightarrow{-\mathrm{id}} & A & \xrightarrow{0} & 0 \end{array}$$

By **TR1**, since the bottom triangle is distinguished, so must be the top. Therefore

$$0 \xrightarrow{0} A[1] \xrightarrow{\mathrm{id}} A[1] \xrightarrow{0} 0[1]$$

is indeed distinguished, proving the claim. □

Lemma 2 *Let* $A \xrightarrow{f} B \xrightarrow{g} C \xrightarrow{h} A[1]$ *be a distinguished triangle. Then* $g \circ f = 0$.

Proof By the last lemma, we have that

$$0 \to C \xrightarrow{id} C \to 0$$

is distinguished.

Consider the following commutative diagram.

$$\begin{array}{ccccccc} A & \xrightarrow{f} & B & \xrightarrow{g} & C & \xrightarrow{\mathrm{h}} & A[1] \\ & & \downarrow{\scriptstyle g} & & \downarrow{\scriptstyle \mathrm{id}} & & \\ 0 & \xrightarrow{0} & C & \xrightarrow{\mathrm{id}} & C & \xrightarrow{0} & 0[1] \end{array}$$

Then by **TR3** there exists a morphism $A \to 0$ making the diagram commute. However the only possible morphism $A \to 0$ is 0.

$$\begin{array}{ccccccc}
A & \xrightarrow{f} & B & \xrightarrow{g} & C & \xrightarrow{h} & A[1] \\
\downarrow{\scriptstyle 0} & & \downarrow{\scriptstyle g} & & \downarrow{\scriptstyle \mathrm{id}} & & \downarrow{\scriptstyle 0} \\
0 & \xrightarrow{0} & C & \xrightarrow{\mathrm{id}} & C & \xrightarrow{0} & 0[1]
\end{array}$$

Therefore by the resulting commutative diagram, we have that $g \circ f = 0$ as required. □

Proposition 3 *Let $\mathcal{D}$ be a triangulated category and $A \xrightarrow{f} B \xrightarrow{g} C \xrightarrow{h} A[1]$ a distinguished triangle in this category. Then for any object X we get the following induced sequences on homomorphism groups.*

$$\mathrm{Hom}(X, A) \xrightarrow{f_*} \mathrm{Hom}(X, B) \xrightarrow{g_*} \mathrm{Hom}(X, C)$$

$$\mathrm{Hom}(A, X) \xleftarrow{f^*} \mathrm{Hom}(B, X) \xleftarrow{g^*} \mathrm{Hom}(C, X)$$

These sequences are exact.

Proof (*for the first sequence*)
Exactness of this sequence means that $\ker g_* = \mathrm{im} f_*$.
First we will prove $\ker g_* \subseteq \mathrm{im} f_*$.
Let $j : X \to B$ be an element of $\ker g_*$. Then $g \circ j = 0$.
By **TR1** we have that the triangle

$$X \xrightarrow{\mathrm{id}} X \to 0 \to X[1]$$

is distinguished.
Consider the following commutative diagram.

$$\begin{array}{ccccccc}
X & \xrightarrow{id} & X & \xrightarrow{0} & 0 & \xrightarrow{0} & X[1] \\
 & & \downarrow{\scriptstyle j} & & \downarrow{\scriptstyle 0} & & \\
A & \xrightarrow{f} & B & \xrightarrow{g} & C & \xrightarrow{h} & A[1]
\end{array}$$

Applying **TR3**, there exists a morphism $\alpha : X \to A$ making the triangle commute.

$$\begin{array}{ccccccc} X & \xrightarrow{id} & X & \xrightarrow{0} & 0 & \xrightarrow{0} & X[1] \\ \downarrow{\scriptstyle \alpha} & & \downarrow{\scriptstyle j} & & \downarrow{\scriptstyle 0} & & \downarrow{\scriptstyle \alpha[1]} \\ A & \xrightarrow{f} & B & \xrightarrow{g} & C & \xrightarrow{h} & A[1] \end{array}$$

Therefore $j \circ \mathrm{id} = j = f \circ \alpha = f_*\alpha$, and j is indeed an element of $\mathrm{im} f_*$.

Next we will prove $\ker g_* \supseteq \mathrm{im} f_*$.

Let $k : X \to A$ be arbitrary. Then $f_*(k) = f \circ k$ is an arbitrary element of $\mathrm{Im}(f_*)$. Then $g_*(f \circ k) = g \circ f \circ k$. By the Lemma 2 $g \circ f = 0$, so $f \circ k$ is an element of $\ker(g_*)$.

The proof for the other sequence is similar.

Corollary 1 *Let $\mathscr{D}$ be a triangulated category and $A \xrightarrow{f} B \xrightarrow{g} C \xrightarrow{h} A[1]$ a distinguished triangle in this category. Then for any object X there is an exact sequence*

$$\cdots \mathrm{Hom}(X, A) \xrightarrow{f_*} \mathrm{Hom}(X, B) \xrightarrow{g_*} \mathrm{Hom}(X, C) \xrightarrow{h_*} \mathrm{Hom}(X, A[1]) \xrightarrow{-f[1]_*} \mathrm{Hom}(X, B[1]) \xrightarrow{-g[1]_*} \cdots .$$

Proof This follows from applying **TR2** to the previous proposition. □

References

1. Atiyah, M.F., Macdonald, I.G.: Introduction to commutative algebra. Addison-Wesley Publishing Co., Reading, Mass.-London-Don Mills, Ont. (1969)
2. Dold, A., Puppe, D.: Homologie nicht-additiver Funktoren. Anwendungen. Ann. Inst. Fourier Grenoble **11**, 201–312 (1961). http://www.numdam.org/item?id=AIF_1961__11__201_0
3. Evans, J.: Tikz code for the octahedral axiom. https://jonathanevans27.wordpress.com/2013/03/01/tikz-code-for-the-octahedral-axiom/. Accessed: 2016, Published: 2013
4. Freyd, P.: Abelian categories. An introduction to the theory of functors. Harper's Series in Modern Mathematics. Harper & Row, Publishers, New York (1964)
5. Hartshorne, R.: Residues and duality. Lecture notes of a seminar on the work of A. Grothendieck, given at Harvard 1963/64. With an appendix by P. Deligne. Lecture Notes in Mathematics, No. 20. Springer-Verlag, Berlin-New York (1966)
6. Holm, T., Jørgensen, P.: Triangulated categories: definitions, properties, and examples. In: Triangulated categories, *London Math. Soc. Lecture Note Ser.*, vol. 375, pp. 1–51. Cambridge Univ. Press, Cambridge (2010). http://dx.doi.org/10.1017/CBO9781139107075.002
7. Huybrechts, D.: Fourier-Mukai transforms in algebraic geometry. Oxford Mathematical Monographs. The Clarendon Press, Oxford University Press, Oxford (2006). http://dx.doi.org/10.1093/acprof:oso/9780199296866.001.0001
8. Krishnan, V.S.: An introduction to category theory. North-Holland Publishing Co., New York-Amsterdam (1981)

9. Neeman, A.: Triangulated categories, *Annals of Mathematics Studies*, vol. 148. Princeton University Press, Princeton, NJ (2001). http://dx.doi.org/10.1515/9781400837212
10. Verdier, J.L.: Des catégories dérivées des catégories abéliennes. Astérisque (239), xii+253 pp. (1997) (1996). With a preface by Luc Illusie, Edited and with a note by Georges Maltsiniotis

Derived Categories and Derived Functors

Nitin Kumar Chidambaram

1 Category of Complexes and Homotopy Category

Given an abelian category $\mathscr{A}$ let us define the category of complexes $\mathrm{Kom}(\mathscr{A})$. Chain complexes, denoted $A^\bullet$, are diagrams of the form

$$\cdots \xrightarrow{d^{i-2}} A^{i-1} \xrightarrow{d^{i-1}} A^i \xrightarrow{d^i} A^{i+1} \xrightarrow{d^{i+1}} \cdots$$

where $A^i \in Obj(\mathscr{A})$ and $d^i \in Mor(\mathscr{A})$ such that $d^i \circ d^{i-1} = 0$, and the morphisms between chain complexes, say $f : A^\bullet \to B^\bullet$, are defined as commutative diagrams of the form

$$\begin{array}{ccccccccc}
\cdots \xrightarrow{d^{i-2}} & A^{i-1} & \xrightarrow{d^{i-1}} & A^i & \xrightarrow{d^i} & A^{i+1} & \xrightarrow{d^{i+1}} \cdots \\
 & \downarrow{\scriptstyle f^{i-1}} & & \downarrow{\scriptstyle f^i} & & \downarrow{\scriptstyle f^{i+1}} & \\
\cdots \xrightarrow{d^{i-2}} & B^{i-1} & \xrightarrow{d^{i-1}} & B^i & \xrightarrow{d^i} & B^{i+1} & \xrightarrow{d^{i+1}} \cdots
\end{array}$$

Definition 1 The category of complexes $\mathrm{Kom}(\mathscr{A})$ of an abelian category $\mathscr{A}$ is the category with objects as chain complexes $A^\bullet$ and morphisms as morphisms between chain complexes.

Remark 1 $\mathrm{Kom}(\mathscr{A})$ is an abelian category. The verification of the existence of kernels and cokernels is straightforward.

N. K. Chidambaram (✉)
Department of Mathematical and Statistical Sciences, University of Alberta, 632 CAB, Edmonton, AB T6G 2G1, Canada
e-mail: chidamba@ualberta.ca

M. Ballard et al. (eds.), *Superschool on Derived Categories and D-branes*, Springer Proceedings in Mathematics & Statistics 240,
https://doi.org/10.1007/978-3-319-91626-2_2

Remark 2 The mapping $A \to (\cdots \longrightarrow 0 \xrightarrow{0} A \xrightarrow{0} 0 \longrightarrow \cdots)$ defines an equivalence of $\mathscr{A}$ with a full subcategory of Kom($\mathscr{A}$).

We shall define two functors on the category of complexes which are important to us:

Definition 2 The *shift* functor T : Kom($\mathscr{A}$) $\to$ Kom($\mathscr{A}$) is defined as the following — $A^\bullet[1] := T(A^\bullet)$ is the complex with $A^\bullet[1]^i := A^{i+1}$ and differential $d^i_{A[1]} := -d^{i+1}_A$. And $f[1] := T(f)$ of the morphism $f : A^\bullet \to B^\bullet$ is the morphism such that $f[1]^i = f^{i+1}$.

Remark 3 The shift functor defines an equivalence of categories.

Remark 4 The shift functor does not give Kom($\mathscr{A}$) the structure of a triangulated category! The problem lies in defining exact triangles.

Definition 3 The *cohomology* functor H^i : Kom($\mathscr{A}$) $\to \mathscr{A}$ is defined as the following on objects — $H^i(A^\bullet) :=$ Ker $(d^i)/$ Im (d^{i-1}). The morphism $f^i : A^i \to B^i$ descend to the cohomology $H^i(f) : H^i(A) \to H^i(B)$.

Remark 5 A complex, $A^\bullet$, is called *acyclic* if $H^i(A^\bullet) = 0 \; \forall \; i \in \mathbb{Z}$.

Remark 6 A short exact sequence in Kom($\mathscr{A}$)

$$0 \longrightarrow A \longrightarrow B \longrightarrow C \longrightarrow 0$$

induces a long exact sequence in cohomology

$$\cdots \longrightarrow H^i(A) \longrightarrow H^i(B) \longrightarrow H^i(C) \longrightarrow H^{i+1}(A) \longrightarrow \cdots$$

Let us define quasi-isomorphisms which will be essential for defining derived categories now.

Definition 4 A morphism of complexes $f : A^\bullet \to B^\bullet$ is called a *quasi-isomorphism* (or qis) if the induced maps on cohomology are isomorphisms, i.e. $H^i(f) : H^i(A^\bullet) \to H^i(B^\bullet)$ is an isomorphism $\forall i \in \mathbb{Z}$.

In the derived category, we wish to make all quasi-isomorphisms invertible (i.e. isomorphisms). In order to do this, we will first pass to the homotopy category of chain complexes which will make a certain class of quasi-isomorphisms invertible (namely the ones that have an inverse up to homotopy).

Definition 5 (*Homotopy of morphisms*) Two morphisms of complexes $f, g : A^\bullet \to B^\bullet$ are said to be homotopic (denoted $f \sim g$) if there exists morphisms $h^i : A^i \to B^{i-1}$, $i \in \mathbb{Z}$ such that

$$f^i - g^i = h^{i+1} \circ d_A^i + d_B^{i-1} \circ h^i$$

$$\begin{array}{ccccccccc}
\cdots & \xrightarrow{d^{i-2}} & A^{i-1} & \xrightarrow{d^{i-1}} & A^i & \xrightarrow{d^i} & A^{i+1} & \xrightarrow{d^{i+1}} & \cdots \\
 & \swarrow^{h^{i-1}} & \downarrow & \swarrow^{h^i} & \downarrow & \swarrow^{h^{i+1}} & \downarrow & & \\
\cdots & \xrightarrow{d^{i-2}} & B^{i-1} & \xrightarrow{d^{i-1}} & B^i & \xrightarrow{d^i} & B^{i+1} & \xrightarrow{d^{i+1}} & \cdots
\end{array}$$

Definition 6 (*Homotopy category of chain complexes*) The *homotopy category of chain complexes*, K($\mathscr{A}$) is the category with the same objects as Kom($\mathscr{A}$) and morphisms defined up to homotopy, i.e. $Hom_{K(\mathscr{A})}(A^\bullet, B^\bullet) = Hom_{Kom(\mathscr{A})}(A^\bullet, B^\bullet)/\sim$.

The verification that this is well defined is fairly straightforward.

We also note that this construction is well defined for any additive category (not necessarily abelian).

Remark 7 If $f \sim g : A^\bullet \to B^\bullet$, $H^i(f) = H^i(g)$.

Remark 8 If $f, g : A^\bullet \to B^\bullet$, such that $f \circ g \sim Id$ and $g \circ f \sim Id$, then f and g are inverses in $K(\mathscr{A})$

Now, we can finally define our object of interest, i.e the derived category.

2 Defining the Derived Category

First of all, let us state the existence result:

Theorem 1 *Let $\mathscr{A}$ be an abelian category and let Kom($\mathscr{A}$) be the category of complexes. Then there exists a category D($\mathscr{A}$), the derived category of $\mathscr{A}$, and a functor*

$$Q : Kom(\mathscr{A}) \to D(\mathscr{A})$$

such that:

- *If f : $A^\bullet \to B^\bullet$ is a quasi-isomorphism, then Q(f) is an isomorphism in D($\mathscr{A}$).*
- *Any functor $F : Kom(\mathscr{A}) \to \mathscr{D}$ satisfying the above condition factorizes uniquely over $Q : Kom(\mathscr{A}) \to D(\mathscr{A})$:*

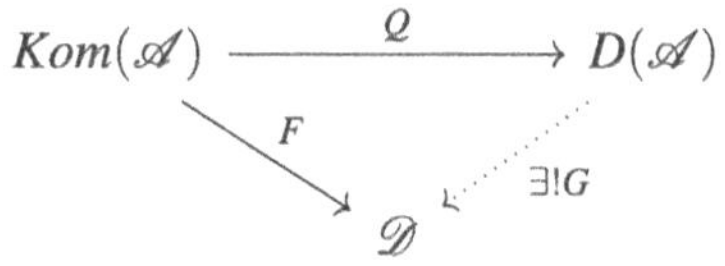

Before we proceed to construct this derived category explicitly, let us make a couple of observations:

Remark 9 The functor Q identifies objects of K($\mathscr{A}$) with the objects of D($\mathscr{A}$).

Remark 10 The cohomology objects $H^i(A^\bullet)$ where $A^\bullet \in D(\mathscr{A})$ are well defined objects of the abelian category $\mathscr{A}$. In other words the cohomology functors H^i factor through Q.

Now we proceed to construct the derived category. The morphisms in the derived category are constructed as follows. We represent morphisms $A^\bullet \to B^\bullet$ in the derived category by equivalence classes of diagrams called *roofs*:

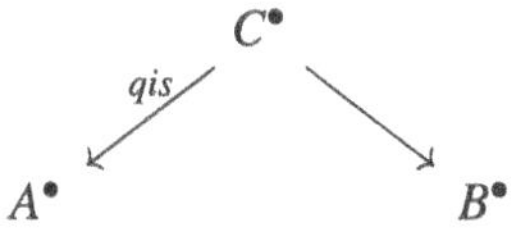

Two roofs representing $A^\bullet \to B^\bullet$ are said to be equivalent if they are dominated in the homotopy category $K(\mathscr{A})$ by a third roof, i.e. there exists a commutative diagram in $K(\mathscr{A})$ of the form:

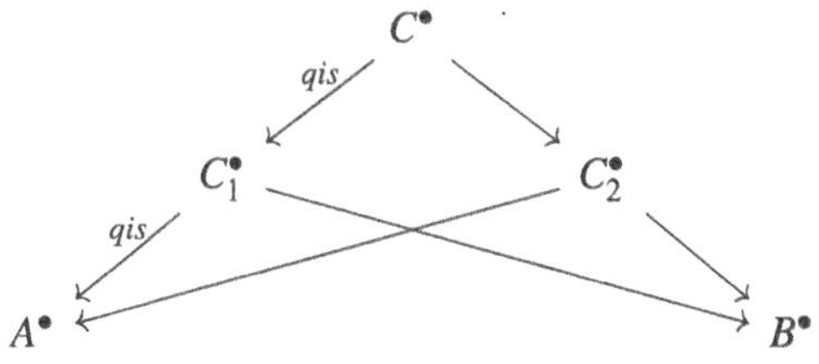

Now we need to define composition of morphism. Say we are given two morphisms:

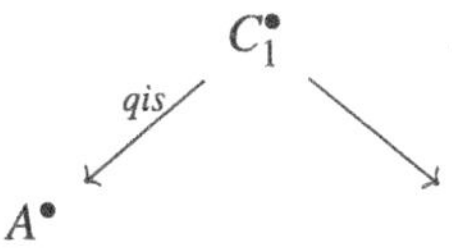

and

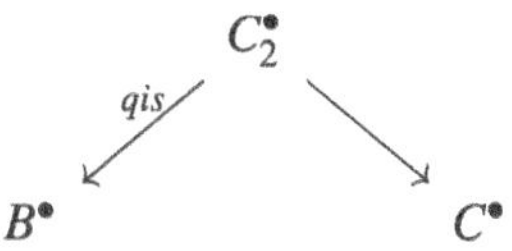

We want the composition to be defined as:

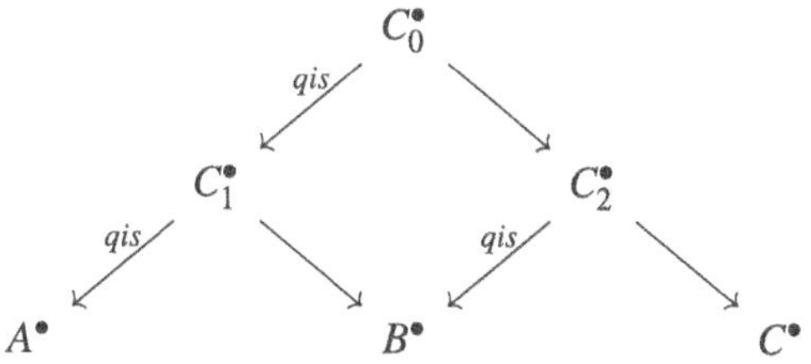

(1)

We will show that this diagram exists and is defined uniquely in the derived category, by introducing a construction called the mapping cone. This mapping cone construction will also tell us how to give the derived category the structure of a triangulated category.

3 Derived Category as a Triangulated Category

Let us define the mapping cone first.

Definition 7 Let $f : A^\bullet \to B^\bullet$ be a morphism in Kom($\mathscr{A}$). The mapping cone C(f) is defined as the complex such that:

$$C(f)^i := A^{i+1} \bigoplus B^i \qquad \text{and}$$

$$d^i_{C(f)} := \begin{pmatrix} -d_A^{i+1} & 0 \\ f^{i+1} & d_B^i \end{pmatrix}$$

Also, there exists two natural morphisms

$$\tau : B^\bullet \to C(f) \qquad \text{and} \qquad \pi : C(f) \to A^\bullet[1]$$

given by the natural injection and surjection respectively.

Remark 11 $B^\bullet \xrightarrow{\iota} C(f) \xrightarrow{\pi} A^\bullet[1]$ is a short exact sequence in Kom($\mathscr{A}$) (We will define this as a distinguished triangle in D($\mathscr{A}$) eventually.)

Proposition 1 (TR3) *A commutative diagram can be completed as follows*

$$\begin{array}{ccccccc} A_1^\bullet & \xrightarrow{f_1} & B_1^\bullet & \longrightarrow & C(f_1) & \longrightarrow & A_1^\bullet[1] \\ \downarrow & & \downarrow & & \vdots\,\exists & & \downarrow \\ A_2^\bullet & \xrightarrow{f_2} & B_2^\bullet & \longrightarrow & C(f_2) & \longrightarrow & A_2^\bullet[1] \end{array}$$

Proof We construct the morphism by sending $A_1^{i+1} \subset C(f_1)^i$ to $A_2^{i+1} \subset C(f_2)^i$ by the given morphism $A_1^\bullet \to A_2^\bullet$ and similarly for $B^i \subset of C(f_1)^i$. By construction the diagram is commutative. □

Proposition 2 (TR2) *Let $f : A^\bullet \to B^\bullet$ be a complex morphism. Then there exists a complex morphism $g : A^\bullet[1] \to C(\tau)$ which is invertible in $K(\mathscr{A})$ that makes the following diagram commutative in $K(\mathscr{A})$:*

$$\begin{array}{ccccccc} B^\bullet & \longrightarrow & C(f) & \longrightarrow & A^\bullet[1] & \xrightarrow{-f} & B^\bullet[1] \\ \downarrow = & & \downarrow = & & \vdots\ \exists g & & \downarrow = \\ B^\bullet & \xrightarrow{\tau} & C(f) & \longrightarrow & C(\tau) & \longrightarrow & B^\bullet[1] \end{array}$$

Proof The morphism $g : A^\bullet[1] \to C(\tau)$ is defined on

$$A^{i+1} \longrightarrow C(\tau)^i = B^{i+1} \bigoplus C(f)^i = B^{i+1} \bigoplus A^{i+1} \bigoplus B^i$$

as $(-f^{i+1}, id, 0)$. See Proposition 2.16 in [1] for details. □

The following proposition will let us define composition in the derived category.

Lemma 1 (Extension lemma) *Given a quasi-isomorphims $f : A^\bullet \to B^\bullet$ and a morphism $g : C^\bullet \to B^\bullet$, there exists the following commutative diagram in $K(\mathscr{A})$:*

$$\begin{array}{ccc} C_0^\bullet & \xrightarrow{qis} & C^\bullet \\ \downarrow & & \downarrow g \\ A^\bullet & \underset{qis}{\xrightarrow{f}} & B^\bullet \end{array}$$

Proof First of all, we use the mapping cone construction on f to get a short exact sequence by Remark 11: $B \to C(f) \to A[1]$. As f is a quasi-isomorphism, passing to the long exact sequence in cohomology gives us that $H^i(C(f)) = 0$ for all $i \in \mathbb{Z}$. Now, using the mapping cone construction on $\tau \circ g$ we know that there exists a morphism of complexes given by

$$C^\bullet \xrightarrow{\tau \circ g} C(f) \longrightarrow C(\tau \circ g)$$

Furthermore, by Remark 11, we know that this is a short exact sequence. So we pass to the long exact sequence in cohomology and using that $H^i(C(f)) = 0$ we get that $C_0^\bullet := C(\tau \circ g)[-1] \to C^\bullet$ is a quasi-isomorphism.

All that remains to be shown is that there exists a map $C_0^\bullet \to A^\bullet$ so that we get a commutative diagram. For this, we use TR3 to show the isomorphism $A^\bullet \simeq C(\tau)$. Then we construct the natural map $C(\tau \circ g) \to C(\tau)$ and use the isomorphism above to get $C_0^\bullet \to A^\bullet$.

$$\begin{array}{ccccccc} C(\tau \circ g)[-1] & \longrightarrow & C^\bullet & \xrightarrow{\tau \circ g} & C(f) & \longrightarrow & C(\tau \circ g) \\ \vdots & & \downarrow g & & \downarrow = & & \vdots \\ A^\bullet & \xrightarrow{f} & B^\bullet & \xrightarrow{\tau} & C(f) & \longrightarrow & A^\bullet[1] \end{array}$$

The commutativity is clear. □

Corollary 1 *The composition of roofs defined by* (1) *is well-defined and unique.*

Proof Apply the extension lemma to the diagram

$$\begin{array}{ccccccc} & & C_1^\bullet & & & C_2^\bullet & \\ & {}^{qis}\swarrow & & \searrow & {}^{qis}\swarrow & & \searrow \\ A^\bullet & & & B^\bullet & & & C^\bullet \end{array}$$

This shows that a roof representing composition exists. Uniqueness can also be proved using the extension lemma multiple times. □

Now that we have defined composition, let us give the homotopy category and the derived category the structure of a triangulated category as follows:

Definition 8 A distinguished triangle in $D(\mathscr{A})$ (or $K(\mathscr{A})$) is any triangle isomorphic to a triangle of the form:

$$A^\bullet \xrightarrow{f} B^\bullet \longrightarrow C(f) \longrightarrow A[1]^\bullet$$

Proposition 3 *Distinguished triangles as defined above along with the shift functor turn both the homotopy category and the derived category into triangulated categories.*

Also the functor $Q_{\mathscr{A}} : K(\mathscr{A}) \to D(\mathscr{A})$ is an exact functor of triangulated categories.

Proof See IV.2 in [2]. □

Remark 12 The general procedure to construct a derived category is called localization, in this case we use the set of quasi-isomorphisms in $K(\mathscr{A})$ as our localizing class of morphisms.

Remark 13 For a semi-simple abelian category, any complex in the derived category is isomorphic to it's cohomology complex.

Remark 14 The distinguished triangles in the derived category should be thought of as a generalization of exact triples in an abelian category. Any distinguished triangle in the derived category, say

$$A^\bullet \longrightarrow B^\bullet \longrightarrow C^\bullet \longrightarrow A[1]^\bullet$$

generates a long exact sequence in cohomology in the abelian category:

$$\cdots \longrightarrow H^i(A) \longrightarrow H^i(B) \longrightarrow H^i(C) \longrightarrow H^{i+1}(A) \longrightarrow \cdots$$

Let us introduce bounded categories now. These will be useful in defining derived functors.

Definition 9 Let $Kom^*(\mathscr{A})$, with $* = +, -$ or b, be the category of complexes $A^\bullet$ such that $A^i = 0$ for $i \ll 0$, $i \gg 0$, or $|i| \gg 0$ respectively.

We can construct bounded homotopy categories and bounded derived categories using $Kom^*(\mathscr{A})$.

The bounded derived category is equivalent to a cohomologically bounded full subcategory of the unbounded derived category. More precisely:

Proposition 4 *The natural inclusion functor $D^*(\mathscr{A}) \to D(\mathscr{A})$, with $* = +, -$ or b, defines an equivalence of $D^*(\mathscr{A})$ with the full triangulated subcategory of objects $A^\bullet \in D(\mathscr{A})$ such that $H^i(A^\bullet) = 0$ for $i \ll 0$, $i \gg 0$, or $|i| \gg 0$ respectively.*

Proof The idea is that the acyclic part of a complex in the derived category can be replaced by 0. See Proposition 2.30 in [1] for details of the construction. □

4 Equivalence Between Homotopy Category of Injectives in $\mathscr{A}$ and the Derived Category $D(\mathscr{A})$

In certain abelian categories (containing enough injectives or projectives), we can avoid working with the derived categories which are harder to understand, but instead work with some homotopy categories which are better understood and easier to work with (in the sense that the morphisms are honest morphisms of complexes).

First of all let us define what injective and projective objects in an abelian category are. There exist multiple equivalent definitions which we shall not state here.

Definition 10 An object $I \in \mathscr{A}$ is called injective if $Hom_{\mathscr{A}}(-, I)$ is exact. Dually, $P \in \mathscr{A}$ is called projective if $Hom_{\mathscr{A}}(P, -)$ is exact.

Definition 11 A category $\mathscr{A}$ is said to have *enough injectives (or projectives)* if there exists an injective morphism $A \to I$ (or a projective morphism $P \to A$) for all objects $A \in \mathscr{A}$.

Remark 15 Not all abelian categories have enough projectives or enough injectives. For example the category of quasicoherent $\mathscr{O}_X$-modules on a scheme X, $QCoh(X)$ has enough injectives but not enough projectives.

Remark 16 An injective resolution is a quasi-isomorphism between a complex $A^\bullet \in Kom(\mathscr{A})$ and a complex $I^\bullet \in Kom(\mathscr{A})$ such that I^i are injective and $I^i = 0$ for $i < 0$. Similarly we define projective resolutions using projective objects.

Proposition 5 *Let $\mathscr{A}$ be an abelian category with enough injectives. For any complex $A^\bullet \in K^+(\mathscr{A})$ there exists a complex $I^\bullet \in K^+(\mathscr{A})$, with I^i injectives, and a quasi-isomorphism $A^\bullet \to I^\bullet$.*

Proof See III.5 in [2]. □

The class of injectives $\mathscr{I}$ forms a full additive subcategory of the abelian category $\mathscr{A}$. We can also define the triangulated category $K^*(\mathscr{I})$ which is the (bounded) homotopy category of this additive category. $K^*(\mathscr{I})$ is called the homotopy category of injectives of $\mathscr{A}$. Dually, we can carry out the same construction (assuming the existence of enough projectives) to get the homotopy category of projectives $K^*(\mathscr{P})$.

Proposition 6 *If $\mathscr{A}$ contains enough injectives, the natural functor*

$$i : K^+(\mathscr{I}) \to D^+(\mathscr{A}) \tag{2}$$

is an exact equivalence of triangulated categories.

Proof See Proposition 2.40 in [1]. □

Remark 17 The above proposition allows us to work with $K^+(\mathscr{I})$ instead of the much more complicated $D^+(\mathscr{A})$. The inverse functor replaces any object in $D^+(\mathscr{A})$ with an injective resolution of this object. We also have the isomorphism

$$\mathrm{Hom}_{D^+(\mathscr{A})}(A, B) \simeq \mathrm{Hom}_{K^+(\mathscr{I})}(i^{-1}(A), i^{-1}(B))$$

which allows us to compute morphisms in the derived category by morphisms of complexes between injective resolutions up to homotopy (which avoids working with roofs).

5 Derived Functors

If we have a functor $F : \mathscr{A} \to \mathscr{B}$ between two abelian categories, a natural question to ask is whether we can define a canonical functor between the derived categories. It turns out that the naive extension only makes sense for an exact functor.

Hence we need to introduce the more complicated idea of a *derived functor* between the derived categories. For a left exact functor $F : \mathscr{A} \to \mathscr{B}$, we define a right derived functor $RF : D^+(\mathscr{A}) \to D^+(\mathscr{B})$. Or given a right exact functor we can define a left derived functor $LF : D^-(\mathscr{A}) \to D^-(\mathscr{B})$.

Now let us define the right derived functor $RF : D^+(\mathscr{A}) \to D^+(\mathscr{B})$ given the left exact functor $F : \mathscr{A} \to \mathscr{B}$, where we also assume that $\mathscr{A}$ has enough injectives. Recall Proposition 6 and consider the following diagram.

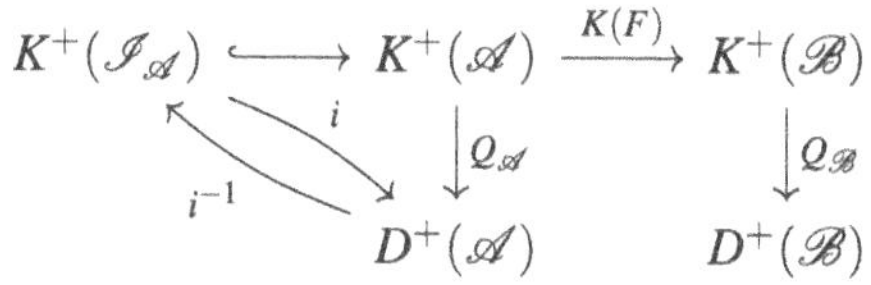

where K(F) is the functor F applied to every object and morphism in the complex (this is well defined for the homotopy categories). Note that this functor is exact as a triangulated functor between triangulated categories.

Definition 12 The right derived functor of F is the functor

$$RF := Q_{\mathscr{B}} \circ K(F) \circ i^{-1} : D^+(\mathscr{A}) \to D^+(\mathscr{B}). \tag{3}$$

Proposition 7 *1. $RF : D^+(\mathscr{A}) \to D^+(\mathscr{B})$ is an exact functor of triangulated categories.*
2. There exists a natural morphism of functors $Q_{\mathscr{B}} \circ K(F) \to RF \circ Q_{\mathscr{A}}$.
3. Suppose $G : D^+(\mathscr{A}) \to D^+(\mathscr{B})$ is an exact functor. Then any functor morphism $Q_{\mathscr{B}} \circ K(F) \to G \circ Q_{\mathscr{A}}$ factors through a unique functor morphism $RF \to G$.

Proof 1. The functor i is exact, hence i^{-1} is exact; and all other functors are exact. For parts 2 and 3, see III6.1 in [2]. □

We also define the higher right derived functors as follows.

Definition 13 Let $RF : D^+(\mathscr{A}) \to D^+(\mathscr{B})$ be the right derived functor of the left exact functor $F : \mathscr{A} \to \mathscr{B}$. Then for any complex $A^\bullet \in D^+(\mathscr{A})$ we define:

$$R^i F(A^\bullet) := H^i(RF(A^\bullet)) \in \mathscr{B}$$

The induced functors

$$R^i F : \mathscr{A} \to \mathscr{B}$$

are called the *higher derived functors* of F.

Remark 18 $R^i F(A) = 0$ for $i < 0$ and $R^0 F(A) = F(A)$ for any $A \in \mathscr{A}$.

Remark 19 An object $A \in \mathscr{A}$ is called *F-acyclic* if $R^i F(A) = 0$ for $i \neq 0$.

The right derived functor RF roughly measures how much the functor F fails to be exact on the right. More precisely,

Proposition 8 *Let $RF : D^+(\mathscr{A}) \to D^+(\mathscr{B})$ be the right derived functor of the left exact functor $F : \mathscr{A} \to \mathscr{B}$. Then any short exact sequence*

$$0 \longrightarrow A \longrightarrow B \longrightarrow C \longrightarrow 0$$

induces a long exact sequence

$$0 \longrightarrow F(A) \longrightarrow F(B) \longrightarrow F(C) \longrightarrow RF^1(A) \longrightarrow \cdots$$

$$\cdots \longrightarrow RF^i(B) \longrightarrow RF^i(C) \longrightarrow RF^{i+1}(A) \longrightarrow \cdots$$

Proof Any short exact sequence in $\mathscr{A}$ gives rise to a distinguished triangle,

$$A^\bullet \longrightarrow B^\bullet \longrightarrow C^\bullet \longrightarrow A[1]$$

in $D(\mathscr{A})$. Applying the exact functor RF, we get another distinguished triangle

$$RF(A^\bullet) \longrightarrow RF(B^\bullet) \longrightarrow RF(C^\bullet) \longrightarrow RF(A^\bullet)[1]$$

This distinguished triangle gives us a long exact sequence in cohomology according to Remark 14. □

Finally, let us look at an example of a derived functor. Consider the left exact functor Hom(A, $-$) : $\mathscr{A} \to Ab$. The familiar Ext functors are the right derived functors of this functor.

Definition 14 If $\mathscr{A}$ has enough injectives, we define

$$Ext^i(A,-) := H^i \circ RHom(A,-)$$

But in fact, these functors have a natural interpretation as just morphisms in the derived category. More precisely,

Proposition 9 *Let A, B be objects of an abelian category $\mathscr{A}$ that has enough injectives. Then there are natural isomorphisms*

$$Ext^i_{\mathscr{A}}(A, B) \simeq Hom_{D(\mathscr{A})}(A, B[i])$$

Proof See Proposition 2.56 in [1].

Dually, the Tor functors are constructed as the left derived functors of the right exact $\otimes$ functor.

References

1. D. Huybrechts, Fourier-Mukai Transforms in Algebraic Geometry, Oxford Science Publications (2006)
2. Sergei I. Gelfand and Yuri I. Manin, Methods of Homological Algebra, Springer-Verlag Berlin Heidelberg (2003)

Introduction to Quivers

Minako Chinen

1 Quivers and Path Algebras

Definition 1 A **quiver** Q is a directed graph consisting of a set of vertices Q_0, a set of arrows Q_1, and maps $h, t : Q_0 \to Q_1$ which specify the head and tail.

$$t(a) \xrightarrow{\quad a \quad} h(a)$$

We assume that both Q_0 and Q_1 are finite sets and that Q is connected.

Definition 2 A **nontrivial path** p in Q of length n from vertex $i \in Q_0$ to vertex $j \in Q_0$ is a sequence of arrows $a_1, \ldots, a_n$.

$$i = t(a_1) \xrightarrow{\quad a_1 \quad} h(a_1) = t(a_2) \xrightarrow{\quad a_2 \quad} \cdots \xrightarrow{\quad a_n \quad} h(a_n) = j$$

with $h(a_k) = t(a_{k+1})$ for $1 \leq k \leq n$; we set $t(p) := t(a_1)$ and $h(p) := h(a_n)$. In addition, each vertex $i \in Q_0$ gives a **trivial path** e_i of length 0 with $h(e_i) = t(e_i) = i$. A **cycle** is a nontrivial path with the same head and tail, and Q is called **acyclic** if it contains no cycles.

Here is an example of a quiver with no cycles. We will revisit this quiver later.

M. Chinen (✉)
University of Alberta, Edmonton, AB T6G 2R3, Canada
e-mail: chinen@ualberta.ca

M. Ballard et al. (eds.), *Superschool on Derived Categories and D-branes*,
Springer Proceedings in Mathematics & Statistics 240,
https://doi.org/10.1007/978-3-319-91626-2_3

Example 1 (Beilinson quiver [2]) The quiver with two vertices and a pair of arrows

$$1 \rightrightarrows 2$$

is called the *Beilinson quiver for* $\mathbb{P}^1$. More generally the Beilinson quiver for $\mathbb{P}^n$ is the quiver

$$v_0 \underset{\longrightarrow}{\overset{\rightrightarrows}{\vdots}} v_1 \underset{\longrightarrow}{\overset{\rightrightarrows}{\vdots}} \cdots \underset{\longrightarrow}{\overset{\rightrightarrows}{\vdots}} v_n$$

consisting of $n+1$ vertices $v_0, \dots v_n$ and there are $n+1$ arrows between vertices.

Definition 3 Let k be a field and Q be a finite connected quiver. The **path algebra** kQ of Q is the associative algebra whose underlying k-vector space is spanned by all paths in Q. The **multiplication** of paths $p = a_1 \dots a_n$ and $q = b_1 \dots b_m$ is defined by the concatenation

$$p \cdot q = \begin{cases} p \cdot q & \text{if } h(p) = t(q) \\ 0 & \text{otherwise} \end{cases}$$

$$t(p) \xrightarrow{a_1} \cdots \xrightarrow{a_n} h(p) = t(q) \xrightarrow{b_1} \cdots \xrightarrow{b_m} h(q)$$

There are several important facts about the path algebra which we don't prove.

Remark 1 [1–3]

1. The path algebra kQ is graded by path length, i.e. $kQ = \bigoplus_{l \in \mathbb{N}} (kQ)_l$ where $(kQ)_l$ denotes the vector space spanned by paths of length l and $(kQ)_l \cdot (kQ)_m \subseteq (kQ)_{l+m}$. The zero graded subring $(kQ)_0 \subset kQ$ spanned by trivial paths e_i is a semisimple ring.
2. kQ is finite dimensional over k if and only if Q is acyclic.
3. The trivial paths e_i are orthogonal idempotents of kQ and $\sum_{i \in Q_0} e_i = 1$; i.e. $e_i^2 = e_i$, $e_i e_j = 0$ if $i \neq j$.
4. Using the decomposition $\sum_{i \in Q_0} e_i = 1$, we get $kQ \cong \bigoplus_{i \in Q_0} kQe_i$
    a. kQe_i is a k-vector space spanned by paths starting at i
    b. Every indecomposable projective kQ-module happens to be one of $\{kQe_i\}_{i \in Q_0}$, and in fact e_i is a primitive idempotent.

Note that because there is a One-to-One correspondence between simple modules and projective indecomposable modules up to isomorphism, being able to decompose kQ into projective indecomposable modules allows us to express kQ using different "building blocks" other than simple modules [4].

2 The Category of Quiver Representations

Definition 4 A **representation of the quiver Q** consists of a k-vector space V_i for each $i \in Q_0$ and a k-linear map $f_a : V_{t(a)} \to V_{h(a)}$ for each arrow $a \in Q_1$. We write the representation of Q as a tuple $V = ((V_i)_{i \in Q_0}, (f_a)_{a \in Q_1})$. A representation is said to be **finite dimensional** if each vector space V_i has finite dimension over k. The **dimension vector of V** is the tuple of nonnegative integers $\underline{dim V} = (dim V_i)_{i \in Q_0}$.

Just to have a better understanding of a representation of a quiver, we will take a look at one simple example. Say we have a quiver given by

$$Q: \qquad 1 \xrightarrow{a} 2 \underset{c}{\overset{b}{\rightrightarrows}} 3$$

The representation of this quiver consists of three vector spaces V_1, V_2, V_3 corresponding to the three vertices and three k-linear maps f_a, f_b, f_c corresponding to the arrows of Q.

$$V: \qquad V_1 \xrightarrow{f_a} V_2 \underset{f_c}{\overset{f_b}{\rightrightarrows}} V_3$$

Definition 5 Suppose that $V = ((V_i)_{i \in Q_0}, (f_a)_{a \in Q_1})$ and $W = ((W_j)_{j \in Q_0}, (g_b)_{b \in Q_1})$ are both representation of a quiver Q. A **morphism** $\phi : V \to W$ between two representations is a collection of k-linear maps $\{\phi_i : V_i \to W_i \mid i \in Q_0\}$ such that the diagram commutes.

$$\begin{array}{ccc} V_{t(a)} & \xrightarrow{f_a} & V_{h(a)} \\ \downarrow{\scriptstyle \phi_{t(a)}} & & \downarrow{\scriptstyle \phi_{h(a)}} \\ W_{t(a)} & \xrightarrow{g_a} & W_{h(a)} \end{array}$$

With the notion of morphism between two representations, we can now define a category of finite dimensional representation of Q.

Definition 6 A **category of finite dimensional representations of the quiver Q** denoted by rep(Q) consists of

- objects: representations V of Q
- morphisms: the morphisms between the finite dimensional representations of Q which we just have defined.

There are two important properties of this category which we will not give proofs.

Remark 2 [2, 3]

1. rep(Q) is equivalent to the category of finitely generated left kQ-modules kQ-mod
2. rep(Q) is an Abelian category

The first remark allows us to obtain a finitely generated left kQ-module from a given representation of a quiver Q and vice versa. In the next example, we will see how to obtain a representation of Q from a left kQ-module.

Example 2 Given a left kQ-module (more precisely a projective indecomposable module) kQe_j where $j \in Q_0$ is fixed, the corresponding representation $((V_i)_{i\in Q_0}, (f_a)_{a\in Q_1})$ consists of the vector spaces

$$V_i = e_i(kQe_j)$$

spanned by paths p starting at $j \in Q_0$ and ending at $i \in Q_0$ and the k-linear map associated to a path $a \in Q_1$

$$f_a : V_{t(a)} \to V_{h(a)}$$

$$p \mapsto pa$$

defined by concatenation of the path a.

Before we move on to another example, let us introduce the notion of a **simple representation** S(i) of Q at a fixed vertex $i \in Q_0$.

$$S(i) = ((S(i)_j)_{j\in Q_0}, (f_a)_{a\in Q_1})$$

$$\text{the vector space} \qquad S(i)_j = \begin{cases} k & \text{if } i = j \\ 0 & \text{if } i \neq j \end{cases}$$

$$\text{the k-linear map} \qquad f_a = 0 \quad \text{for any } a \in Q_1$$

The following shows two simple representations at each vertex of a quiver Q consisting of a single arrow going from a vertex 1 to vertex 2.

$$\text{If we have} \quad Q: \quad 1 \xrightarrow{a} 2 \;\; \text{then}$$

$$S(1): \qquad k \xrightarrow{0} 0$$

$$S(2): \qquad 0 \xrightarrow{0} k$$

What we did to obtain the simple representations, is put a 1 dimensional vector space k at the vertex we fixed in the beginning and then put 0 for every other vertex.

Moreover all the k-linear maps are zero maps. Note that any simple representation of a quiver Q is isomorphic to $S(i)$ for a unique $i \in Q_0$ [1].

The following example shows how to determine a kQ-module associated to a given representation of Q.

Example 3 Given the simple representation $S(i)$ for some $i \in Q_0$, the corresponding kQ-module is a quotient module

$$kQe_i/\langle \text{nontrivial paths starting at } i \rangle = k$$

where kQe_i is the vector space generated by all paths starting at the vertex i. This is a 1-dimensional vector space spanned by the trivial path e_i, and its kQ-module structure is defined by

$$\begin{cases} e_t m = m & \text{if } t = i \\ e_t m = 0 & \text{if } t \neq i \\ am = 0 & \text{for any } m \in k \text{ and } a \in Q_1 \end{cases}.$$

3 Bound Quivers

Definition 7 A **relation** in Q is a linear combination of paths

$$\sum_{i=1}^{l} c_i p_i$$

where $c_i \in k$ and p_i is a path such that $t(p_1) = \cdots = t(p_l)$ and $h(p_1) = \cdots = h(p_l)$. A quiver Q with a set R of relations is called a **bound quiver** (or a **quiver with relations**) denoted by (Q, R).

Like before, we can define a representation of a bound quiver (Q, R). The only difference is that we now have to incorporate the relations.

Definition 8 A **representation** of a bound quiver (Q, R) consists of a k-vector space V_i for each $i \in Q_0$ and a k-linear map $f_a : V_{t(a)} \to V_{h(a)}$ for each $a \in Q_1$ such that if $r \in R$ then $f_r = 0$.

For example, if $r = ac - 2bc \in R$ where a, b, c are arrows of Q, then the corresponding k-linear map is $f_r = f_a f_c - 2 f_b f_c = 0$.

Note that any finite set R of relations in Q generates a two sided ideal $I = \langle R \rangle$ in the path algebra kQ. Therefore the **path algebra for the bounded quiver** (Q, R) is kQ/I.

We have seen the Beilinson quiver in the beginning of the talk. Now we are going to add some relations to it. In general, the Beilinson quiver for $\mathbb{P}^n$ with a set of relations has geometric significance.

Example 4 (Bound Beilinson quiver for $\mathbb{P}^2$ [2]) Consider the case when $n = 2$. We have $Q_0 = \{0, 1, 2\}$ and $Q_1 = \{a_1, a_2, a_3, b_1, b_2, b_3\}$.

$$0 \;\substack{\xrightarrow{a_0}\\ \xrightarrow{a_1}\\ \xrightarrow{a_2}}\; 1 \;\substack{\xrightarrow{b_0}\\ \xrightarrow{b_1}\\ \xrightarrow{b_2}}\; 2$$

Set $R = \{a_0b_1 - a_1b_0, a_1b_2 - a_2b_1, a_2b_0 - a_0b_2\}$. Then the quotient algebra $kQ/\langle R\rangle$ is isomorphic to the endomorphism algebra of the vector bundle $\mathcal{O}_{\mathbb{P}^2} \oplus \mathcal{O}_{\mathbb{P}^2}(1) \oplus \mathcal{O}_{\mathbb{P}^2}(2)$ [2].

Let us make two final remarks before we end.

Remark 3 [2]

1. Finite dimensional representations of (Q,R) form a Abelian category denoted by rep(Q,R)
2. rep(Q,R) is equivalent to the category of finitely generated kQ/I-modules

References

1. M. Brion, *Representations of Quivers*, 2008, Available at: https://www-fourier.ujf-grenoble.fr/~mbrion/notes_quivers_rev.pdf
2. A. Craw, *Explicit Methods for Derived Categories of Sheaves*, 2007, available at: http://www.math.utah.edu/dc/tilting.pdf
3. A. Craw, *Quiver Representations in Toric Geometry*, 2008, available at: https://arxiv.org/pdf/0807.2191v1.pdf
4. T. Leinster, *The bijection between projective indecomposable and simple modules*, 2014, available at: https://arxiv.org/pdf/1410.3671.pdf

Semi-orthogonal Decompositions of Derived Categories

Yijia Liu

1 Preliminaries

We start with some definitions. The first question to answer is: what is orthogonality and why do we prefer "semi" to "full" orthogonality?

Definition 1 Let $\mathscr{A}$ be a full subcategory of a triangulated category $\mathscr{T}$:

1. the *left orthogonal of* $\mathscr{A}$ is

$$^{\perp}\mathscr{A} := \{T \in \mathscr{T} \mid \mathrm{Hom}(T, A[l]) = 0, \forall l \in \mathbb{Z}, \forall A \in \mathscr{A}\}; \tag{1}$$

2. the *right orthogonal of* $\mathscr{A}$ is

$$\mathscr{A}^{\perp} := \{T \in \mathscr{T} \mid \mathrm{Hom}(A[l], T) = 0, \forall l \in \mathbb{Z}, \forall A \in \mathscr{A}\}. \tag{2}$$

Definition 2 Let $\mathscr{A}$, $\mathscr{B}$ be full triangulated subcategories of a triangulated category $\mathscr{T}$. We say that $\mathscr{T}$ has an *orthogonal decomposition (OD)*, if

1. $\mathrm{Hom}(\mathscr{A}, \mathscr{B}) = 0 = \mathrm{Hom}(\mathscr{B}, \mathscr{A})$
2. Any object $T \in \mathscr{T}$ fits in a distinguished triangle (d.t.)

$$A \to T \to B \to \cdot \tag{3}$$

   with $A \in \mathscr{A}$, $B \in \mathscr{B}$.

We write $\mathscr{T} = \mathscr{A} \oplus \mathscr{B}$.

However, this is a too strong condition since we have:

Y. Liu (✉)
McGill University, Montreal, QC, Canada
e-mail: yijia.liu@mail.mcgill.ca

M. Ballard et al. (eds.), *Superschool on Derived Categories and D-branes*,
Springer Proceedings in Mathematics & Statistics 240,
https://doi.org/10.1007/978-3-319-91626-2_4

Remark 1 $D^b(X)$ has a nontrivial OD $\Leftrightarrow$ X is disconnected.

Proof ($\Leftarrow$) Suppose that X has two disjoint components Y and Z with corresponding embeddings i, j. Then for any $T \in D^b(X)$, the split exact sequence:

$$0 \to i_* i^* T \to T \to j_* j^* T \to 0 \tag{4}$$

induces the desired distinguished triangle.

($\Rightarrow$) It suffices to split $\mathcal{O}_X = \mathcal{O}_Y \oplus \mathcal{O}_Z$. This can be done using e.g. Lemma 4.10 of [16]. □

So OD detects whether X is connected and that is it. We have to weaken the definition.

Definition 3 Let $\mathscr{A}$, $\mathscr{B}$ be full triangulated subcategories of a triangulated category $\mathscr{T}$. We say that $\mathscr{T}$ has a *semiorthogonal decomposition (SOD)*, if

1. $\mathrm{Hom}(\mathscr{A}, \mathscr{B}) = 0$
2. Any object $T \in \mathscr{T}$ fits in a d.t.

$$A \to T \to B \to \cdot \tag{5}$$

with $A \in \mathscr{A}$, $B \in \mathscr{B}$.

We write $\mathscr{T} =< \mathscr{B}, \mathscr{A} >$. (Be careful with the order).

In general:

Definition 4 Let $\mathscr{A}_1, \mathscr{A}_2, \ldots, \mathscr{A}_n$ be full triangulated subcategories of a triangulated category $\mathscr{T}$. We say that $\mathscr{T}$ has a *semiorthogonal decomposition (SOD)*, if

1. $\mathrm{Hom}(\mathscr{A}_i, \mathscr{A}_j) = 0, i > j$
2. $\mathscr{A}_1, \mathscr{A}_2, \ldots, \mathscr{A}_n$ generate $\mathscr{T}$. i.e. For any object $T \in \mathscr{T}$, we have $T_i \in \mathscr{T}, i = 0, 1, \ldots, n$ and a sequence:

$$0 = T_n \to T_{n-1} \to \cdots \to T_1 \to T_0 = T \tag{6}$$

with $\mathrm{cone}(T_i \to T_{i-1}) \in \mathscr{A}_i$. (We can use $n - 1$ d.t. to generate any given object.)

We write $\mathscr{T} =< \mathscr{A}_1, \mathscr{A}_2, \ldots, \mathscr{A}_n >$.

The easiest way to build a SOD is to find an admissible subcategory.

Recall: A full triangulated subcategory $\mathscr{A} \subset \mathscr{T}$ is *right admissible* if the embedding functor $i : \mathscr{A} \to \mathscr{T}$ has a right adjoint $i^! : \mathscr{T} \to \mathscr{A}$. It is equivalent to: any $T \in \mathscr{T}$ fits in a d.t.

$$A \to T \to B \to \cdot \tag{7}$$

with $A \in \mathscr{A}$, $B \in \mathscr{A}^\perp$. (In this case $i^!(T) = A$ and is well-defined.)

There is a dual definition for left admissible. In fact, if $\mathscr{T}$ admits a Serre functor, then these two notions are equivalent, by Bondal and Kapranov [3]. This is true for our first example ($S = (\_ \otimes \omega_X)[n]$) and for derived categories of modules over finite dimensional finite global dimensional algebra (e.g. path algebra of any finite acyclic quiver).

We immediately get:

Remark 2 $\mathscr{A}$ is admissible $\Rightarrow {}^{\perp}\mathscr{A}, \mathscr{A}^{\perp}$ are full triangulated subcategories of $\mathscr{T}$ (actually admissible with the existence of a Serre functor), which implies the existence of two SODs of $\mathscr{T}$:

$$\mathscr{T} = < \mathscr{A}, {}^{\perp}\mathscr{A} > \tag{8}$$

and

$$\mathscr{T} = < \mathscr{A}^{\perp}, \mathscr{A} > . \tag{9}$$

Moreover:

Remark 3 $\mathscr{A}_1, \ldots, \mathscr{A}_n$ semiorthogonal admissible subcategories $\Rightarrow < \mathscr{A}_1, \ldots, \mathscr{A}_l >$ is admissible and can be extended to a SOD of $\mathscr{T}$, for each $0 \leq l \leq n$:

$$\mathscr{T} = < \mathscr{A}_1, \ldots, \mathscr{A}_l, {}^{\perp} < \mathscr{A}_1, \ldots, \mathscr{A}_l > \cap < \mathscr{A}_{l+1}, \ldots, \mathscr{A}_n >^{\perp}, \mathscr{A}_{l+1}, \ldots, \mathscr{A}_n > . \tag{10}$$

The simplest admissible subcategory is equivalent to $\mathrm{D}^{\mathrm{b}}(\mathrm{Vec}_k^{\mathrm{f.d.}})$, which is generated by an exceptional object. Let's make a dumb observation. Every functor

$$\mathrm{D}^{\mathrm{b}}(\mathrm{Vec}_k^{\mathrm{f.d.}}) \to \mathscr{T} \tag{11}$$

is determined by the image of the 0-complex k, say $E \in \mathscr{T}$, denoted by φ_E. Note $\varphi_E(V^\bullet) = V^\bullet \otimes E$. By tensor-hom adjunction, its right adjoint is given by

$$\varphi_E^!(F) = \mathrm{Hom}^\bullet(E, F) = \oplus \mathrm{Hom}(E, F[k])[-k], \tag{12}$$

with trivial differentials. Therefore φ_E is a fully faithful embedding $\Leftrightarrow \varphi_E^! \varphi_E \cong Id \Leftrightarrow \mathrm{Hom}^\bullet(E, E) = k$ as 0-complex, since $\varphi_E^! \varphi_E(V^\bullet) = \mathrm{Hom}^\bullet(E, V^\bullet \otimes E) = \mathrm{Hom}^\bullet(E, E) \otimes V^\bullet$.

In this spirit, we define:

Definition 5 An object E is *exceptional* if

$$\mathrm{Hom}(E, E[l]) = \begin{cases} k \text{ if } l = 0 \\ 0 \text{ if } l \neq 0 \end{cases} \tag{13}$$

Adding semiorthogonality, we define:

Definition 6 A collection of exceptional objects $E_\bullet = E_1, \ldots, E_r$ is *exceptional of length r* if

$$\mathrm{Hom}(E_i, E_j[l]) = 0, \text{ for } i > j. \tag{14}$$

- It is *strong* if in addition

$$\mathrm{Hom}(E_i, E_j[l]) = 0, \text{ for } l \neq 0. \tag{15}$$

Recall: For an exceptional collection we have a SOD:

$$\mathscr{T} =< \mathscr{A}, E_1, \ldots, E_r >, \tag{16}$$

where $\mathscr{A} =< E_1, \ldots, E_r >^{\perp}$.
- It is *full* if $\mathscr{A} = 0$.

2 Two Classes of Examples

Let us return to our two classes of examples. We will start with the second one since that is easier.

2.1 *Derived Categories of Bound Quivers* $\mathbf{D}^{\mathrm{b}}(Q, I)$

Consider the case when Q is finite, ordered: $Q_0 = \{1, 2, \ldots, n\}$ and $s(a) < t(a)$, for any $a \in Q_1$. In this case, Q is acyclic and $A = kQ/I$ is called a *quiver algebra*. Let e_l be the path of length zero at vertex l. As a right A-module over itself, A can be decomposed to

$$A = \bigoplus_{q \in Q_0} P_q, \tag{17}$$

with $P_q = e_q A$, where P_q are indecomposable projective modules. We have for any right A-module M a natural isomorphism:

$$\mathrm{Hom}(P_q, M) \cong M_q := M e_q. \tag{18}$$

Indeed, $P_1, P_2, \ldots, P_n$ is a full strong exceptional collection of $\mathrm{D}^{\mathrm{b}}(Q, I)$.

2.2 *Derived Categories of Coherent Sheaves* $\mathbf{D}^{\mathrm{b}}(X)$

Let us return back to the first class of examples. In general, it is hard to find even an exceptional object, let alone a full strong exceptional collection.

Remark 4 • $\mathscr{O}_X$ is exceptional $\Leftrightarrow h^{i,0}(X) = 0$ for $i > 0$. In this case, any line bundle is exceptional. $\Rightarrow$ Every Fano has a SOD.

On the contrary,

- No CY has a SOD. Indeed, suppose it has one as $< \mathscr{A}, \mathscr{B} >$, with $\mathrm{Hom}(\mathscr{B}, \mathscr{A}) = 0$. Since its Serre functor is $(\_ \otimes \omega_X)[n] = [n]$, $\mathrm{Hom}(\mathscr{A}, \mathscr{B}) = \mathrm{Hom}(\mathscr{B}, \mathscr{A}[n])^\star = 0$ and therefore it is an OD $A \oplus B$.

Having a full exceptional collection is even more restrictive. For curves, only $\mathbb{P}^1$ has SODs (and a full exceptional collection). For surfaces, King [11] showed that all rational surfaces admit full exceptional collections and they are conjecturally the only surfaces doing so. Kawamata [10] proved any toric variety has a full exceptional collection. The classification is still beyond touch at the moment.

Some constraints for X to admit a full exceptional collection can be seen in the following properties.

Proposition 1 *If X admits a full exceptional collection $E_\bullet = E_1, E_2, \ldots, E_r$ of length r, then:*

1. *$\mathrm{H}^{p,q}(X) = 0$, $\forall p \neq q$ and $\chi(X) = \sum h^{p,p} = r$. In particular, $p_g = q = 0$ for surfaces.*
2. *The Grothendieck group $\mathrm{K}_0(X)$ is free abelian of rank r with basis the isomorphism classes of exceptional objects $[E_i]$.*

Proof 1. We apply Hochschild–Konstant–Rosenberg's theorem:

$$\mathrm{HH}_\bullet(X) = \bigoplus_{p-q=i} \mathrm{H}^{p,q}(X) \tag{19}$$

and used the fact that Hochschild homology is additive w.r.t. SOD (see [12]). The result follows the fact that each piece is isomorphic to $\mathrm{D^b}(pt)$.

2. Easy to prove that K_0 is additive w.r.t. SOD. Then notice that

$$\mathrm{K}_0(< E_i >) = \mathrm{K}_0(\mathrm{D^b}(pt)) = \mathbb{Z}\,. \tag{20}$$

□

Remark 5 These properties are necessary conditions, however they are not sufficient. For instance, it was shown by Böhning, Graf von Bothmer, Katzarkov and Sosna [7] that Barlow surface admits a exceptional collection of length 11 which gives rise to a basis of K_0. But it is not full since its orthogonal is a phantom category!

Well, how do we determine when an exceptional collection is full or not? We can follow the ideal of Beilinson to find a resolution of the diagonal, more precisely:

Lemma 1 *Let $E_\bullet = E_1, E_2, \ldots, E_n$ be an exceptional collection of sheaves on X. Assume there exists a resolution of the diagonal $\mathscr{O}_\Delta$ on $X \times X$:*

$$0 \to E_1 \boxtimes F_1 \to \cdots \to E_n \boxtimes F_n \to \mathscr{O}_\Delta \to 0. \tag{21}$$

(Here $F_\bullet$ is not necessarily exceptional.) Then $E_\bullet$ is full.

Recall that:

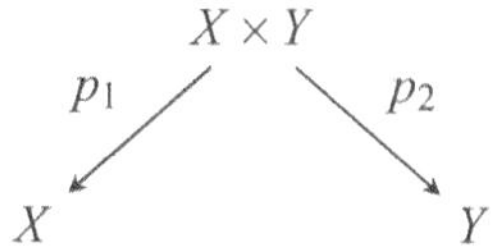

- Exterior tensor product:

$$E \boxtimes F := p_1^* E \otimes p_2^* F. \tag{22}$$

- Fourier–Mukai transform w.r.t. $\mathscr{F}$ on $X \times Y$:

$$\mathrm{FM}_{\mathscr{F}} : \mathrm{D^b}(Y) \to \mathrm{D^b}(X) \tag{23}$$

$$\mathscr{E} \mapsto p_{1*}(\mathscr{F} \otimes p_2^*(\mathscr{E})). \tag{24}$$

Proof This resolution is equivalent to a list of short exact sequences:

$$0 \to H_{k-1} \to E_k \boxtimes F_k \to H_k \to 0, \qquad k = 1, \dots, n, \tag{25}$$

with H_i sheaves on $X \times X$, $H_0 = 0$, $H_n = \mathscr{O}_\Delta$. We use an induction on k to show that $\forall C \in \mathrm{D^b}(X)$, $\mathrm{FM}_{H_k}(C) \in < E_1, \dots, E_k >$. When $k = 0$ it is trivial. In the inductive step, the above short exact sequence induces a distinguished triangle

$$\mathrm{FM}_{H_{k-1}}(C) \to \mathrm{FM}_{E_k \boxtimes F_k}(C) \to \mathrm{FM}_{H_k}(C) \to \cdot, \tag{26}$$

where $\mathrm{FM}_{H_{k-1}}(C) \in < E_1, \dots, E_{k-1} >$ by inductive assumption and $\mathrm{FM}_{E_k \boxtimes F_k}(C) = E_k \otimes \Gamma(F_k \otimes C)$. Therefore $\mathrm{FM}_{H_k}(C) \in < E_1, \dots, E_k >$. In particular when $k = n$, $\mathrm{FM}_{H_n} = \mathrm{FM}_{\mathscr{O}_n} = Id_{\mathrm{D^b}(X)} \Rightarrow C \in < E_1, E_2, \dots, E_n >$. Since C is arbitrary, $E_\bullet$ is full, as desired. □

We have the following theorem for projective spaces:

Theorem 1 (Beilinson [2]) *For $\mathscr{F} = \mathscr{O}(-1) \boxtimes \Omega^1(1)$ on $\mathbb{P}^n \times \mathbb{P}^n$, there exists a resolution of $\mathscr{O}_\Delta$:*

$$0 \to \wedge^n \mathscr{F} \to \cdots \wedge^2 \mathscr{F} \to \mathscr{F} \to \mathscr{O} \boxtimes \mathscr{O} \to \mathscr{O}_\Delta \to 0. \tag{27}$$

Proof Find a global section of $\mathscr{O}(1) \boxtimes T(-1)$ ($s = \sum_{i=0}^{n} x_i \boxtimes \frac{\partial}{\partial y_i}$) with zero locus Δ. Then apply the Koszul resolution. □

Note that $\wedge^k \mathscr{F} = \wedge^k(\mathscr{O}(-1) \boxtimes \Omega^1(1)) = \mathscr{O}(-k) \boxtimes \Omega^k(k)$, here $\Omega^k(k) := \wedge^k(\Omega^1(1))$. Therefore as a corollary we have:

Corollary 1 *$\mathcal{O}(-n), \mathcal{O}(-n+1), \ldots, \mathcal{O}(1), \mathcal{O}$ is a full strong exceptional collection for $D^b(\mathbb{P}^n)$.*

Proof It suffices to check it is a strong exceptional collection. Let $E_i = \mathcal{O}(i-1-n), i = 1, \ldots, n+1$.

$$\begin{aligned}
\mathrm{Hom}_D(E_i, E_j[l]) &= \mathrm{Hom}_D(\mathcal{O}(i-1-n), \mathcal{O}(j-1-n)[l]),\\
&= \mathrm{Ext}^l_{\mathcal{O}}(\mathcal{O}(i-1-n), \mathcal{O}(j-1-n)),\\
&= R^l\,\mathrm{Hom}_{\mathcal{O}}(\mathcal{O}(i-1-n), \mathcal{O}(j-1-n)),\\
&= R^l\Gamma(\mathbb{P}^n, \mathcal{O}(j-i)),\\
&= \mathrm{H}^l(\mathbb{P}^n, \mathcal{O}(j-i)), \qquad (28)\\
&= \begin{cases} 0 & \text{if } l \neq 0\\ \Gamma(\mathbb{P}^n, \mathcal{O}(j-i)) = Sym^{j-i}V^\star & \text{if } l = 0,\end{cases}\\
&= \begin{cases} k \text{ if } l = 0 \text{ and } i = j\\ 0 \text{ if } l = 0, i > j \text{ or } l \neq 0.\end{cases}
\end{aligned}$$

□

Remark 6 One can check similarly its dual: $\Omega^n(n), \Omega^{n-1}(n-1), \ldots, \Omega^1(1), \mathcal{O}$ is also a full strong exceptionally collection for $D^b(\mathbb{P}^n)$.

2.3 Actions on SODs

We can see from above remark that a derived category, admitting a full exceptional collection, may have many other full exceptional collections, or more generally many SODs. Precisely speaking, there are two groups acting on the set of SODs of a triangulated category $\mathcal{T}$: Aut $\mathcal{T}$ and a braid group of so-called mutations.

Let us now define a mutation. Roughly a mutation on a SOD removes an object in it and add a new object at some other place. Consider an admissible subcategory $\mathcal{B}$ of $\mathcal{T}$. Recall that we have a pair of SOD of $\mathcal{T}$: $\mathcal{T} =< \mathcal{B}^\perp, \mathcal{B} >=< \mathcal{B}, {}^\perp\mathcal{B} >$. We need to define an action, denoted $R_{\mathcal{B}}$, that sends the first SOD to the second one, namely an equivalence of categories $R_{\mathcal{B}} : \mathcal{B}^\perp \to {}^\perp\mathcal{B}$. We denote $\mathcal{A} := \mathcal{B}^\perp$ and $R_{\mathcal{B}}\mathcal{A} := {}^\perp\mathcal{B}$. Since $\mathcal{B}$ is admissible, so is $R_{\mathcal{B}}\mathcal{A}$. Therefore the embedding functor $i : R_{\mathcal{B}}\mathcal{A} \to \mathcal{T}$ has a right adjoint $i^! : \mathcal{T} \to R_{\mathcal{B}}\mathcal{A}$. Restrict it to $\mathcal{A}$, we define:

$$R_{\mathcal{B}} := i^!|_{\mathcal{A}} : \mathcal{A} \to R_{\mathcal{B}}\mathcal{A}. \qquad (29)$$

Remark 7 $R_{\mathcal{B}}$ is an equivalence of categories. Indeed its inverse is given by the left adjoint of the embedding functor $j : \mathcal{A} \to \mathcal{T}$ restricted to $R_{\mathcal{B}}\mathcal{A}$:

$$j^*|_{R_{\mathcal{B}}\mathcal{A}} : R_{\mathcal{B}}\mathcal{A} \to \mathcal{A}. \qquad (30)$$

Definition 7 The admissible subcategory $R_{\mathscr{B}}\mathscr{A}$ together with the functor $R_{\mathscr{B}}$ realizes a transformation from $\mathscr{B}^{\perp}$ to ${}^{\perp}\mathscr{B}$ and is called the *right mutation of* $\mathscr{A}$ *through* $\mathscr{B}$.

Remark 8 $R_{\mathscr{B}}$ acting on the SOD $< \mathscr{A}, \mathscr{B} >$ is the SOD $< \mathscr{B}, R_{\mathscr{B}}\mathscr{A} >$.

Remark 9 We can similarly define the left mutation of $\mathscr{B}$ through $\mathscr{A}$. The explicit image of an object under the mutation functor is given by finding a distinguished triangle.

Now we can define the mutations on exceptional collections.

Definition 8 Let $E_\bullet = E_1, \ldots, E_r$ be an exceptional collection. For each $i = 1, \ldots, r-1$, we define the *left/right ith mutation of* $E_\bullet$ by:

$$L_i(E_\bullet) = E_1, \ldots, E_{i-1}, L_{E_i}E_{i+1}, E_i, E_{i+2}, \ldots, E_n, \tag{31}$$
$$R_i(E_\bullet) = E_1, \ldots, E_{i-1}, E_{i+1}, R_{E_{i+1}}E_i, E_{i+2}, \ldots, E_n. \tag{32}$$

Remark 10 One can check the following properties:

- $L_i(E_\bullet)$, $R_i(E_\bullet)$ are exceptional collections. They are full if $E_\bullet$ is.
- The mutations have relations:

$$L_iR_i = R_iL_i = Id, \tag{33}$$
$$L_{i+1}L_iL_{i+1} = L_iL_{i+1}L_i, \tag{34}$$
$$R_{i+1}R_iR_{i+1} = R_iR_{i+1}R_i, \tag{35}$$
$$L_iL_j = L_jL_i, \ R_iR_j = R_jR_i, \text{ for } |i-j| \geqslant 2. \tag{36}$$

Therefore generators L_i and R_i with these relations form a braid group.

Return back to our two classes of examples.

Example 1 ($\mathrm{D^b}(Q, I)$) We have seen that $P_1, \ldots, P_n$ is a full strong exceptional collection of $\mathrm{D^b}(Q, I)$. Let $\mathscr{A} =< P_1, \ldots, P_{n-1} >$. Then $\mathrm{D^b}(Q, I) =< \mathscr{A}, P_n >$ under the action of $L_{\mathscr{A}}$ is $< L_{\mathscr{A}}P_n, \mathscr{A} >$, where

$$L_{\mathscr{A}} = i^*|_{{}^{\perp}\mathscr{A}} : {}^{\perp}\mathscr{A} \to \mathscr{A}^{\perp}, \quad i : \mathscr{A}^{\perp} \hookrightarrow \mathrm{D^b}(Q, I). \tag{37}$$

We claim that $L_{\mathscr{A}}P_n = i^*P_n = S_n$. To define i^*, we consider the distinguished triangle:

$$0 \to \oplus_{k=1}^{n-1}(P_n)_k \to P_n \to S_n \to 0. \tag{38}$$

Note that $\oplus_{k=1}^{n-1}(P_n)_k \in \mathscr{A} \subset {}^{\perp}(\mathscr{A}^{\perp})$, it suffices to check that $S_n \in \mathscr{A}^{\perp}$. Indeed

$$\mathrm{Hom}(P_i, S_n[l]) = \begin{cases} \mathrm{Hom}(P_i, S_n) \cong (S_n)_i = e_nAe_ne_i = 0 \text{ if } l = 0 \\ \mathrm{Ext}^l(P_i, S_n) = 0 \text{ since } \mathrm{P}_i \text{is projective if } l \neq 0 \end{cases} \tag{39}$$

for $i < n$ and $\forall\, l \in \mathbb{Z}$.

Therefore we have a new full exceptional collection $S_n, P_1, \ldots, P_{n-1}$. However it is no longer strong, since $\mathrm{Hom}(P_1, S_n[1]) = \mathrm{Hom}(S_1, S_n[1]) = \mathrm{Ext}^1(S_1, S_n) =$ the number of arrows from 1 to $n \neq 0$. So strongness is not preserved by mutations! Continuing the mutations, we have a series of full (not strong) exceptional collections:

$$S_n, S_{n-1}, P_1, \ldots, P_{n-2} \tag{40}$$

$$S_n, S_{n-1}, S_{n-2}, P_1, \ldots, P_{n-3} \tag{41}$$

$$\ldots\ldots \tag{42}$$

$$S_n, S_{n-1}, \ldots, S_1 \tag{43}$$

Example 2 $(\mathrm{D}^{\mathrm{b}}(\mathbb{P}^n))$ Start with the full strong exceptional collection

$$\mathscr{O}, \mathscr{O}(1), \ldots, \mathscr{O}(n), \tag{44}$$

one could check that

$$L_{<\mathscr{O},\ldots,\mathscr{O}(k)>}\mathscr{O}(k+1) \cong \Omega^{k+1}(k+1)[k+1], \text{ for } k = 0, \ldots, n-1. \tag{45}$$

Thus step by step we have another full (and actually strong) exceptional collection

$$\Omega^n(n)[n], \ldots, \Omega^1(1)[1], \mathscr{O}. \tag{46}$$

In fact it is proved that any full exceptional collection of $\mathrm{D}^{\mathrm{b}}(\mathbb{P}^n)$ arises in this way.

2.4 More Examples

From the well-known SODs of $\mathrm{D}^{\mathrm{b}}(X)$, we can also construct SODs of some new categories, such as the derived categories of its projective bundle and birational transforms. Roughly their SODs have the following correspondence:

- Projective bundle of rank $n \leftrightarrow$ "add" n-copies of the base under twists.
- Birational transforms $\begin{cases} \text{Blowup} \leftrightarrow \text{"add" items} \\ \text{Flip} \quad \leftrightarrow \text{"remove" items} \\ \text{Flop} \quad \leftrightarrow \text{equivalent categories} \end{cases}$

Consider a vector bundle of rank $n+1$, $\pi : E \to X$. We associate it with a projective bundle $p : P := \mathbb{P}(E) \to X$. We have:

Theorem 2 (Orlov [14]) *The pullback functor* $p^* : \mathrm{D}^{\mathrm{b}}(X) \to \mathrm{D}^{\mathrm{b}}(P)$ *is fully faithful and there exists a SOD*

$$\mathrm{D}^{\mathrm{b}}(P) = < \mathrm{D}^{\mathrm{b}}(X)_{-n}, \mathrm{D}^{\mathrm{b}}(X)_{-n+1}, \ldots, \mathrm{D}^{\mathrm{b}}(X)_{-1}, \mathrm{D}^{\mathrm{b}}(X)_0 > . \tag{47}$$

Here $\mathrm{D^b}(X)_0 = p^* \mathrm{D^b}(X)$, $\mathrm{D^b}(X)_{-k} = p^* \mathrm{D^b}(X) \otimes \mathcal{O}_P(k)$ *for* $k = 1, \ldots, n$.
In particular, if X *admits a full exceptional collection, so does* P.

Applying this theorem on trivial bundles we have:

Corollary 2 *If* X, Y *are smooth projective varieties admitting full exceptional collections* $E_1, \ldots, E_m$ *and* $F_1, \ldots, F_n$ *respectively. Then* $\{E_i \boxtimes F_j\}$ *is a full exceptional collection on* $X \times Y$ *with compatible order.*

Example 3 $\mathbb{P}^n \times \mathbb{P}^n$ admits a full exceptional collection $\mathcal{O}(i, j) := \mathcal{O}(i) \boxtimes \mathcal{O}(j)$, $0 \leqslant i, j \leqslant n$. In particular, $\mathrm{D^b}(\mathbb{P}^1 \times \mathbb{P}^1) =< \mathcal{O}, \mathcal{O}(0, 1), \mathcal{O}(1, 0), \mathcal{O}(1, 1) >$.

Now we turn to the case of blowups. Let $i : Y \hookrightarrow X$ denote a closed smooth subvariety of codimension $c = \mathrm{codim}(Y|X)$ in a smooth projective variety X. The blowup of X in Y is a fiber square:

$$\begin{array}{ccc} \mathbb{P}(\mathcal{N}_{Y|X}) = \widetilde{Y} & \overset{j}{\hookrightarrow} & \widetilde{X} \\ p \downarrow & & \downarrow \pi \\ Y & \overset{i}{\hookrightarrow} & X \end{array}$$

Here i, j are embeddings of smooth varieties, $p : \mathbb{P}(\mathcal{N}_{Y|X}) \to Y$ is a projective bundle of rank $c - 1$ and $\pi : \widetilde{X} \to X$ is a natural projective morphism.

Let $\mathcal{O}(1)$ be the canonical line bundle on $\widetilde{Y}$. By Theorem 2:

$$\mathrm{D^b}(\widetilde{Y}) =< \mathrm{D^b}(Y)_{-(c-1)}, \ldots, \mathrm{D^b}(Y)_{-1}, \mathrm{D^b}(Y)_0 >, \tag{48}$$

where $\mathrm{D^b}(Y)_0 = p^* \mathrm{D^b}(Y)$, $\mathrm{D^b}(Y)_k = p^* \mathrm{D^b}(Y) \otimes \mathcal{O}(k)$. The embedding functor $j_* : \mathrm{D^b}(\widetilde{Y}) \to \mathrm{D^b}(\widetilde{X})$ is not full, however we restrict it to $\mathrm{D^b}(Y)_k$, then it is full. We have the following Blowup formula:

Theorem 3 (Orlov [14])

1. $\pi^* : \mathrm{D^b}(X) \to \mathrm{D^b}(\widetilde{X})$ *is a fully faithful embedding.*
2. $j_* : \mathrm{D^b}(\widetilde{Y}) \to \mathrm{D^b}(\widetilde{X})$ *restricted to* $\mathrm{D^b}(Y)_k$ *is a fully faithful embedding.*
3. $\mathrm{D^b}(\widetilde{X}) =< \widetilde{D^bY}_{-(c-1)}, \ldots, \widetilde{D^bY}_{-1}, \mathrm{D^b}(X)_0 >$,
   where $\mathrm{D^b}(X)_0 = \pi^* \mathrm{D^b}(X)$, $\widetilde{D^bY}_{-k} = j_* \mathrm{D^b}(Y)_k$.

In particular, if X *and* Y *both admit full exceptional collections, so does* $\widetilde{X}$.

Example 4 ($\widetilde{\mathbb{P}^2_p}$) We consider $\mathbb{P}^2$ blown up at one point ($c = 2$). Let E be the exceptional divisor. Then

$$\mathrm{D^b}(\widetilde{\mathbb{P}^2_p}) \cong < \mathcal{O}_E(-1), \mathcal{O}_{\mathbb{P}^2}(-2), \mathcal{O}_{\mathbb{P}^2}(-1), \mathcal{O}_{\mathbb{P}^2} > . \tag{49}$$

3 Connections

Let us now try to establish a connection between our two classes of examples. For a smooth projective variety X, every full strong exceptional collection $E_\bullet$ defines so-called tilting sheaf $T = \bigoplus E_i$. Let $A = \mathrm{End}_{\mathcal{O}_X}(T)$, then we can understand $\mathrm{D^b}(X)$ via $\mathrm{D^b}(\text{mod-}A)$, which is a significant simplification. Because it has finite length and can be visualized via the corresponding bound quiver (Q, I).

First, we define:

Definition 9 $T \in \mathrm{Coh}(X)$ is a *tilting sheaf* if

1. $A := \mathrm{End}_{\mathcal{O}_X}(T)$, called the *tilting algebra*, has finite global dimension (i.e. maximal projective dimension of any object in mod- A is finite).
2. $\mathrm{Ext}^k_{\mathcal{O}_X}(T, T) = 0, \forall\, k > 0$.
3. T generates $\mathrm{D^b}(X)$.

If T is locally free, then it is called a *tilting bundle*.

Remark 11 Every full strong exceptional collection $E_\bullet = E_1, \ldots, E_r$ of vector bundles defines a tilting bundle $T = \bigoplus_{i=1}^r E_i$. Indeed, (2) and (3) are trivial. To see (1), notice that

$$\mathrm{Hom}(E_i, E_j) = H^0(X, E_j \otimes E_i^{-1}) \tag{50}$$

$$= \begin{cases} k^d & \text{if } i < j \\ k & \text{if } i = j \\ 0 & \text{if } i > j \end{cases} \tag{51}$$

$\Rightarrow A \cong$ an algebra of lower triangle matrix.

$\Rightarrow A$ is a finite dimensional and finite global dimensional algebra.

The correspondence lies in the following theorem, due independently to Baer and Bondal.

Theorem 4 (Baer [1], Bondal [5]) *The functor* $\mathrm{Hom}_{\mathcal{O}_X}(T, \_) : \mathrm{Coh}(X) \to$ *mod-A induces an equivalence of triangulated categories*

$$R\,\mathrm{Hom}(T, \_) : \mathrm{D^b}(X) \to \mathrm{D^b}(\textit{mod-}\,A) \tag{52}$$

with quasi-inverse

$$\_ \otimes^L_A T : \mathrm{D^b}(\textit{mod-}\,A) \to \mathrm{D^b}(X). \tag{53}$$

In addition, since the tilting algebra is finite dimensional over k, we have:

Proposition 2 *When A is a tilting algebra, we have an isomorphism* $A \cong kQ/ < I >$, *a path algebra of a bound quiver* (Q, I), *where*

1. *The vertices of Q correspond to* E_i.

2. *The edges from i to j correspond to a basis of the vector space* $\mathrm{Hom}_{\mathcal{O}_X}(E_i, E_j)$.
3. *Two paths are equal (their different is in I) if the corresponding morphisms are equal.*

Remark 12 • Q constructed above is finite and ordered.
• $\mathrm{D^b}(X) \cong \mathrm{D^b}(\text{mod-} A) \cong \mathrm{D^b}(\text{mod-} kQ/ < I >) \cong \mathrm{D^b}(\mathrm{rep}(Q, I)^{op})$.

Example 5 ($\mathbb{P}^1$) $\mathrm{D^b}(\mathbb{P}^1) = < E_1, E_2 > = < \mathcal{O}, \mathcal{O}(1) >$. $\mathrm{Hom}_{\mathcal{O}}(E_1, E_2) = \Gamma(\mathcal{O}(1)) = kx + ky$. Therefore

$$A = \begin{bmatrix} k & 0 \\ k^2 & k \end{bmatrix} \tag{54}$$

and we obtain the Kronecker quiver:

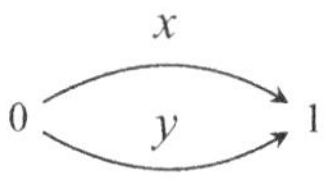

Example 6 ($\mathbb{P}^2$) $\mathrm{D^b}(\mathbb{P}^2) = < E_1, E_2, E_3 > = < \mathcal{O}, \mathcal{O}(1), \mathcal{O}(2) >$. $\mathrm{Hom}_{\mathcal{O}}(E_1, E_2) = \mathrm{Hom}_{\mathcal{O}}(E_2, E_3) = \Gamma(\mathcal{O}(1)) = kx + ky + kz$. $\mathrm{Hom}_{\mathcal{O}}(E_1, E_3) = \Gamma(\mathcal{O}(2)) = kx^2 + ky^2 + kz^2 + kxy + kxz + kyz$. Therefore

$$A = \begin{bmatrix} k & 0 & 0 \\ k^3 & k & 0 \\ k^6 & k^3 & k \end{bmatrix} \tag{55}$$

and we obtain the quiver

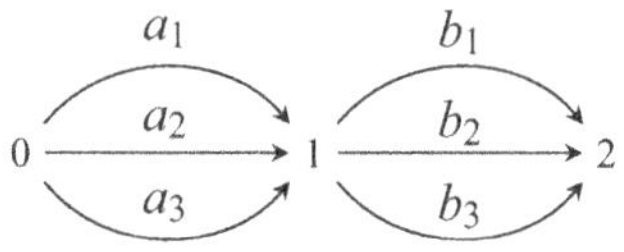

with relations $a_i b_j = a_j b_i$, $i, j \in \{1, 2, 3\}$.

More generally:

Example 7 ($\mathbb{P}^n$) We obtain the Beilinson quiver:

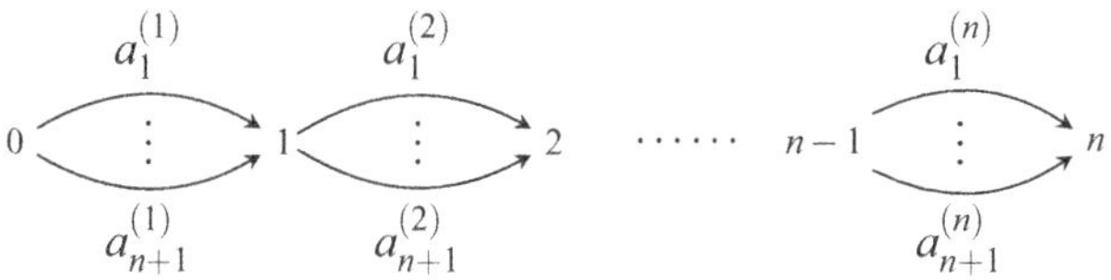

with relations $a_i^{(l)} a_j^{(l+1)} = a_j^{(l)} a_i^{(l+1)}$, $i, j \in \{1, 2, \ldots, n+1\}$, $l \in \{1, \ldots, n-1\}$.

Example 8 ($\mathbb{P}^1 \times \mathbb{P}^1$) $\quad \mathrm{D}^{\mathrm{b}}(\mathbb{P}^1 \times \mathbb{P}^1) = < E_1, E_2, E_3, E_4 > = < \mathcal{O}, \mathcal{O}(0,1), \mathcal{O}(1,0), \mathcal{O}(1,1) >$.

$$\mathrm{Hom}(E_2, E_3) = 0, \tag{56}$$

$$\mathrm{Hom}(E_1, E_2) = kx_1 + ky_1 = \mathrm{Hom}(E_3, E_4), \tag{57}$$

$$\mathrm{Hom}(E_1, E_3) = kx_2 + ky_2 = \mathrm{Hom}(E_2, E_4), \tag{58}$$

$$\mathrm{Hom}(E_1, E_4) = kx_1y_1 + kx_1y_2 + kx_2y_1 + kx_2y_2. \tag{59}$$

$$\Rightarrow A = \begin{bmatrix} k & 0 & 0 & 0 \\ k^2 & k & 0 & 0 \\ k^2 & 0 & k & 0 \\ k^4 & k^2 & k^2 & k \end{bmatrix} \tag{60}$$

and the bound quiver

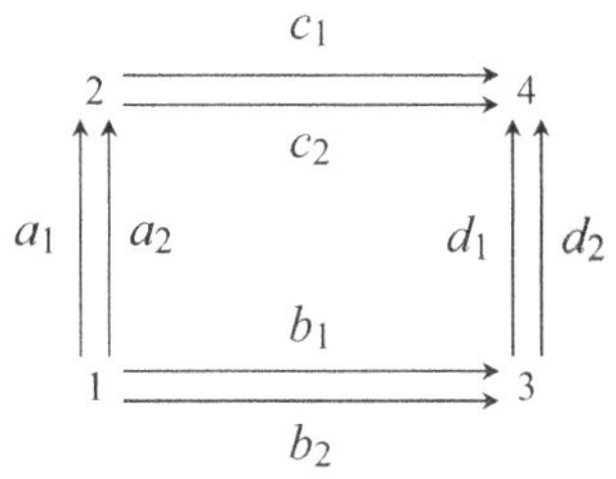

with relations $a_i c_j = b_j d_i$, $i, j \in \{1, 2\}$.

Example 9 ($\widetilde{\mathbb{P}^2_p}$) $\quad X = \widetilde{\mathbb{P}^2_p} = \mathbb{P}(\mathcal{O}_{\mathbb{P}^1} \oplus \mathcal{O}_{\mathbb{P}^1}(1))$. For $(k, l) \in \mathbb{Z}^2$, $\mathcal{O}_X(k, l) := \mathcal{O}_X(kD_1 + lD_4) \in \mathrm{Pic}\, X$. Then $\mathrm{D}^{\mathrm{b}}(X) = < \mathcal{O}, \mathcal{O}(1,0), \mathcal{O}(0,1), \mathcal{O}(1,1) >$.

$$\Rightarrow A = \begin{bmatrix} k & 0 & 0 & 0 \\ k^2 & k & 0 & 0 \\ k^3 & k & k & 0 \\ k^6 & k^3 & k^2 & k \end{bmatrix} \tag{61}$$

and the bound quiver

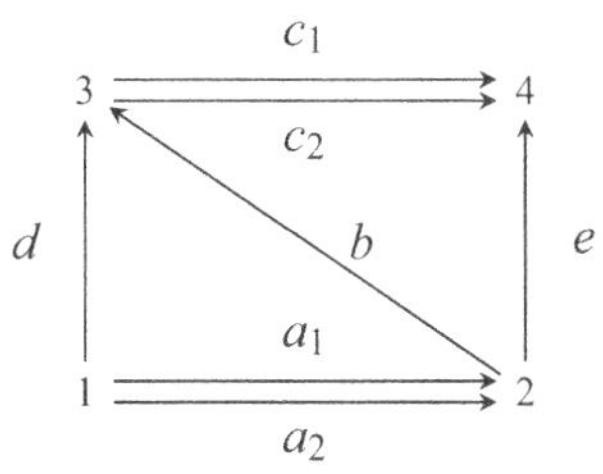

with relations $dc_i = a_i e$, $i \in \{1, 2\}$, $a_2 b c_1 = a_1 b c_2$.

Acknowledgements The author thanks the University of Alberta and the organizers for the hospitality and excellent opportunity to participate in the superschool. For detailed treatments of this survey, see [4, 6, 8, 9, 13, 15, 17].

References

1. Dagmar Baer. Tilting sheaves in representation theory of algebras. Manuscripta Math., 60(3):323–347, 1988.
2. A.A. Beilinson. Coherent sheaves on $\mathbb{P}_n$ and problems of linear algebra. *Functional Analysis and Its Applications 12 (1978), no. 3, 214–216.*, 1978.
3. A. I. Bondal and M. M. Kapranov. Representable functors, Serre functors, and reconstructions. *Izv. Akad. Nauk SSSR Ser. Mat.*, 53(6):1183–1205, 1337, 1989.
4. Alexei Bondal and Dmitri Orlov. Semiorthogonal decomposition for algebraic varieties. *arXiv preprint alg-geom/9506012*, 1995.
5. A. I. Bondal. Helices, representations of quivers and Koszul algebras. In *Helices and vector bundles*, volume 148 of *London Math. Soc. Lecture Note Ser.*, pages 75–95. Cambridge Univ. Press, Cambridge, 1990.
6. Alexei I Bondal. Representation of associative algebras and coherent sheaves. *Mathematics of the USSR-Izvestiya*, 34(1):23, 1990.
7. Christian Böhning, Hans-Christian Graf von Bothmer, Ludmil Katzarkov, and Pawel Sosna. Determinantal Barlow surfaces and phantom categories. *arXiv preprint* arXiv:1210.0343, 2012.
8. Alastair Craw. Quiver representations in toric geometry. *arXiv preprint* arXiv:0807.2191, 2008.
9. D. Huybrechts. Fourier-Mukai transforms in algebraic geometry. Oxford Mathematical Monographs. The Clarendon Press, Oxford University Press, Oxford, 2006.
10. Yujiro Kawamata. Derived categories of toric varieties. Michigan Math. J., 54(3):517–535, 2006.
11. Alastair King. Tilting bundles on some rational surfaces. *Unpublished manuscript*, 1997.
12. Alexander Kuznetsov. Hochschild homology and semiorthogonal decompositions. *arXiv preprint* arXiv:0904.4330, 2009.
13. Alexander Kuznetsov. Semiorthogonal decompositions in algebraic geometry. *arXiv preprint* arXiv:1404.3143, 2014.
14. Dmitri Olegovich Orlov. Projective bundles, monoidal transformations, and derived categories of coherent sheaves. Russian Academy of Sciences. Izvestiya Mathematics, 41(1):133, 1993.
15. Dmitri O Orlov. Geometric realizations of quiver algebras. Proceedings of the Steklov Institute of Mathematics, 290(1):70–83, 2015.
16. Raphaël Rouquier. Derived categories and algebraic geometry. In *Triangulated categories*, volume 375 of *London Math. Soc. Lecture Note Ser.*, pages 351–370. Cambridge Univ. Press, Cambridge, 2010.
17. Evgeny Shinder. Derived categories of coherent sheaves and phantoms. *Lectures given at the University of Mainz*, 2013.

Introduction to Stability Conditions

Rebecca Tramel

1 Motivation

Let X be a smooth projective Calabi–Yau variety over $\mathbb{C}$. Then $\mathcal{D}^b(X)$, the derived category of coherent sheaves on X, is equivalent to the category of D-branes on X [9]. In [10], Douglas defined a notion of stability for D-branes on X called Π-stability. This notion of stability was meant to pick out BPS-branes on X. In [7], Bridgeland aimed to define a notion of stability directly for objects in $\mathcal{D}^b(X)$ which would correspond to Π-stability for D-branes. Bridgeland's stability can be defined on any triangulated category, and hence has been studied in other cases, such as for varieties which are not Calabi–Yau.

2 Definition of Stability

2.1 Example: $\mathbb{P}^1$

Consider the example of Coh($\mathbb{P}^1$), the category of coherent sheaves on $\mathbb{P}^1$. The objects in this category are all direct sums of the following building blocks:

1. Line bundles $\mathcal{O}(n)$, $n \in \mathbb{Z}$.
2. Torsion sheaves $\mathcal{O}_{nx}$, $x \in \mathbb{P}^1$.

There are two invariants which can be assigned to each type of sheaf. First, there is the rank of the sheaf. The rank of a line bundle is 1, and the rank of a skyscraper sheaf is 0. Further, there is the degree of the sheaf. The degree of the line bundle $\mathcal{O}(n)$ is n, and the degree of the torsion sheaf $\mathcal{O}_{nx}$ is n.

Both the rank and degree functions can be defined more generally for any sheaf on $\mathbb{P}^1$. Both invariants are additive on short exact sequences. So, for example, the

R. Tramel (✉)
Mount Holyoke College, South Hadley, MA, USA
e-mail: rtramel@illinois.edu

M. Ballard et al. (eds.), *Superschool on Derived Categories and D-branes*,
Springer Proceedings in Mathematics & Statistics 240,
https://doi.org/10.1007/978-3-319-91626-2_5

rank and degree of $\mathcal{O}(2) \oplus \mathcal{O}(4) \oplus \mathcal{O}_x$ are 2 and 7 respectively. Similarly, all objects in $\mathcal{D}^b(\mathrm{Coh}(\mathbb{P}^1))$ are extensions or shifts of line bundles and torsion sheaves, hence we could define the degree and rank functions for any objects in $\mathcal{D}^b(\mathrm{Coh}(\mathbb{P}^1))$.

We can define a group homomorphism $Z\colon K(\mathcal{D}^b(\mathrm{Coh}(\mathbb{P}^1))) \to \mathbb{C}$ as

$$Z(E^{\cdot}) = -\mathrm{degree}(E^{\cdot}) + i\,\mathrm{rank}(E^{\cdot})$$

for $E^{\cdot} \in \mathcal{D}^b(\mathrm{Coh}(\mathbb{P}^1))$. This is well-defined since degree and rank are additive on short exact sequences. Further, if we consider the subcategory $\mathrm{Coh}(\mathbb{P}^1)$ inside $\mathcal{D}^b(\mathrm{Coh}(\mathbb{P}^1))$, the image is the upper half plane.

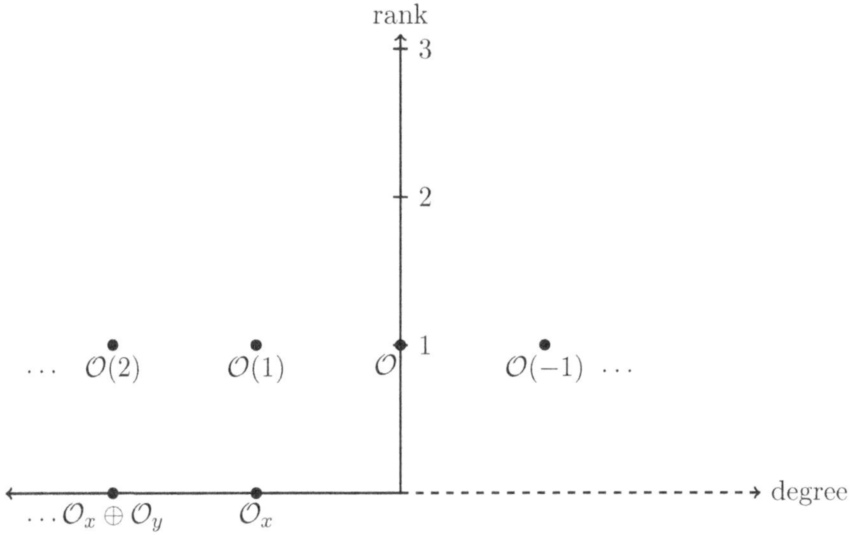

For each $E^{\cdot} \in \mathcal{D}^b(\mathrm{Coh}(\mathbb{P}^1))$ we can write $Z(E^{\cdot}) = m(E^{\cdot})e^{\pi i\phi(E)}$ for some $m(E^{\cdot}) > 0$. We call $m(E^{\cdot})$ the mass of $E^{\cdot}$ and $\phi(E^{\cdot})$ the phase of $E^{\cdot}$. Note that for objects E in $\mathrm{Coh}(\mathbb{P}^1)$, the phase lies in the range $0 < \phi(E) \leq 1$.

For $E \in \mathrm{Coh}(\mathbb{P}^1)$, we say E is Z-stable if for all subsheaves $F \subsetneq E$, $\phi(F) < \phi(E)$. We say E is semistable if for all subsheaves $F \subsetneq E$, $\phi(F) \leq \phi(E)$. It is easy to check that the only stable sheaves are line bundles and skyscraper sheaves, and that a sheaf is semistable if and only if it is either a direct sum of skyscraper sheaves or a direct sum of line bundles all of the same degree.

We can use this fact to construct a filtration of a sheaf $E \in \mathrm{Coh}(\mathbb{P}^1)$ whose successive quotients are semistable sheaves of strictly decreasing phase as follows. We write $E = \bigoplus_{x_i \in \mathbb{P}^1} \mathcal{O}_x \oplus \bigoplus_{j=1}^{s} \mathcal{O}(n_j)$ for a collection of points $x_i \in \mathbb{P}^1$ and $n_1 \geq n_2 \geq \cdots \geq n_s$. Then we can construct a filtration

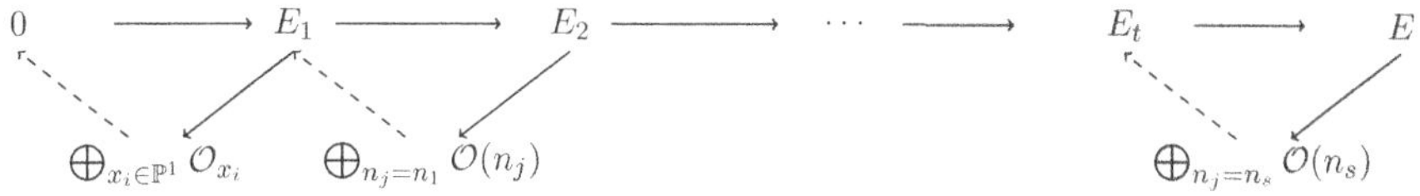

by building E out of its summands, one type at a time. Such a filtration is called a Harder-Narasimhan, or HN, filtration.

2.2 Definition

Definition 2.1 Let $\mathcal{D}$ be a triangulated category. A heart of a bounded t-structure is a full additive subcategory $\mathcal{A}$ of $\mathcal{D}$ satisfying

1. $\mathrm{Hom}^i(A, B) = 0$ for $i < 0$ and $A, B \in \mathcal{A}$.
2. Objects in $\mathcal{D}^b(X)$ have filtrations by cohomology objects in $\mathcal{A}$. That is, for all nonzero $E^{\cdot} \in \mathcal{D}^b(X)$, there is a sequence of exact triangles

such that $A_i[-k_i] \in \mathcal{A}$ for integers $k_1 > \cdots > k_n$.

Definition 2.2 ([7, Proposition 5.3]) A Bridgeland stability condition is a pair $\sigma = (Z, \mathcal{A})$ where $Z\colon K_0(\mathcal{D}^b(X)) \to \mathbb{C}$ is a group homomorphism and $\mathcal{A}$ is a heart of a bounded t-structure. The pair must further satisfy that

1. $Z(\mathcal{A} \setminus \{0\}) \subseteq \{re^{i\pi\phi} \mid r > 0,\ 0 < \phi \leq 1\}$. Define the phase of $0 \neq E \in \mathcal{A}$ to be $\phi(E) := \phi$. We say $E \in \mathcal{A}$ is Z-semistable if for all nonzero subobjects $F \in \mathcal{A}$ of E, $\phi(F) \leq \phi(E)$. E is Z-stable if for all nonzero subobjects $F \in \mathcal{A}$ of E, $\phi(F) < \phi(E)$.
2. The objects of $\mathcal{A}$ have Harder-Narasimhan filtrations with respect to Z. That is, for every $E \in \mathcal{A}$ there is a unique sequence of inclusions

$$0 = E_0 \subseteq E_1 \subseteq \cdots \subseteq E_{n-1} \subseteq E_n = E$$

such that the successive quotients E_i/E_{i-1} are Z-semistable, and the phases $\phi(E_1/E_0) > \phi(E_2/E_1) > \cdots > \phi(E_{n-1}/E_{n-2}) > \phi(E_n/E_{n-1})$.

There is an alternate definition of a Bridgeland stability condition, given in [7, Definition 5.1]. I will give this definition as well. First, we must define a slicing of a triangulated category.

Definition 2.3 A slicing $\mathcal{P}$ of a triangulated category $\mathcal{D}$ consists of full additive subcategories $\mathcal{P}(\phi)$ for each $\phi \in \mathbb{R}$ satisfying

1. For all $\phi \in \mathbb{R}$, $\mathcal{P}(\phi + 1) = \mathcal{P}(\phi)[1]$.
2. If $\phi_1 > \phi_2$, $A_1 \in \mathcal{P}(\phi_1)$, and $A_2 \in \mathcal{P}(\phi_2)$, then $\mathrm{Hom}(A_1, A_2) = 0$.
3. For every $E \in \mathcal{D}$, there is a finite sequence of real numbers

$$\phi_1 > \phi_2 > \cdots > \phi_n$$

so that there is a sequence of exact triangles

such that $A_i \in \mathcal{P}(\phi_i)$ for each $i = 1, \ldots, n$.

Definition 2.4 A stability condition $\sigma = (Z, \mathcal{P})$ on $\mathcal{D}$ consists of a group homomorphism $Z\colon K(\mathcal{D}) \to C$ and a slicing $\mathcal{P}$ such that if $0 \neq E^{\cdot} \in \mathcal{P}(\phi)$, then $Z(E^{\cdot}) = m(E)e^{\pi i \phi(E^{\cdot})}$ for some $m(E^{\cdot}) \in \mathbb{R}_{>0}$.

In this definition, the semistable objects of phase ϕ are defined to be the objects of $\mathcal{P}(\phi)$. Note that the phase of an arbitrary $E^{\cdot} \in \mathcal{D}$ is not well-defined, only the objects of slicings $\mathcal{P}(\phi)$ have well-defined phase.

This definition is equivalent to the previous definition. The heart $\mathcal{A}$ is replaced by the category $\mathcal{P}(0, 1]$, the extension closure of the collection of objects in $\mathcal{P}(\phi)$ for $0 < \phi \leq 1$. That this category is necessarily abelian is shown in [7, Proposition 5.3]. In fact, one can show that all the subcategories $\mathcal{P}(\phi)$ are abelian [7, Lemma 5.2].

3 Examples

3.1 Curves

For a smooth projective curve C of genus g, stability conditions can be constructed of the type described for $\mathbb{P}^1$, with heart Coh(S) and central charge $Z = -\text{degree} + i\,\text{rank}$. Note that for $g > 0$, sheaves are more complicated, and vector bundles are no longer necessarily direct sums of line bundles. Hence HN filtrations must be constructed more carefully.

There is an action of $\widetilde{\mathrm{GL}}^+(2, \mathbb{R})$ [7, Lemma 8.2] on the space of stability conditions on C (or on any smooth projective variety). If we consider an element of this group to be a pair (T, f) where T is a linear transformation from $\mathbb{R}^2$ to $\mathbb{R}^2$ which is an orientation preserving isomorphism, and $f\colon \mathbb{R} \to \mathbb{R}$ is increasing, and satisfies

$f(\phi + 1) = f(\phi) + 1$ for all $\phi \in \mathbb{R}$, then it acts on a stability condition $(Z, \mathcal{P})$ by replacing Z with $T^{-1}Z$, and replacing $\mathcal{P}(\phi)$ with $\mathcal{P}(f(\phi))$.

In fact, if $g > 0$, then up to the action of $\widetilde{\mathrm{GL}}^+(2, \mathbb{R})$ the previous construction of stability conditions on a curve in terms of rank and degree gives all possible stability conditions on C [11]. There are other possible stability conditions on $\mathbb{P}^1$ described in [5, 12].

3.2 *Quivers*

First, consider the following quiver, Q.

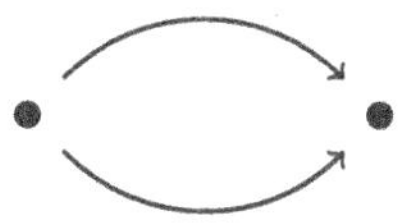

A representation $\overline{V}$ of this quiver consists of a choice of two vector spaces, V_1 and V_2, and two linear maps x and y from V_1 to V_2.

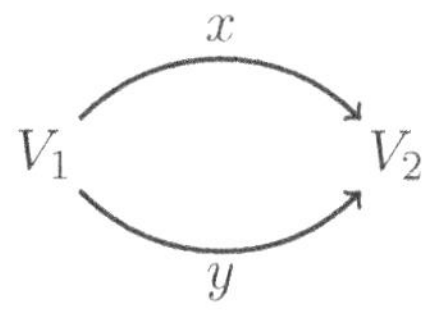

Suppose we wish to define a stability condition on Q. We may start with the abelian category Rep(Q). If we pick any two numbers $z_1, z_2 \in \mathbb{C}$ which lie in the upper half plane or along the negative real axis, we can define a central charge

$$Z(\overline{V}) = z_1 \dim(V_1) + z_2 \dim(V_2).$$

In other words, we choose the images of the two simple representations, S_1 and S_2, pictured below.

S_1 :

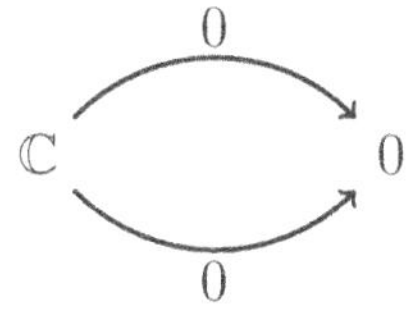

S_2 :

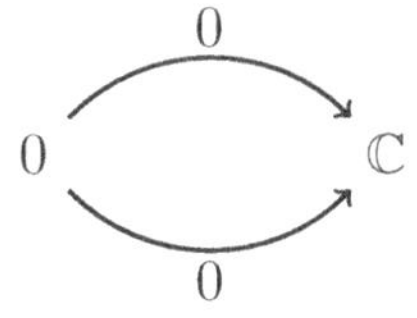

We can then extend our central charge to complexes of representations by requiring it to be additive on exact triangles. We claim that the pair $(\text{Rep}(Q), Z)$ is a Bridgeland stability condition on $\mathcal{D}^b(\text{Rep}(Q))$.

The fact that the image of $\text{Rep}(Q)$ lies in the upper half plane follows from the fact that dimensions are positive, and from the choice of z_1 and z_2. It remains to show that each representation $\overline{V}$ of Q has an HN filtration. This is argued nicely in [2, Theorem 2.1.6]

It is interesting to note in this example how the choice of z_1 and z_2 controls which representations are stable. Suppose first that we pick z_2 so that its phase is larger than z_1. S_2 is a subobject of any representation for which $V_2 \neq 0$, and S_1 is a quotient of any representation for which $V_1 \neq 0$. Hence no object can be stable besides the simple representations.

On the other hand, suppose we choose z_1 so that its phase is larger than z_2. Then again, S_1 and S_2 are necessarily stable. Now, however, so is any representation for which V_1 and V_2 are one-dimensional. Hence, these stable objects are parameterized by the choice of linear maps x, y. Up to scaling, we can suppose $x = 1$. In this way, we see a one-to-one correspondence between stable representations of Q and points of $\mathbb{P}^1$.

Reference [6] shows that there is an equivalence of categories $\mathcal{D}^b(\text{Rep}(Q)) \cong \mathcal{D}^b(\text{Coh}(\mathbb{P}^1))$. This equivalence is given explicitly by the functor $R\text{Hom}(\mathcal{O} \oplus \mathcal{O}(1), -)\colon \mathcal{D}^b(\text{Coh}(\mathbb{P}^1)) \to \mathcal{D}^{\text{b}}(\text{Rep(Q)})$. Such an equivalence always sends a heart of a bounded t-structure to a heart of a bounded t-structure. Hence if we consider the stability conditions we have constructed here on $\text{Rep}(Q)$, there should be corresponding stability conditions on a heart in $\mathcal{D}^b(\text{Coh}(\mathbb{P}^1))$. Note that the inverse image of $\text{Rep}(Q)$ under this equivalence is not $\text{Coh}(\mathbb{P}^1)$, so this already gives an example of a stability condition on $\mathbb{P}^1$ with a heart that is not $\text{Coh}(\mathbb{P}^1)$. The heart on $\mathbb{P}^1$ we get via this map can also be constructed by the process of tilting, described below.

3.3 Surfaces, Threefolds, and Higher Dimensional Varieties

Let X be a smooth projective variety of dimension n. In order to define a central charge, we may wish to start with the example of curves and generalize the ideas of degree and rank. In order to do this, we may choose an ample divisor ω on X, and use the Chern characters of sheaves on X to define the central charge. This is convenient, since these quantities are once again additive on short exact sequences.

For a sheaf $E \in \mathrm{Coh(X)}$ (or an object $E^{\cdot} \in \mathcal{D}^b(\mathrm{Coh(X)})$, our numerical invariants are now $\omega^n \cdot \mathrm{ch}_0(\mathrm{E}^{\cdot}), \omega^{n-1} \cdot \mathrm{ch}_1(\mathrm{E}^{\cdot}), \ldots, \omega^0 \mathrm{ch}_n(\mathrm{E}^{\cdot})$. Note that if $n = 1$, this does not depend on the choice of ω, and gives us exactly the rank and degree of E.

However, it is not possible to define a stability condition on the heart Coh(X) with central charge in terms of these quantities for $n > 1$. Hence we must find a different choice of abelian subcategory of $\mathcal{D}^b(\mathrm{Coh(X)})$. One technique for constructing new hearts inside $\mathcal{D}^b(\mathrm{Coh(X)})$ is called tilting. In order to perform the process of tilting, one chooses two full additive subcategories $\mathcal{T}$ and $\mathcal{F}$ in Coh(X) which form what is called a torsion pair.

Definition 3.1 A torsion pair in a heart $\mathcal{A}$ is a pair $(\mathcal{T}, \mathcal{F})$ of full additive subcategories of $\mathcal{A}$ such that

1. If $T \in \mathcal{T}$ and $F \in \mathcal{F}$, then $\mathrm{Hom}(T, F) = 0$.
2. For all $E \in \mathcal{A}$ there is an object $T \in \mathcal{T}$ and $F \in \mathcal{F}$ so that the sequence $0 \to T \to E \to F \to 0$ is exact.

We then replace our category Coh(X) with the tilt

$$\mathcal{A}^{\#} = \{E^{\cdot} \in \mathcal{D}^b(X) \mid H^0_{\mathcal{A}}(E^{\cdot}) \in \mathcal{T},\ H^{-1}_{\mathcal{A}}(E^{\cdot}) \in \mathcal{F},\ H^i_{\mathcal{A}}(E^{\cdot}) = 0 \text{ for } i \neq 0, -1\}$$

whose elements are 2-term complexes with restrictions on cohomology. This process can then be repeated to construct more hearts.

If X is a surface, it is shown in [1, 8] that this process can be used to construct stability conditions on X. In particular, we choose another class $B \in \mathrm{NS}_{\mathbb{R}}(X)$, and then can write the central charge formula explicitly as

$$Z(E^{\cdot}) = -\int_X e^{B+i\omega}\mathrm{ch}(\mathrm{E}^{\cdot}).$$

In particular, [1] shows that this central charge, paired with a heart which is a single tilt of Coh(X) given explicitly in terms of ω and B, give a stability condition on X.

For $n > 2$, one might hope a similar process might work. We might hope that the same central charge formula, and a heart constructed in terms of ω and B by tilting Coh(X) perhaps $n - 1$ times could give a stability condition on X. Unfortunately, this has been difficult to prove. It is conjectured true for threefolds in [4], with the heart given explicitly, although the exact conjecture in [4] has been shown not to hold for certain threefolds in [14]. It is shown only for certain threefolds, in [3, 4, 13].

References

1. Daniele Arcara and Aaron Bertram. Bridgeland-stable moduli spaces for K-trivial surfaces. *J. Eur. Math. Soc. (JEMS)*, 15(1):1–38, 2013. With an appendix by Max Lieblich.
2. A. Bayer. A tour to stability conditions on derived categories.

3. Arend Bayer, Emanuele Macrì, and Paolo Stellari. The space of stability conditions on abelian threefolds, and on some Calabi-Yau threefolds. 2014.
4. Arend Bayer, Emanuele Macrì, and Yukinobu Toda. Bridgeland stability conditions on threefolds I: Bogomolov-Gieseker type inequalities. J. Algebraic Geom., 23(1):117–163, 2014.
5. A. Bertram, S. Marcus, and J. Wang. The stability manifolds of $\mathbb{P}^1$ and local $\mathbb{P}^1$.
6. A. I. Bondal. Representations of associative algebras and coherent sheaves. Izv. Akad. Nauk SSSR Ser. Mat., 53(1):25–44, 1989.
7. Tom Bridgeland. Stability conditions on triangulated categories. *Ann. of Math. (2)*, 166(2):317–345, 2007.
8. Tom Bridgeland. Stability conditions on K3 surfaces. Duke Math. J., 141(2):241–291, 2008.
9. Michael R. Douglas, Bartomeu Fiol, and Christian Römelsberger. Stability and BPS branes. *J. High Energy Phys.*, (9):006, 15, 2005.
10. Michael R. Douglas. Dirichlet branes, homological mirror symmetry, and stability. In *Proceedings of the International Congress of Mathematicians, Vol. III (Beijing, 2002)*, pages 395–408. Higher Ed. Press, Beijing, 2002.
11. Emanuele Macrì. Stability conditions on curves. may 2007.
12. S. Okada. Stability manifold of $\mathbb{P}^1$. mar 2005.
13. B. Schmidt. A generalized Bogomolov-Gieseker inequality for the smooth quadric threefold. September 2013.
14. B. Schmidt. Counterexample to the generalized Bogomolov-Gieseker inequality for threefolds. February 2016.

A Brief Introduction to Geometric Invariant Theory

Nathan Grieve

Mathematics Subject Classification (2010) Primary 14L24 · Secondary 13A50

1 Introduction

We work over the complex numbers $\mathbb{C}$ and our aim is to describe some aspects of Geometric Invariant Theory (GIT). Everything we describe here is well-known and we follow the excellent treatments given in a number of works, notably [2–4, 6–11], closely. We include no proofs for the results that we state. Instead, here we content ourselves with stating and describing some of the main theorems of GIT, while, at the same time, still providing detailed references as to where accessible proofs can be found in the literature. Further, we illustrate aspects of the general theory by considering important instructive examples.

Broadly speaking, GIT concerns questions related to a *reductive group* G acting on an algebraic variety X. It is a technique for forming quotient spaces in algebraic geometry and provides a fundamental method for the construction and study of moduli spaces of projective varieties. To begin with, one issue in forming quotients in algebraic geometry is that the orbit space X/G does not exist, in general, in the category of separated algebraic varieties. Instead, to form quotients of X by G, the idea of GIT is to first choose a G-*linearization* of an ample line bundle L on X and then construct a *good categorical quotient* $X /\!\!/ \mathrm{G}$. This quotient depends, in general,

N. Grieve (✉)
Department of Mathematics, Michigan State University, Lansing 48824, MI, UK
e-mail: n.grieve@unb.ca

N. Grieve
Department of Mathematics and Statistics, University of New Brunswick, Fredericton, NB, Canada

M. Ballard et al. (eds.), *Superschool on Derived Categories and D-branes*, Springer Proceedings in Mathematics & Statistics 240,
https://doi.org/10.1007/978-3-319-91626-2_6

on the choice of G-linearization, and it is an interesting question to understand the extent to which the quotient is independent both of the G-linearization of the line bundle L and also of the line bundle L itself.

As mentioned, the purpose of these notes is to give a brief introduction and summary of the most basic concepts and questions related to GIT. It is hoped that the interested reader will consult the references for more details and further study.

2 Reductive Groups

Since GIT concerns taking quotients of an algebraic variety acted on by a *reductive algebraic group*, we start by briefly explaining some of these concepts. To this end, first we recall that an *algebraic group* is a group G together with the structure of an algebraic variety such that the maps $\mathrm{G} \times \mathrm{G} \to \mathrm{G}$, $(g, g') \mapsto gg'$ and $\mathrm{G} \to \mathrm{G}$, $g \mapsto g^{-1}$ are morphisms of algebraic varieties. Next we recall that a *linear algebraic group* is an algebraic group G which can be realized as a Zariski closed subset of the general linear group

$$\mathrm{GL}_n(\mathbb{C}) := \operatorname{Spec} \mathbb{C}[T_{11}, T_{12}, \ldots, T_{nn}, D^{-1}],$$

where $D = \det(T_{ij})$, for some n.

Example 2.1 The following are examples of linear algebraic groups:

(a) $\mathrm{SL}_n(\mathbb{C}) = \operatorname{Spec} \mathbb{C}[T_{11}, T_{12}, \ldots, T_{nn}]/(\det(T_{ij}) - 1)$ (Type A_{n-1});
(b) $\mathrm{Sp}_{2n}(\mathbb{C}) = \{g \in \mathrm{GL}_{2n}(\mathbb{C}) : {}^t g J g = J\}$, for $J = \begin{pmatrix} 0 & I_n \\ -I_n & 0 \end{pmatrix}$ (Type C_n);
(c) the other classical and exceptional simple groups (i.e., those of Types B_n, D_n, F_4, G_2, E_6, E_7, E_8); and
(d) algebraic tori $\mathrm{G}_m^n = \underbrace{\mathrm{GL}_1(\mathbb{C}) \times \cdots \times \mathrm{GL}_1(\mathbb{C})}_{n\text{-times}} = (\mathbb{C}^\times)^n$.

The concept of a *reductive group* is deeply tied to the concept of *complete reducibility* of *rational actions* of linear algebraic groups. In this direction, we recall that a linear algebraic group is *reductive* if its *radical* (i.e., its unique maximal connected solvable normal subgroup) is isomorphic to a direct product copies of $\mathbb{C}^\times$, see for example [1, p. 158] and [2, p. 42].

Example 2.2 All of the groups described in Example 2.1 are reductive while the additive group G_a is not reductive, see for instance [2, Exercise 4.1, p. 62].

Next, we say that a *rational (right) action* of a linear algebraic group G on a finitely generated $\mathbb{C}$-algebra R is a map

$$R \times \mathrm{G} \to R, \text{ defined by } (s, g) \mapsto s^g$$

with the properties that:

(a) $s^{gg'} = (s^g)^{g'}$ and $s^e = s$ for all $s \in R$, $g, g' \in \mathrm{G}$ and e the identity of G;

(b) the map $s \mapsto s^g$ is a $\mathbb{C}$-algebra automorphism of R for all $g \in \mathrm{G}$; and
(c) every element of R is contained in a finite dimensional subspace V which is invariant under G and on which G acts by a *rational representation*, i.e., by a homomorphism of algebraic groups:

$$\rho : \mathrm{G} \to \mathrm{GL}(V) \simeq \mathrm{GL}_n(\mathbb{C}),$$

for $n = \dim V$, [9, p. 47].

Given a rational action of a linear algebraic group G on a finitely generated $\mathbb{C}$-algebra R, one formulation of Hilbert's 14th problem asks if the G-invariant subalgebra

$$R^{\mathrm{G}} = \{s \in R : s^g = s, \text{ for all } g \in \mathrm{G}\}$$

is finitely generated, see for instance [9, p. 47] or [2, Chap. 4] for more details.

Furthermore, a related question concerns the concept of *complete reducibility* which is essentially the question of as to whether or not a given rational representation decomposes into a direct sum of representations each of which possesses no proper invariant subspaces, [9, p. 48].

Two important theorems related to reductive groups are:

Theorem 2.3 ([9, Remark 3.2, p. 48]) *A (complex) linear algebraic group* G *is reductive if and only if every rational representation is completely reducible. Further, if a (complex) reductive group* G *acts linearly on* $\mathbb{C}^n$, *then for every invariant point* $0 \neq v \in \mathbb{C}^n$, *there exists an invariant homogeneous polynomial* s *of positive degree with* $s(v) \neq 0$.

Theorem 2.4 ([9, Theorem 3.4, p. 49], [2, Theorem 3.3, p. 41]) *If a (complex) reductive group* G *acts rationally on a finitely generated* $\mathbb{C}$*-algebra* R*, then* R^{G} *is finitely generated.*

Remark 2.5 Similar more technical theorems hold true over fields of positive characteristic. In that setting, one needs to distinguish between the concepts of a linear algebraic group being reductive, geometrically reductive, and/or linearly reductive, see [9, Remark 3.2, p. 48], for instance, for a more detailed discussion.

Example 2.6 A rational action of G_m on a finitely generated $\mathbb{C}$-algebra R is equivalent to giving a $\mathbb{Z}$*-grading*, that is a decomposition of the form $R = \bigoplus_{i \in \mathbb{Z}} R_i$ with $R_i R_j \subset R_{i+j}$ and $\dim_{\mathbb{C}} R_i < \infty$. For a more detailed explanation, see [2, Example 3.1, p. 38].

3 Group Quotients

In this section, we discuss concepts related to taking group quotients within the category of algebraic varieties. To this end, let G be an algebraic group acting on an algebraic variety X. In particular, we are given a morphism $\sigma : \mathrm{G} \times X \to X$ of algebraic varieties such that $\sigma(g, \sigma(g', x)) = \sigma(gg', x)$ and $\sigma(e, x) = x$, for all $g, g' \in \mathrm{G}$, all $x \in X$, and e the identity of G, [9, p. 43].

We state a couple of definitions.

Definition 3.1 ([10, Definition 1.4, p. 515]) A morphism $f : X \to Y$ of algebraic varieties is a *categorical quotient* if the following properties hold true:

(a) the morphism f is G-*invariant*, i.e., for the trivial action of G on Y, f is a G-morphism; and
(b) for all G-morphisms $g : X \to Z$, there exists a unique morphism $h : Y \to Z$ such that $g = h \circ f$.

Remark 3.2 A categorical quotient is uniquely determined up to isomorphism.

Definition 3.3 ([10, Definition 1.5, p. 516]) Assume that G is linear. A G-morphism $f : X \to Y$ of algebraic varieties is said to be a *good quotient* if the following properties hold true:

(a) f is a surjective, affine G-invariant morphism;
(b) $f_*(\mathcal{O}_X)^{\mathrm{G}} = \mathcal{O}_Y$; and
(c) if Z is a closed G-stable subset of X, then $f(Z)$ is closed in Y. Further, if Z_1 and Z_2 are two closed G-stable subsets of X such that $Z_1 \cap Z_2 = \varnothing$, then $f(Z_1) \cap f(Z_2) = \varnothing$.

Definition 3.4 ([10, Definition 1.6, p. 516]) Assume that G is linear. A G-morphism $f : X \to Y$ of algebraic varieties is said to be a *geometric quotient* if the following properties hold true:

(a) f is a good quotient; and
(b) for all $x \in X$, the G-orbit $\mathrm{O}(x)$ through x is closed in X.

In general we have the implications:

$$\text{Geometric Quotient} \Rightarrow \text{Good Quotient} \Rightarrow \text{Categorical Quotient};$$

the first two implications are clear from the definitions and we refer to [10, p. 516] for details concerning the fact that a good quotient is a categorical quotient.

4 Linearization of an Invertible Sheaf

Here, we discuss the concept of linearizing a line bundle with respect to a group action. Throughout this section, we let X be an irreducible algebraic variety over $\mathbb{C}$, we let L be a line bundle on X, we let G be a reductive group acting on X via $\sigma : \mathrm{G} \times X \to X$ and we denote the group law on G by $\mu : \mathrm{G} \times \mathrm{G} \to \mathrm{G}$. The following definition is fundamental to GIT.

Definition 4.1 ([8, Definition 1.6, p. 30], [2, p. 104]) A G-*linearization* of L with respect to σ is an isomorphism of $\mathscr{O}_{G\times X}$-modules

$$\phi : \sigma^* L \xrightarrow{\sim} p_2^* L$$

satisfying the *cocycle condition*, that is the diagram, of maps of line bundles on $G \times G \times X$,

$$\begin{array}{ccc} (\sigma \circ (1_G \times \sigma))^* L & \xrightarrow{(1_G \times \sigma)^* \phi} & (p_2 \circ (1_G \times \sigma))^* L \\ \| & & \downarrow \\ (\sigma \circ (\mu \times 1_X))^* L & \xrightarrow{(\mu \times 1_X)^* \phi} & (p_2 \circ (\mu \times 1_X))^* L \end{array}$$

commutes.

The concept of a G-linearization of L with respect to σ can be made more concrete in a number of ways. To begin with, if $a \in G(\mathbb{C})$, then let $\tau_a : X \to X$ denote the automorphism defined by $x \mapsto a \cdot x$. Then if L admits a G-linearization with respect to σ, the isomorphism ϕ restricts to an isomorphism

$$\phi_a : \tau_a^* L \xrightarrow{\sim} L.$$

The cocycle condition then implies that

$$\phi_{ab} = \phi_b \circ \tau_b^* \phi_a,$$

for all $a, b \in G(\mathbb{C})$; equivalently the diagram

$$\begin{array}{ccc} \tau_{ab}^* L & \xrightarrow{\phi_{ab}} & L \\ & \searrow^{\tau_b^* \phi_a} & \uparrow \phi_b \\ & & \tau_b^* L \end{array}$$

commutes, [8, p. 31].

Next let $\mathscr{L} = \operatorname{Spec}\left(\bigoplus_{m \leqslant 0} L^{\otimes m}\right)$ be the total space of L and $\pi : \mathscr{L} \to X$ the projection; then $L = \mathscr{S}(\mathscr{L}/X)$ the sheaf of sections of π, [5, Exercise II.5.18]. A G-linearization

$$\phi : \sigma^* L \xrightarrow{\sim} p_2^* L$$

corresponds canonically to a bundle isomorphism

$$(G \times X) \times_X \mathscr{L} \xleftarrow{\sim} (G \times X) \times_X \mathscr{L} : \Phi$$

which corresponds canonically to a G-bundle action

$$\Sigma = p_2 \circ \Phi : \mathrm{G} \times \mathscr{L} \to \mathscr{L}$$

of G on $\mathscr{L}$ which covers σ. In particular, the diagram

$$\begin{array}{ccc} \mathrm{G} \times \mathscr{L} & \xrightarrow{\Sigma} & \mathscr{L} \\ {\scriptstyle 1_{\mathrm{G}} \times \pi} \downarrow & & \downarrow {\scriptstyle \pi} \\ \mathrm{G} \times X & \xrightarrow[\sigma]{} & X \end{array}$$

commutes. Conversely, every G-bundle action

$$\Sigma : \mathrm{G} \times \mathscr{L} \to \mathscr{L}$$

which covers σ determines a G-linearization ϕ of L with respect to σ, [8, p. 31].

Since the concept of G-linearization can be a source of confusion, we provide more details concerning the natural bijective correspondence between G-linearizations of L and those of $\mathscr{L}$. To this end, if $a \in \mathrm{G}(\mathbb{C})$, then let $\tau_a^* \mathscr{L}$ denote the total space of $\tau_a^* L$. In particular, $\tau_a^* \mathscr{L} = X \times_X \mathscr{L}$ and given a G-linearization Φ of $\mathscr{L}$, we obtain, for each $a \in \mathrm{G}(\mathbb{C})$, a (linear) automorphism of $\mathscr{L}$ covering τ_a. By the universal property of Cartesian squares, Φ_a induces a (linear) isomorphism

$$\widetilde{\phi}_a : \mathscr{L} \xrightarrow{\sim} \tau_a^* \mathscr{L}$$

over X. Setting $\phi_a = \widetilde{\phi}_a^{-1}$, we naturally obtain an $\mathscr{O}_X$-module isomorphism

$$\phi_a : \tau_a^* L \xrightarrow{\sim} L.$$

The collection of such isomorphisms ϕ_a, for $a \in \mathrm{G}(\mathbb{C})$, determine the G-linearization

$$\phi : \sigma^* L \xrightarrow{\sim} p_2^* L$$

corresponding to Φ.

Another important feature of G-linearizations is that they allow for the study of $\mathrm{H}^0(X, L)^{\mathrm{G}}$ the space of G-invariant sections of L with respect to ϕ, [8, p. 32] compare also with [2, Sect. 7.3, p. 110]. In particular, given a G-linearization ϕ of L, we obtain a representation of G on $\mathrm{H}^0(X, L)$. This representation is defined, for each $a \in \mathrm{G}(\mathbb{C})$, by:

$$\mathrm{H}^0(X, L) \xrightarrow{\widetilde{\phi}_a} \mathrm{H}^0(X, \tau_a^* L) \xrightarrow{\tau_{-a}^*} \mathrm{H}^0(X, L). \tag{1}$$

Alternatively, but equivalently, we can define $\mathrm{H}^0(X, L)^{\mathrm{G}}$ as the invariants for the dual action of G on $\mathrm{H}^0(X, L)$. This (dual) action is defined, for each $a \in \mathrm{G}(\mathbb{C})$, by:

$$\mathrm{H}^0(X, L) \xrightarrow{\tau_a^*} \mathrm{H}^0(X, \tau_a^* L) \xrightarrow{\phi_a} \mathrm{H}^0(X, L); \tag{2}$$

the representation (1) is related to the dual action (2) via:

$$\phi_a \circ \tau_a^* = (\tau_{-a}^* \circ \widetilde{\phi}_a)^{-1}.$$

It is not difficult to check that the tensor product of two G-linearized line bundles and the inverse of two G-linearized line bundles are naturally G-linearized. In particular, the collection of G-linearized line bundles modulo isomorphism form a group which we denote by $\mathrm{Pic}^{\mathrm{G}}(X)$. Furthermore, if $f : X \to Y$ is a G-linear morphism of algebraic varieties over $\mathbb{C}$, then there is an induced homomorphism

$$f^* : \mathrm{Pic}^{\mathrm{G}}(Y) \to \mathrm{Pic}^{\mathrm{G}}(X),$$

[8, p. 32]. When (Y, f) is a geometric quotient of X by G and if the action of G on X is free, then a consequence of descent theory is that

$$\mathrm{Pic}^{\mathrm{G}}(X) \simeq \mathrm{Pic}(Y),$$

see [8, p. 32]. As one final related comment, we note that if G is connected, if X is normal and if L is a line bundle on X, then some tensor power of L is always linearizable, [8, Corollary 1.6, p. 35].

Example 4.2 (*Compare with* [2, Example 8.4, p. 123]) Let the multiplicative group G_m act on $\mathbb{A}^n := \mathrm{Spec}\,\mathbb{C}[x_1, \ldots, x_n]$ by

$$t \cdot x = t \cdot (x_1, \ldots, x_n) := (t^{r_1} x_1, \ldots, t^{r_n} x_n),$$

for integers $r_1, \ldots, r_n$. We then have that $\mathrm{Pic}(\mathbb{A}^n) = 0$ while, by contrast,

$$\mathrm{Pic}^{\mathrm{G}_m}(\mathbb{A}^n) = \mathbb{Z}.$$

To see this, since $\mathrm{Pic}(\mathbb{A}^n) = 0$, every line bundle on $\mathbb{A}^n$ is isomorphic to the trivial line bundle $L = \mathscr{O}_{\mathbb{A}^n}$ with total space $\mathscr{L} = \mathbb{A}^n \times \mathbb{A}^1$. As one consequence of this fact, it follows that the collection of isomorphism classes of G_m-linearizations of L with respect to our given G_m-action on $\mathbb{A}^n$ is in bijection with the collection of those determined by the formula

$$t \cdot (x, v) = (t \cdot x, t^a v),$$

for some $a \in \mathbb{Z}$ and $x = (x_1, \ldots, x_n) \in \mathbb{A}^n$, $v \in \mathbb{A}^1$. In particular, if L_a denotes the trivial line bundle $L = \mathscr{O}_{\mathbb{A}^n}$ with G_m-linearization determined by $a \in \mathbb{Z}$, then the isomorphism

$$\mathrm{Pic}^{\mathrm{G}_m}(\mathbb{A}^n) \xrightarrow{\sim} \mathbb{Z}$$

is set-up by $a \mapsto L_a$.

We now consider $\mathrm{H}^0(\mathbb{A}^n, L^{\otimes d})^{\mathrm{G}_m}$ the space of G_m-invariant sections for the G_m-linearization on $L^{\otimes d}$ determined by $a \in \mathbb{Z}$. To this end, sections $s \in \mathrm{H}^0(\mathbb{A}^n, L^{\otimes d})$ are in bijection with morphisms

$$s : \mathbb{A}^n \to \mathscr{L}^{\otimes d} = \mathbb{A}^n \times \mathbb{A}^1$$

given by

$$s(x) = (x, F(x)),$$

for some polynomial $F(x) \in \mathbb{C}[x_1, \dots, x_n]$. Using this bijection, the sections $s \in \mathrm{H}^0(\mathbb{A}^n, L^{\otimes d})$ fixed by the dual action

$$\mathrm{H}^0(\mathbb{A}^n, L^{\otimes d}) \xrightarrow{\tau_t^*} \mathrm{H}^0(\mathbb{A}^n, \tau_t^* L^{\otimes d}) \xrightarrow{\phi_t} \mathrm{H}^0(\mathbb{A}^n, L^{\otimes d})$$

are of the form $s(x) = (x, F(x))$, where $F \in \mathbb{C}[x_1, \dots, x_n]$ has the property that

$$F(t \cdot x) = t^{da} F(x), \tag{3}$$

for each $t \in \mathrm{G}_m$ and each $x \in \mathbb{A}^n$. Indeed, to see that (3) holds true, we simply note, as in [2, p. 123], that the G_m-action given by (1), applied to our given G_m-linearization on $L^{\otimes d}$, takes the form:

$$t \cdot s(x) = (x, t^{da} F(t^{-1} \cdot x)),$$

for each $x \in \mathbb{A}^n$ and each $t \in \mathrm{G}_m$. In particular, $\mathrm{H}^0(\mathbb{A}^n, L^{\otimes d})^{\mathrm{G}_m} = \mathbb{C}[x_1, \dots, x_n]_{da}$ where the grading on $\mathbb{C}[x_1, \dots, x_n]$ is given by $\deg(x_i) = r_i$.

5 Semi-Stability and the First Main Theorem of GIT

Let X be an algebraic variety and G a reductive group acting on X. In this section, we discuss the concept of semi-stability for X with respect to G and then state, in Theorem 5.1, the first main theorem of GIT. This theorem pertains to existence of good categorical quotients.

First of all, recall that a line bundle L on X is said to be *ample* if there exists a morphism $f : X \to \mathbb{P}^n$, for some n, such that f maps X isomorphically onto a quasi-projective variety in $\mathbb{P}^n$ and $f^*\mathscr{O}_{\mathbb{P}^n}(1) \simeq L^{\otimes d}$ for some $d \in \mathbb{Z}_{>0}$. Next, let L be a G-linearized ample line bundle on X. The subsets of *semistable*, *stable* and *unstable* points of X, with respect to the G-linearized ample line bundle L, are described respectively by:

$$X^{ss}(L) := \{x \in X : \text{ there exists } s \in \mathrm{H}^0(X, L^{\otimes d})^{\mathrm{G}} \text{ such that } s(x) \neq 0\},$$

$$X^s(L) := \{x \in X^{ss}(L) : \mathrm{G} \cdot x \text{ is closed in } X^{ss}(L) \text{ and the stabilizer } \mathrm{G}_x \text{ is finite}\}$$

and

$$X^{us}(L) := X \setminus X^{ss}(L),$$

see for instance [3, p. 6].

Having defined the concepts of semi-stability and stability, we can state the first main theorem of GIT:

Theorem 5.1 ([9, Theorem 3.21, p. 84], [8, Theorem 1.10, p. 38], [10, Theorem 1.1, p. 517]) *Let L be a* G*-linearized ample line bundle on X. There exists a good (categorical) quotient*

$$\pi : X^{ss}(L) \to X^{ss}(L)/\!\!/\mathrm{G}.$$

Further, there exists an open subset $U \subseteq X^{ss}(L)/\!\!/\mathrm{G}$ such that $X^s(L) = \pi^{-1}(U)$ and such that $(U, \pi|_{X^s(L)})$ is a geometric quotient of $X^s(L)$ by G*. Finally, there exists an ample line bundle M on $X^{ss}(L)/\!\!/\mathrm{G}$ such that $\pi^*(M) \simeq L^{\otimes d}|_{X^{ss}(L)}$, for some $d > 0$. In particular, $X^{ss}(L)/\!\!/\mathrm{G}$ is a quasi-projective variety.*

In the setting of Theorem 5.1, we make the following remarks.

Remark 5.2 (a) The sets $X^{ss}(L)$, $X^s(L)$, $X^{us}(L)$ and the quotient $X^{ss}(L)/\!\!/\mathrm{G}$ remain unchanged if we replace L by $L^{\otimes d}$ for some $d > 0$.

(b) The sets $X^{ss}(L)$, $X^s(L)$, $X^{us}(L)$ and the quotient $X^{ss}(L)/\!\!/\mathrm{G}$ are not in general independent of the choice of G-linearization.

(c) The set $X^{ss}(L)$ and the quotient $X^{ss}(L)/\!\!/\mathrm{G}$ are independent of the G-algebraic equivalence class of L, [11, Proposition 2.1].

An important special case of Theorem 5.1 reads:

Proposition 5.3 ([10, Theorem 1.1 B]) *In addition to the assumptions of Theorem 5.1, assume that X is projective and let*

$$R = \bigoplus_{d \geqslant 0} \mathrm{H}^0(X, L^{\otimes d}).$$

Then

$$X^{ss}(L)/\!\!/\mathrm{G} \simeq \mathrm{Proj}\left(R^{\mathrm{G}}\right).$$

In particular, $X^{ss}(L)/\!\!/\mathrm{G}$ is a projective variety.

Example 5.4 ([11, Example 1.16, p. 699], [2, Example 8.6, p. 125]) We now consider the case of $\mathrm{G} = \mathrm{G}_m$ acting on $X = \mathbb{A}^4$ by

$$t \cdot (x_1, \dots, x_4) = (tx_1, tx_2, t^{-1}x_3, t^{-1}x_4)$$

and we consider GIT quotients with respect to the G-linear line bundle L_a, for $a \in \mathbb{Z}$. To this end, let $R = \mathbb{C}[x_1, x_2, x_3, x_4]$ with grading defined by

$$\deg(x_1) = \deg(x_2) = 1 \text{ and } \deg(x_3) = \deg(x_4) = -1$$

and put $R = \bigoplus_{d \geqslant 0} R_d$. Recall that we then have that

$$\mathrm{H}^0(X, L^{\otimes d})^{\mathrm{G}} = \mathrm{H}^0(X, L^{\otimes d})^{\mathrm{G}_m} = \mathrm{H}^0(X, L_a^{\otimes d})^{\mathrm{G}_m} = R_{da}.$$

Consider first the case that $a = 0$. In this case, we have for all $d > 0$, that

$$\mathrm{H}^0(X, L^{\otimes d})^{\mathrm{G}} = R_0$$

and since $1 \in R_0$, it follows that $X = X^{ss}(L)$. Also

$$R_0 = R^{\mathrm{G}} = \mathbb{C}[x_1x_3, x_1x_4, x_2x_3, x_2x_4]$$

which is isomorphic to

$$R^{\mathrm{G}} \simeq \mathbb{C}[T_1, T_2, T_3, T_4]/\langle T_1T_4 - T_2T_3 \rangle.$$

A consequence of this is that, when $a = 0$, the quotient admits an isomorphism

$$X^{ss}(L) /\!\!/ \mathrm{G}_m \simeq Y_0 \subseteq \mathbb{A}^4,$$

with Y_0 defined by $T_1T_4 - T_2T_3 = 0$.

Next consider the case $a > 0$. Without loss of generality we assume that $a = 1$, see for example [11, p. 694] or [2, Exercise 8.3, p. 127] for more details. Then

$$\bigoplus_{d>0} \mathrm{H}^0(X, L^{\otimes d})^{\mathrm{G}} = \bigoplus_{d>0} R_d = R_{>0} = x_1 R_{\geqslant 0} + x_2 R_{\geqslant 0}$$

and we deduce that

$$X^{ss}(L) = \mathbb{A}^4 \setminus V(x_1, x_2)$$

which is covered by the affine open subsets U_i defined by the conditions that $x_i \neq 0$, for $i = 1, 2$. Further

$$\mathscr{O}_X(U_1)^{\mathrm{G}} = R_{(x_1)} = R_0[x_2/x_1],$$

$$\mathscr{O}_X(U_2)^{\mathrm{G}} = R_{(x_2)} = R_0[x_1/x_2],$$

and it follows, since $R_0 = \mathbb{C}[x_1x_3, x_1x_4, x_2x_3, x_2x_4]$, that

$$Y_+ := X^{ss}(L) /\!\!/ \mathrm{G}_m$$

is isomorphic to the closed subvariety of $\mathbb{A}^4 \times \mathbb{P}^1$ given by:

$$Y_+ = V(T_1X_2 - T_3X_1, T_2X_2 - T_4X_1, T_1T_4 - T_2T_3) \subseteq \mathbb{A}^4 \times \mathbb{P}^1.$$

Indeed, if Y_i' is the subset of Y_+ given by the condition that $X_i \neq 0$, for $i = 1, 2$, then

$$\mathcal{O}_{Y_+}(Y_i') \simeq \mathcal{O}_X(U_i)^{\mathrm{G}}$$

and we check that these isomorphism glue. In addition, there exists a canonical morphism $f_+ : Y_+ \to Y_0$ which is given by the inclusion of rings $R_0 \hookrightarrow R_{(x_i)}$ and, in fact,

$$Y_+ = \mathrm{BL}_{T_1=T_3=0}(Y_0)$$

and $f_+ : Y_+ \to Y_0$ is a *small resolution* because the exceptional set is of codimension >1.

Next consider the case that $a < 0$. Again without loss of generality we may assume that $a = -1$. In this case, we can show:

$$X^{ss}(L)/\!\!/\mathrm{G}_m \simeq Y_{-1} = V(T_1X_4 - T_2X_3, T_3X_4 - T_4X_3, T_1T_4 - T_2T_3) \subseteq \mathbb{A}^4 \times \mathbb{P}^1.$$

In this setting, the variety Y_{-1} admits a canonical morphism $f_- : Y_- \to Y_0$ and the birational morphisms f_+ and f_- fit into a diagram:

$$\begin{array}{ccc} Y_+ & & Y_- \\ & {\scriptstyle f_+}\searrow \quad \swarrow{\scriptstyle f_-} & \\ & Y_0 & \end{array}$$

which is called a *flip*. The varieties Y_+ and Y_- are not isomorphic but they are isomorphic outside the fibres $f_\pm^{-1}(0) \simeq \mathbb{P}^1$.

Example 5.5 (*Toric varieties*, [2, Chap. 12]) Let $\mathrm{T} = \mathrm{G}_m^r$ act linearly on $\mathbb{A}^n$ by the formula

$$(t_1, \ldots, t_r) \cdot (x_1, \ldots, x_n) = (t^{\mathbf{a}_1}x_1, \ldots, t^{\mathbf{a}_n}x_n)$$

where

$$\mathbf{a}_j = (a_{1j}, \ldots, a_{rj}) \in \mathbb{Z}^r, \text{ and } t^{\mathbf{a}_j} = t_1^{a_{1j}} \ldots t_r^{a_{rj}}, \text{ for } j = 1, \ldots, n.$$

We view the $\mathbf{a}_j$ as elements of $\mathfrak{X}(\mathrm{T}) = \mathbb{Z}^r$ the character group of T. The group of T-linearized line bundles on $\mathbb{A}^n$ has the form

$$\mathrm{Pic}^{\mathrm{T}}(\mathbb{A}^n) \simeq \mathfrak{X}(\mathrm{T}) \simeq \mathbb{Z}^r.$$

Fix $\mathbf{a} = (\alpha_1, \ldots, \alpha_r) \in \mathbb{Z}^r$ and denote by $L_{\mathbf{a}}$ the linearized line bundle with total space $\mathscr{L}_{\mathbf{a}}$ linearized by:

$$t \cdot (x, w) = (t \cdot x, t^{\mathbf{a}} w) = (t \cdot x, t_1^{\alpha_1} \ldots t_r^{\alpha_r} w).$$

We then have, as in Example 4.2, that $F \in \mathrm{H}^0(\mathbb{A}^n, L^{\otimes d}) = \mathbb{C}[x_1, \ldots, x_n]$, for $d \geqslant 0$, is an element of $\mathrm{H}^0(\mathbb{A}^n, L^{\otimes d})^{\mathrm{T}}$, the space of invariant sections for the linearization of $L^{\otimes d}$ determined by $\mathbf{a}$, if and only if

$$F(t^{\mathbf{a}_1} x_1, \ldots, t^{\mathbf{a}_n} x_n) = t^{d\mathbf{a}} F(x_1, \ldots, x_n).$$

Equivalently, F is a linear combination of monomials

$$\mathbf{x}^{\mathbf{m}} := x_1^{m_1} \cdot \cdots \cdot x_n^{m_n}, \text{ for } \mathbf{m} = (m_1, \ldots, m_n) \in \mathbb{Z}^n_{\geqslant 0},$$

such that

$$m_1 \mathbf{a}_1 + \cdots + m_n \mathbf{a}_n = d\mathbf{a};$$

equivalently

$$\begin{pmatrix} a_{11} & \ldots & a_{1n} \\ \vdots & & \vdots \\ a_{r1} & \ldots & a_{rn} \end{pmatrix} \begin{pmatrix} m_1 \\ \vdots \\ m_n \end{pmatrix} = A \cdot \mathbf{m} = d\mathbf{a}, \tag{4}$$

for $A = (a_{ij})$.

Let S_d be the set of non-negative integral solutions to (4) and $\mathbb{C}[S_d]$ the $\mathbb{C}$-vector space that it determines. We then have that $\Gamma(\mathbb{A}^n, L^{\otimes d})^{\mathrm{T}} = \mathbb{C}[S_d]$ and

$$\bigoplus_{d \geqslant 0} \Gamma(\mathbb{A}^n, L^{\otimes d})^{\mathrm{T}} \simeq \mathbb{C}[S] := \bigoplus_{d \geqslant 0} \mathbb{C}[S_d] = \bigoplus_{d \geqslant 0} \mathbb{C}[S]_d.$$

Next, fix a minimal set of monomial generators $\mathbf{x}^{\mathbf{m}_1}, \ldots, \mathbf{x}^{\mathbf{m}_\ell}$ for the ideal

$$\mathbb{C}[S]_{>0} := \bigoplus_{d > 0} \mathbb{C}[S_d] = \bigoplus_{d > 0} \mathbb{C}[S]_d$$

and for each $\mathbf{m}_j = (m_{1j}, \ldots, m_{nj})$, for $j = 1, \ldots, \ell$, let $I_j := \{i : m_{ij} \neq 0\}$ and for each subset $I \subseteq \{1, \ldots, n\}$ let $\mathbf{x}_I = \prod_{i \in I} x_i$. We then have:

$$(\mathbb{A}^n)^{ss}(L) = \bigcup_{j=1}^{\ell} D(\mathbf{x}_{I_j})$$

where $D(\mathbf{x}_{I_j}) = \mathbb{A}^n \setminus \{\mathbf{x}_{I_j} = 0\}$.

Finally, let $\mathbb{Z}^n \to \mathbb{Z}^r$ be the map given by the matrix A, let $M := \ker A$ denote its kernel, let N be the image of $(\mathbb{Z}^n)^\vee \to M^\vee$, the dual map of the inclusion $M \hookrightarrow \mathbb{Z}^n$, and put $q := n - \operatorname{rank}(A)$. Further let Σ be the N-fan formed by the ℓ convex cones

$$\sigma_j \subseteq N_\mathbb{R} = N \otimes \mathbb{R} \simeq \mathbb{R}^q$$

spanned by the vectors $\overline{e}_i^\vee$, $i \notin I_j$; here $e_1^\vee, \ldots, e_n^\vee$ is the dual basis of the standard basis $e_1, \ldots, e_n$ of $\mathbb{Z}^n$ and $\overline{e}_i^\vee$ denotes the image of $e_i^\vee$ in $M^\vee$. In particular, Σ is a finite collection of rational convex polyhedral cones $\{\sigma_i\}$ in $\mathbb{R}^q$ with the property that $\sigma_i \bigcap \sigma_j$ is a common face of σ_i and σ_j.

The fan Σ determines a toric variety X_Σ. In more detail, for each cone $\sigma \in \Sigma$, put

$$\sigma^\vee = \{y \in \mathbb{R}^q : x \cdot y \geqslant 0, \text{ for all } x \in \sigma\}$$

and let $A_\sigma := \mathbb{C}[\sigma^\vee \cap M]$ be the semi-group algebra determined by $\sigma^\vee \cap M$. That the A_σ are finitely generated $\mathbb{C}$-algebras follows from Gordan's lemma, see for example [2, Lemma 12.1, p. 189]. In addition, the affine varieties $X_\sigma := \operatorname{Spec} A_\sigma$, for $\sigma \in \Sigma$, glue together to form the toric variety X_Σ and the following theorem expresses this toric variety as a GIT quotient for the line bundle $L = \mathscr{O}_{\mathbb{A}^n}$ linearized by $\mathbf{a} \in \mathbb{Z}^r$.

Theorem 5.6 ([2, Theorem 12.1, p. 192]) *In the above setting,*

$$(\mathbb{A}^n)^{ss}(L) /\!\!/ \mathrm{T} \simeq X_\Sigma.$$

6 The Numerical Criterion

Let G be a reductive group acting on an irreducible projective variety X and $\mathscr{L}$ the total space of a G-linearized ample line bundle $L \in \operatorname{Pic}^{\mathrm{G}}(X)$. In this section, we describe, in Theorem 6.1, the numerical criterion for stability, which, perhaps, can be seen as the second main theorem of GIT.

With this in mind, for every $x \in X$ and every 1-parameter subgroup

$$\lambda(t) : \mathrm{G}_m = \mathbb{C}^\times \to \mathrm{G},$$

the subgroup $\lambda(t)$ acts on the fibre $\mathscr{L}|_{x_0}$ over the point

$$x_0 := \lim_{t \to 0} \lambda(t) \cdot x$$

via the character

$$t \mapsto t^{m^L(x,\lambda)}$$

for some integer $m^L(x, \lambda)$; put

$$\mu^L(x, \lambda) := -m^L(x, \lambda),$$

compare with [8, Definition 2.2, p. 49]. In what follows we refer to the integer $m^L(x, \lambda)$ as the λ-*weight* of x. Furthermore, we remark that the definition of x_0 is given by the valuative criterion for properness of X over $\mathbb{C}$, see [8, p. 49] for a more detailed explanation.

The second main theorem of GIT expresses the conditions for semi-stability and stability for $x \in X$ with respect to L in terms of the functions $\mu^L(x, \lambda)$.

Theorem 6.1 ([8, Theorem 2.1, p. 49], [9, Theorem 4.9, p. 105], [3, p. 10]) *In the above setting, the following assertions hold true:*

(a) $x \in X^{ss}(L)$ *if and only if* $\mu^L(x, \lambda) \geqslant 0$ *for all* 1*-parameter subgroups* λ*; and*
(b) $x \in X^{s}(L)$ *if and only if* $\mu^L(x, \lambda) > 0$ *for all* 1*-parameter subgroups* λ*.*

In what follows, motivated by the numerical criterion Theorem 6.1, given a 1-parameter subgroup λ, we say that $x \in X$ is λ-*semi-stable* if $\mu^L(x, \lambda) \geqslant 0$ and that $x \in X$ is λ-*unstable* if $\mu^L(x, \lambda) < 0$.

As explained in [3, Sect. 1.1.5], we can use the *Weight polytope* to give a combinatorial description of the numerical criteria for the case of a linear action of the torus. To this end, we consider the torus $\mathrm{T} = \mathrm{G}_m^n$ acting linearly on an irreducible variety X in $\mathbb{P}(V^*)$. Then $\mathfrak{X}(\mathrm{T}) \simeq \mathbb{Z}^n$ via the isomorphism that associates to every $(r_1, \ldots, r_n) \in \mathbb{Z}^n$, the homomorphism $\chi : \mathrm{T} \to \mathrm{G}_m$ defined by the formula

$$\chi(t_1, \ldots, t_n) = t_1^{r_1} \ldots t_n^{r_n}.$$

Furthermore, every 1-parameter subgroup $\lambda(t) : \mathrm{G}_m \to \mathrm{T}$ is given by the formula

$$\lambda(t) = (t^{r_1}, \ldots, t^{r_n}),$$

for some $(r_1, \ldots, r_n) \in \mathbb{Z}^n$. Thus, in this way, we can identify the set of 1-parameter subgroups, which we denote by $\mathfrak{X}_*(\mathrm{T})$, of T with the group $\mathbb{Z}^n$.

Now, let $\lambda \in \mathfrak{X}_*(\mathrm{T})$ and $\chi \in \mathfrak{X}(\mathrm{T})$. The composition $\chi \circ \lambda$ is a homomorphism

$$\chi \circ \lambda : \mathrm{G}_m \to \mathrm{G}_m$$

and hence is defined by an integer which we denote by $\langle \lambda, \chi \rangle$. In addition, the pairing

$$\mathfrak{X}_*(\mathrm{T}) \times \mathfrak{X}(\mathrm{T}) \to \mathbb{Z},$$

defined by

$$(\lambda, \chi) \mapsto \langle \lambda, \chi \rangle,$$

is isomorphic to the natural dot product pairing

$$\mathbb{Z}^n \times \mathbb{Z}^n \to \mathbb{Z}.$$

Thus, if we now let

$$V^* = \bigoplus_{\chi \in \mathfrak{X}(\mathrm{T})} V_\chi,$$

where

$$V_\chi = \{v \in V^* : t \cdot v = \chi(t) \cdot v\},$$

then we can write every $v \in V^*$ as

$$v = \sum_\chi v_\chi,$$

where $v_\chi \in V_\chi$. The group T acts on the vector v by the formula

$$t \cdot v = \sum_\chi \chi(t) \cdot v_\chi,$$

for $t \in \mathrm{T}$, and we set

$$\mathrm{wt}(V^*) = \{\chi \in \mathfrak{X}(\mathrm{T}) : V_\chi \neq \{0\}\}.$$

This is a finite subset of $\mathbb{Z}^n$ and its convex hull in $\mathbb{R}^n$ is called the *weight polytope* and is denoted by $\overline{\mathrm{wt}(V^*)}$.

Concretely, let $x \in \mathbb{P}(V^*) = \mathrm{Proj}\,\mathrm{Sym}^\bullet(V)$ be represented by a vector

$$v = \sum_\chi v_\chi \in V^*.$$

We set

$$\mathrm{wt}(x) = \{\chi : v_\chi \neq 0\};$$

this is the *weight set* of x and we define the *weight polytope* of x by setting:

$$\overline{\mathrm{wt}(x)} = \text{convex hull of } \mathrm{wt}(x) \text{ in } \mathfrak{X}(\mathrm{T}) \otimes \mathbb{R}.$$

In this setting, we have:

$$-\mu^L(x, \lambda) = m^L(x, \lambda) = \min_{\chi \in \mathrm{wt}(x)} \langle \lambda, \chi \rangle,$$

for $L = \mathscr{O}_{\mathbb{P}(V^*)}(1)|_X$, compare with [4, Definition 4.16].

Example 6.2 ([4, p. 202] or [7, Sect. 1.9]) Consider now the case of the 1-parameter subgroup

$$\lambda(t) = \operatorname{diag}(t^a, t^b, t^c), \text{ with } a + b + c = 0,$$

of the diagonal torus $\mathrm{T} \subseteq \mathrm{SL}_3(\mathbb{C})$, acting on $\mathbb{P}^{N_d} = \mathbb{P}(V^*)$, for $N_d = \binom{d+2}{d} - 1$, the projective space of plane curves of degree d.

In this case, $\overline{\mathrm{wt}(V^*)}$ is the 3-simplex with barycentre $(d/3, d/3, d/3)$. In what follows, we identify the top vertex of the simplex with the monomial x^d, the bottom left vertex of the simplex with the monomial y^d and the bottom right vertex of the simplex with the monomial z^d. We also view λ as determining a line $ai + bj + ck = 0$ passing through the point $(d/3, d/3, d/3)$. This line cuts the simplex in two. Monomials in the top half (including those on the line) are semi-stable while monomials in the bottom half (excluding those on the line) are unstable.

Further, if

$$s(x, y, z) = \sum_{i+j+k=d} c_{ijk} x^i y^j z^k,$$

then

$$\mathrm{wt}(s) = \{(i, j, k) : c_{ijk} \neq 0\}.$$

We can also consider the case of cubic curves in more detail. For instance suppose we fix the 1-parameter subgroup

$$\lambda(t) = \operatorname{diag}(t^{-5}, t^1, t^4).$$

We then can make the table:

Monomial	x^3	x^2y	x^2z	xy^2	xyz	xz^2	y^3	y^2z	yz^2	z^3
λ-weight	-15	-9	-6	-3	0	3	3	6	9	12

from which it is clear that the λ-semi-stable monomials are

$$x^3, x^2y, x^2z, xy^2, xyz$$

while the λ-unstable monomials are

$$xz^2, y^3, y^2z, yz^2, z^3.$$

Furthermore, if, for example,

$$s = x^3 + x^2y + z^3,$$

then

$$m^{\mathcal{O}_{\mathbb{P}^{N_d}}(1)}(s, \lambda) = \min\{-15, -9, 12\} = -15$$

so that s is λ-semi-stable. On the other hand, if, for example, $s = y^3 + y^2 z$, then

$$m^{\mathcal{O}_{\mathbb{P}^{N_d}}(1)}(s, \lambda) = \min\{3, 6\} = 3$$

and so s is λ-unstable.

It turns out that the stability of a degree 3 form $s(x, y, z)$ is related to the singularities of the curve it defines. For instance, $s(x, y, z)$ will be λ-stable if and only if the curve it defines is smooth, while $s(x, y, z)$ will have at worst nodes if and only if the curve it defines is λ-semi-stable, see [4, p. 204] for a more complete discussion.

7 Stratifying the Unstable Locus

In this final section, we make some remarks about stratifying the unstable locus. For the most part we follow [3, Sect. 1.3]. To this end, let G be a reductive group acting on a projective variety X and let $L \in \mathrm{Pic}^{\mathrm{G}}(X)$ be an ample G-linearized line bundle on X. Furthermore, let T be a maximal torus of G and $\mathrm{W} = \mathrm{N}_{\mathrm{G}}(\mathrm{T})/\mathrm{T}$ its Weyl group. Let $\mathfrak{X}_*(\mathrm{T})$ denote the set of 1-parameter subgroups of T and $\mathfrak{X}_*(\mathrm{G})$ the set of 1-parameter subgroups of G. We then have

$$\mathfrak{X}_*(\mathrm{T}) \otimes \mathbb{R} \simeq \mathbb{R}^n,$$

for $n = \dim \mathrm{T}$. Next, fix a W-invariant norm $||\cdot||$ on $\mathbb{R}^n$ and for every 1-parameter subgroup $\lambda(t)$ of G define

$$||\lambda|| := ||\,\mathrm{Int}(g) \circ \lambda||$$

for $\mathrm{Int}(g)$ an inner automorphism of G such that $\mathrm{Int}(g) \circ \lambda \in \mathfrak{X}_*(\mathrm{T})$. For each $x \in X$, put

$$\overline{m}^L(x, \lambda) := \frac{m^L(x, \lambda)}{||\lambda||}$$

and

$$M^L(x) := \sup_{\lambda \in \mathfrak{X}_*(\mathrm{G})} \overline{m}^L(x, \lambda).$$

In what follows, we say that a 1-parameter subgroup $\lambda(t) \in \mathfrak{X}_*(\mathrm{G})$ is *adapted* to x, with respect to L, if

$$M^L(x) = \frac{m^L(x, \lambda)}{||\lambda||},$$

and we denote by $\Lambda^L(x)$, the set of *primitive* (i.e., not divisible by a positive integer) adapted 1-parameter subgroups.

For each integer $d > 0$ and each conjugacy class $\langle\lambda\rangle$ of a 1-parameter subgroup $\lambda \in \mathfrak{X}_*(\mathrm{G})$ put

$$\mathrm{S}^L_{d,\langle\lambda\rangle} := \{x \in X : M^L(x) = d \text{ and there exists } g \in \mathrm{G} \text{ such that } \mathrm{Int}(g) \circ \lambda \in \Lambda^L(x)\}.$$

Then, if $\mathscr{C}$ denotes the set of conjugacy classes of 1-parameter subgroups of G, we can write

$$X = X^{ss}(L) \bigcup \left(\bigcup_{d>0, \langle\lambda\rangle \in \mathscr{C}} \mathrm{S}^L_{d,\langle\tau\rangle} \right).$$

This is a finite stratification of X into Zariski locally closed G-invariant subvarieties of X.

In fact, more is true:

Theorem 7.1 ([3, Theorem 1.3.9, p. 16]) *In the setting of this section, the following assertions hold true:*

(a) *the set of locally closed subvarieties* S *of* X *which can be realized as the stratum* $\mathrm{S}^L_{d,\langle\tau\rangle}$ *for some ample* $L \in \mathrm{Pic}^{\mathrm{G}}(X)$, $d > 0$ *and* $\lambda \in \mathfrak{X}_*(\mathrm{G})$ *is finite; and*
(b) *the set of possible open subsets of* X *which can be realized as the set of semistable points with respect to some ample* G*-linearized line bundle is finite.*

As a final comment we note that, inside of $\mathrm{NS}^{\mathrm{G}}(X) \otimes \mathbb{R}$, the G-linearized real Néron-Severi space of X, for $\mathrm{NS}^{\mathrm{G}}(X)$, the Néron-Severi group of G-linearized ample line bundles modulo homological equivalence, we have the G-ample cone $\mathrm{C}^{\mathrm{G}}(X)$. This is the convex cone spanned by the classes of G-linearized ample line bundles on X.

In this direction, the starting point to the study of $\mathrm{C}^{\mathrm{G}}(X)$ is:

Theorem 7.2 ([3, Theorem 0.2.3, p. 8], [11, Theorem 2.3, p. 701]) *The following assertions hold true:*

(a) *there are only finitely many chambers, walls and cells inside of* $\mathrm{C}^{\mathrm{G}}(X)$*;*
(b) *each wall of* $\mathrm{C}^{\mathrm{G}}(X)$ *is a closed convex cone in* $\mathrm{C}^{\mathrm{G}}(X)$*; and*
(c) *the closure of a chamber of* $\mathrm{C}^{\mathrm{G}}(X)$ *is a rational polyhedral cone inside of* $\mathrm{C}^{\mathrm{G}}(X)$.

Acknowledgements The author was financially supported by an AARMS postdoctoral fellowship at the University of New Brunswick. The author also thanks the organizers of the PIMS Superschool on derived categories and D-branes for asking him to give an introductory lecture about GIT and also for suggesting a possible list of topics to cover.

References

1. A. Borel, Linear algebraic groups, Springer-Verlag, New York, 1991.
2. I. Dolgachev, Lectures on invariant theory, Cambridge University Press, Cambridge, 2003.
3. I. V. Dolgachev and Y. Hu, *Variation of geometric invariant theory quotients*, Inst. Hautes Études Sci. Publ. Math. (1998), no. 87, 5–56.
4. J. Harris and I. Morrison, Moduli of curves, Springer-Verlag, New York, 1998.
5. R. Hartshorne, Algebraic geometry, Springer-Verlag, New York, 1977.

6. F. C. Kirwan, Cohomology of quotients in symplectic and algebraic geometry, Mathematical Notes, Princeton University Press, Princeton, NJ, 1984.
7. D. Mumford, *Stability of projective varieties*, Enseignement Math. (2) **23** (1977), no. 1-2, 39–110.
8. D. Mumford, *Geometric invariant theory*, Springer-Verlag, 1982.
9. P. E. Newstead, Introduction to moduli problems and orbit spaces, Tata Institute of Fundamental Research, Bombay; by the Narosa Publishing House, New Delhi, 1978.
10. C. S. Seshadri, *Quotient spaces modulo reductive algebraic groups*, Ann. of Math. (2) **95** (1972), 511–556; errata, ibid. (2) 96 (1972), 599.
11. M. Thaddeus, Geometric invariant theory and flips, J. Amer. Math. Soc. 9 (1996), no. 3, 691–723.

Birational Geometry and Derived Categories

Colin Diemer

The aim of these notes is to describe some relations between the birational geometry of algebraic varieties and their associated derived categories. This is a large subject with several diverging paths, so we'll restrict our focus to the realm of topics discussed at the Alberta superschool. Being lecture notes, the discussion here is somewhat informal, and we only attempt a general overview, and referring the reader to the relevant original research articles for details. Given the background of the participants of the Alberta superschool, these notes take the somewhat unorthodox approach of assuming that the reader has modest familiarity with derived and triangulated categories, but is perhaps not as familiar with the more cabalistic aspects of birational geometry.

We'll start with some of the basic (read: Hartshorne level) notions of birational geometry, along with some elementary observations about how these might interact with derived categories of coherent sheaves. The default reference for the general theory of birational geometry is the book of Kollár and Mori [17], although the more elementary Bulletin article by Kollár [16] is also a classic and highly recommended. Many of the results on derived categories discussed in these notes are covered in detail in Huybrecht's book [12], and specifically in chapters "A Brief Introduction to Geometric Invariant, Introduction to Mirror Symmetry, Differential Graded Categories and this chapter". Historically these results on derived categories emerged from the pioneering work of Bondal and Orlov, for which we refer to their ICM address [4].

Let X be an algebraic variety; for ease in exposition we will only work over the field $k = \mathbb{C}$ in these notes. Associated to X we have another field, namely the field of rational functions on X, denoted $\mathbb{C}(X)$. If two varieties X and Y are such that

C. Diemer (✉)
University of Alberta Summer Superschool 2016, Edmonton, AB, Canada
e-mail: cdiemer@gmail.com

M. Ballard et al. (eds.), *Superschool on Derived Categories and D-branes*, Springer Proceedings in Mathematics & Statistics 240,
https://doi.org/10.1007/978-3-319-91626-2_7

$\mathbb{C}(X) \cong \mathbb{C}(Y)$ as fields, we say that X and Y are **birational**. Varieties are birational exactly when they are isomorphic generically, i.e when the local rings of their generic points are isomorphic. Chow's lemma says that any variety is birational to a projective variety, so we will often tacitly assume that any variety in this article is projective or at least quasi-projective. Hironaka's big theorem says that any variety is birational to a smooth variety; however, for many purposes it is best to *not* assume X is smooth as there may be spaces birational to X with mild singularities, but with some desirable properties, as we will see.

Example: Rational varieties. Let $X = \mathbb{P}^n$. Then

$$\mathbb{C}(X) \cong \mathbb{C}(x_1, \ldots, x_n)$$

is the usual field of rational functions. Varieties whose function fields are isomorphic to $\mathbb{C}(\mathbb{P}^n)$ for some n are dubbed **rational**. For example, products of projective spaces are rational, but they not actually isomorphic to a single projective space. Explicit criteria for determining the rationality of a given variety are often quite delicate. We will not purse the matter here, but see Kuznetsov's article [18] for a survey of how methods from derived categories apply to problems in rationality.

In practice we are interested in the more geometric manifestations of birationality. Two varieties X and Y are birational if and only if there exist Zariski open subsets $U_X \subseteq X$ and $U_Y \subseteq Y$ and an honest isomorphism of varieties $f : U_X \xrightarrow{\sim} U_Y$. When X and Y are birational we write

$$f : X \dashrightarrow Y.$$

Of course, the direction of the arrow is rather arbitrary as we also have $f^{-1} : Y \dashrightarrow X$. If f is actually defined on all of X, i.e. f extends to an honest morphism of varieties we say, not surprisingly, that we have a **birational morphism**. If $f : X \to Y$ is a birational morphism, the locus in X where f is not a local isomorphism is called the **exceptional locus**, denoted $\mathrm{Exc}(f) \subset X$. If X is smooth or had mild singularities,[1] then $\mathrm{Exc}(f)$ is codimension one, i.e. a divisor ("van der Waerden purity").

Remark: In his thesis [29], Usnich observed a very direct relation between birationality and derived categories (see also [19]). Let X be smooth and let $\mathrm{D}^b\mathrm{coh}(X)$ be the usual bounded derived category of coherent sheaves on X. Let $\mathrm{coh}^1(X)$ be the full subcategory of $\mathrm{coh}(X)$ consisting of sheaves having support of codimension ≥ 1. By standard methods, one can form the quotient category

$$\mathrm{coh}_1(X) := \mathrm{coh}(X)/\mathrm{coh}^1(X).$$

It is not terribly hard to see that $\mathrm{coh}_1(X)$ is equivalent to the (abelian) category of finite dimensional vector spaces over the function field $\mathbb{C}(X)$ (and is thus a birational

[1] Here "mild" actually means normal and $\mathbb{Q}$-factorial. The author apologizes for occasionally taking the liberty to be relaxed about types of singularities in this article, despite their fundamental role in birational geometry.

invariant). One can also form the derived category

$$D_1^b(X) := D^b(\mathrm{coh}_1(X)),$$

and it is likewise not too hard to verify that X and Y are birational if and only if $D_1^b(X)$ and $D_1^b(Y)$ are equivalent as triangulated categories. Unfortunately, beyond the articles mentioned above, the structure of these quotient categories does not appear to have been much studied (Usnich was able to make some progress on its structure for $X = \mathbb{P}^2$). In the remainder of these notes we will thus instead look at $D^b\mathrm{coh}(X)$ directly.

The subject of birational geometry studies equivalence classes of varieties up to birational equivalence. In particular, it asks for criteria for when two varieties are or are not birational, and aims to find canonical representatives of a given birational equivalence class. Restricting to smooth varieties, one may ask (and indeed, many have) the analogous questions for the derived categories of coherent sheaves of varieties up to equivalences of triangulated categories.

Caution: A thoughtful inspection of the definition of birationality suggests heuristically that relating the derived categories of birational varieties will, in general, be a hopeless endeavor. If we view the derived category of coherent sheaves $D^b\mathrm{coh}(X)$ as naively consisting of (complexes) of things like vector bundles or structure sheaves of subvarieties, then such objects have no a priori reason to behave well under birational morphisms. For example, if we take the structure sheaf of a point $\mathcal{O}_p$ where $p \in \mathrm{Exc}(f)$ and view it as a complex concentrated in degree zero, it is not at all obvious what type of object it should correspond to in $D^b\mathrm{coh}(Y)$ under a hypothetical correspondence between the derived categories. The situation is even worse if f is not a morphism but only birational, so that the value of f on sine $p \in X$ need not even be well-defined. In these notes we'll try to get a feeling for some situations where one actually can make sense of such issues.

Before proceeding, let's review some more basic facts and constructions.

Example: algebraic curves. Here one has the following two well-known theorems:

1. If $C \dashrightarrow C'$ are smooth projective curves which are birational, then C and C' are actually isomorphic.
2. If C is any curve, then it is birational to a smooth projective curve (which is unique by the above) via normalization.

So if one restricts to smooth curves, birational geometry is fundamentally uninteresting in dimension one. This matches up well with the situation for derived categories: two smooth curves have equivalent derived categories of coherent sheaves if and only if the curves are isomorphic. That said, the cautious reader may note that the proof[2] of this is slightly non-trivial, as the case of elliptic curves needs to be handled separately. Recall that elliptic curves are in particular curves whose canonical

[2]See e.g. Theorem 6.15 in [12].

bundle K_C is trivial. We should remark that the other curves (i.e. curves with genus >0 having K_C ample, and rational curves having $-K_C$ ample) can be handled by a special case of the following important theorem of Bondal-Orlov.

Theorem: Let X and Y be smooth projective varieties of any dimension and such that $D^b\mathrm{coh}(X) \cong D^b\mathrm{coh}(Y)$ as triangulated categories. If the canonical divisor K_X is either ample or anti-ample,[3] then X and Y are isomorphic.

The theorem states that, up to isomorphism the space X can be recovered completely from its derived category. The hypotheses on the canonical bundle are in very rough analogy with the general classification theory in birational geometry, as we'll see later on.

Remark: The Bondal-Orlov theorem suggests that varieties with K_X trivial (i.e. Calabi–Yau) may have auto-equivalences which do not come from automorphisms of X (or even other obvious homological operations such as degree shifts or twists by invertible objects). Indeed, such exotic auto-equivalences can and do exist, and this is a rich and ongoing topic of study, going under the title "**spherical functors**." We will regretfully not discuss this topic here, and defer only to the original article of Seidel and Thomas where they were first studied [26]. Likewise, when K_X is neither ample nor anti-ample, it is possible to have another variety Y and a derived equivalence

$$D^b\mathrm{coh}(X) \cong D^b\mathrm{coh}(Y),$$

but X and Y are *not* birational. The famous example is the so-called Pfaffian-Grassmannian derived equivalence, for which we refer to the work of Borisov–Căldăraru [5]. (A more elementary example is Mukai's observation that the derived categories of an abelian variety and its dual variety are equivalent.) Generalizations of the Pfaffian-Grassmannian equivalence and the general phenomenon of non-birational derived equivalences is still a very active line of research. For these notes, we simply take their existence as further evidence that one needs to tread carefully when relating the two subjects.

With such cautions in mind, let's move on to the most basic situation where birational morphisms do actually induce easily understood functors between derived categories.

Example: blow-ups. Let X be a smooth projective variety of dimension ≥ 2. If $Z \subset X$ is any smooth subvariety of codimension $r \geq 2$, we can form the blow-up $\mathrm{Bl}_Z(X)$, which is itself smooth projective, and the blow-down map is a birational morphism $\pi : \mathrm{Bl}_Z(X) \to X$. For derived categories, one has the famous semiorthogonal decomposition of Bondal Orlov:

$$D^b\mathrm{coh}(\mathrm{Bl}_Z(X)) = \langle \underbrace{D^b\mathrm{coh}(Z), \ldots D^b\mathrm{coh}(Z)}_{r-\text{copies}}, D^b\mathrm{coh}(X) \rangle.$$

[3]I.e. $-K_X$ is ample.

In particular, the derived pull-back

$$\pi^* : \mathrm{D}^b\mathrm{coh}(X) \to \mathrm{D}^b\mathrm{coh}(\mathrm{Bl}_Z(X))$$

is a fully-faithful embedding.

It is desirable to understand how this semi-orthogonal decomposition can be generalized. We suggested above that it is unreasonable to expect such a decomposition for arbitrary birational morphisms. The proposed class of preferred birational morphisms will indeed come from the main structure theories in higher dimensional birational geometry, which we will now attempt to survey. Once that is in place we can state the main conjectures and results connecting this theory with derived categories and functors between them.

The main structure theories alluded to above all center around the question of trying to find a preferred or distinguished birational model for a given variety X. The existence of blow-ups means that in dimension at least two there are always infinitely many smooth varieties in the same birational equivalence class. This suggests that finding a unique or at least distinguished representative within a birational equivalence class is not so trivial. The theory developed to accomplish this is called the minimal model program (**MMP**) or the **Mori program**. As a warm up, let's review how this works for surfaces.

Example: surfaces and blow-ups. If $f : \mathrm{Bl}_p(S) \to S$ is the blow-up of a smooth surface at a point, then $\mathrm{Exc}(f) = C \cong \mathbb{P}^1$ is a rational curve and the self-intersection is -1, i.e. is a -1 curve. A theorem of Castelnuovo says that the converse is true as well: any -1 curve on a smooth surface can be blown-down, and the resulting surface is smooth. By adjunction, having $C^2 = -1$ and C rational is equivalent to $K_S \cdot C = -1$. Thus, for surfaces, blowing-down can be controlled by looking at intersections of rational curves with the canonical divisor.

Remark: The fact that $\mathrm{Exc}(f) \cong \mathbb{P}^1$ for blowing up smooth points on surfaces generalizes in a certain way to higher dimensions. Namely, if $f : X \to Y$ is a birational morphism with X smooth and f not an isomorphism, then through the general point of any component of $\mathrm{Exc}(f)$ there is a rational curve contracted by f, i.e. $\mathrm{Exc}(f)$ is "**uniruled**." Unfortunately, the higher dimensional analogue of Castelnuovo's contractibility theorem is a much more delicate issue and one of the many issues that the MMP tackles.

The MMP for surfaces: Here the MMP is not so bad, although requires some non-trivial work to fill in all the details in the proofs. It was basically known to Zariski, although is now usually terminologically re-packaged a bit after Mori's work.

Theorem (baby MMP for surfaces): A smooth projective surface is birational to either:

1. $\mathbb{P}^1 \times C$ for some curve C, or
2. A smooth surface S' which the property that $K_{S'} \cdot C \geq 0$ for any curve $C \subset S'$ (i.e. S' is a "**minimal model**").

Note that rational surfaces (including $\mathbb{P}^2$ itself) are subsumed in the first case as they are birational to $\mathbb{P}^1 \times \mathbb{P}^1$. The reader may know that there are further sub-cases and refinements one can include in the above theorem via certain numerical invariants, and the above version was formulated for brevity's sake.

A sketch of a proof of this classification is as follows: first you want to contract all -1 curves. You can do this iteratively by Castelnuovo. A surface may actually have infinitely many -1 curves (the blow up of $\mathbb{P}^2$ at nine general points is the standard example), but blowing down lowers the rank of H_2 (or the Picard number) by one, so the process actually terminates finitely. So we need to show that a smooth surface which is not birational to a product and which has no -1 curves is a minimal model. Suppose for contradiction that one has a curve C on such a surface which is such that $K_S \cdot C < 0$. A theorem of Mori ("**bend and break**") says that if a curve C is such that $K_S \cdot C < 0$, then you actually have a rational curve with the same property, i.e. we may assume that $C \cong \mathbb{P}^1$. Since we are not birational to a product, our curve does not deform, so $C^2 \leq 0$ (recall the self-intersection is the degree of the normal bundle). A quick inspection of the adjunction formula shows that C is forced to be a -1 curve, a contradiction.

Remark: It is important to understand the extent to which the above is a reasonable classification. The first case in the theorem looks like a good output for a classification: products of curves are easy to understand, and we've effectively reduced our study a problem in a lower dimension. But why would we be satisfied with having minimal models as an output as in the second case? The definition does not shed much light on what such an S' actually looks like. There are deep conceptual answers to this (key words: "abundance conjecture" and "canonical models"). A more heuristic argument is the following easy fact.

Proposition: Let $f : X \to Y$ a birational morphism (here X may be of any dimension) and suppose that X contains no rational curves. Then f is an isomorphism.

The proposition follows trivially from the remark above about the uniruledness of exceptional loci. Intuitively, the proposition says that a birational model free of rational curves is minimal in the sense that it cannot have any non-trivial birational morphisms coming out of it. (Such models are sometimes called "absolute minimal models" in the literature, though terminology differs). A related observation, which follows immediately from the bend and break theorem, is that if X contains no rational curves, then K_X is nef. So morally we take the definition of minimal model to be simultaneously a weakening of the property of having no rational curves and of the property of being an absolute minimal model. A main goal of the MMP is to establish the existence of minimal models in higher dimensions via a process very roughly analogous to blowing down -1 curves.

Remark: We haven't said anything about derived categories for some time. For now, let's only remark briefly that smooth varieties with K_X ample (i.e. canonically polarized varieties) are in particular minimal models and thus potential and desirable outputs of the MMP. Above, we saw that the structure of their derived categories are particular nice in that the derived category completely determines the variety. On the other hand, varieties with K_X trivial (i.e. Calabi–Yau) are also minimal models, and as noted above their derived categories can have exotic auto-equivalences. In a very rough sense, we see that the structure of the derived category reflects how complicated the variety is from the perspective of the MMP.

The MMP in higher dimensions. One can mimic the surface case discussed above, but essentially every aspect requires technical improvements to even state properly, and monstrous amounts of historical efforts to prove. For our purposes, let's focus on one aspect: how to generalize the process of blowing down a -1 curve. The following is a succinct version of what are known as the Cone Theorem and Contraction Theorem of Mori.

Theorem: Suppose X is a smooth variety which is not a minimal model. Then there is an extremal curve[4] C such that $K_X \cdot C < 0$ and there exists a normal variety Y equipped with a birational morphism $f : X \to Y$ which obeys:

- $f(C)$ is a point.
- $\rho(X) - \rho(Y) = 1$ where ρ denotes Picard number.
- $-K_X$ is f-ample.

The birational morphisms $f : X \to Y$ supplied by the theorem are called extremal contractions. The third condition is actually automatic if you instead carefully define which curves besides C are to be contracted.

There end up being only three main sub-cases to study depending on the behavior of f. First, f could be a fibration in which case the situation is similar to that of a ruled surface - this situation is called the **Mori Fiber Space** case. Or f could contract a divisor containing C, in which case it is similar to a blow-down - this situation is called a **divisorial contraction**. The last case is when f contracts something of smaller codimension, this is called a **small extremal contraction**. They are the hardest case to deal with, as we will soon see. The idea of the MMP is then to produce a minimal model by iteratively running extremal contractions.

Caution: If one starts with $X = \mathbb{P}^n$ and tries to run the above process, you'll observe that the output would be the trivial Mori Fiber Space $X \to \{\text{point}\}$. This is fine, and reminds us that when K_X is sufficiently negative (e.g. Fano), the birational model one gets as an output of the MMP will have a (possibly trivial) fibration structure with Fano fibers. Thus, the birational classification of varieties in the first sub-case is a topic a bit detached from the MMP.

The Problem. So why are things so much harder in higher dimensions than for surfaces, even granted the above theorem? The obstruction is that one may not be able iterate the contraction theorem for a subtle reason: the above theorem did not guarantee Y has to be smooth (or anything except normal) even if X was smooth. In practice, for the small extremal contractions, Y is so singular[5] that K_Y will not be

[4]I won't define "extremal" except to say that it's a suitable generalization of being a -1 curve on a surface. The definition is not hard, but formulating it is a bit orthogonal to our purposes in these notes.

[5]I.e. is Y is not $\mathbb{Q}$-factorial.

well behaved for the purposes of computing intersection numbers, and such a Y will be too singular to apply the contraction theorem to for iteration.

The fix: flips. Let $f : X \to Y$ be a small extremal contraction. Let X^+ be a variety with mild singularities which is not isomorphic to X, but itself admits a small birational morphism $f^+ : X^+ \to Y$. This is called a **flip** of f.

There is some structure hidden in the definition of a flip. Recall that $-K_X$ is f-ample. One can argue that K_{X^+} has to be f^+-ample (note the sign change). In practice we can't require X^+ smooth, but will allow singularities which are are mild enough to still be $\mathbb{Q}$-factorial (so-called terminal singularities). The following result was due to Mori in dimension three [20], and Birkar–Cascini–Hackon–McKernan [3] in all dimensions.

Huge Theorem: flips exist.

There's a variant of the definition of a flip where you require instead that K_X is f-trivial (and thus K_{X^+} is f^+-trivial). This is a **flop**. These don't show up directly in the MMP. However, instead of "minimizing" within a birational equivalence class, they reflect the ambiguity in choices of minimal models. Namely, we have the following theorem of Kawamata [16]:

Theorem: If X and Y are minimal models which are birational, then they are birational via a sequence of flops.

Let's see an example of flips/flops which avoids messy singularities.

Example: Atiyah flips/flops. Let X be smooth and suppose there is a subvariety $Y \subset X$ such that $Y \cong \mathbb{P}^k$ and the normal bundle obeys

$$N_{Y|X} \cong \mathcal{O}_Y(-1)^{\oplus l+1}$$

(so $k + l + 1 = \dim(X)$). Notice that the condition on the normal bundle is very similar to the contractibility criterion for rational curves in the surface case. If we blow up X along Y it's easy to see that the exceptional locus is now

$$\mathbb{P}^k \times \mathbb{P}^l \subset \mathrm{Bl}_Y(X).$$

It's also easy to see that we can blow down "in the other direction" giving a smooth variety X^+ containing $Y^+ \cong \mathbb{P}^l$. The composition of the inverse of the first blow-up with the second blow down is a birational map which is a flip when $k < l$ and a flop when $k = l$. When $k = l$ we actually have that X is isomorphic to X^+ as spaces, but the composed birational map is not the identity.

The general philosophy is that the distinguished birational maps coming from the MMP are exactly the types of birational morphisms for which one should hope for a corresponding functor of derived categories (we will review some precise conjectures below). We've already seen an example of this with the blow-up and the fully-faithful functor $\mathrm{D}^b\mathrm{coh}(X) \to \mathrm{D}^b\mathrm{coh}(\mathrm{Bl}_Y(X))$. So at least in simple cases where the singularities are under-control, divisorial contractions give a fully-faithful functor. Likewise, in the nicest cases, the Mori fiber spaces appearing the MMP will be actual fiber bundles, and then induce fully-faithful functors due to standard results on derived categories of bundles. To emphasize the core idea being suggested, we quote Bondal and Orlov:

"*This suggests the idea that the minimal model program of the birational geometry can be viewed as a minimization of the derived category* $\mathrm{D}^b\mathrm{coh}(X)$ *in a given birational class of X.*"

Here the process of minimization is implemented by fully-faithful functors. We state in particular, the following conjecture of Bondal and Orlov which has attracted a fair amount of attention.

Conjecture: Let $f : X \to X^+$ be either a flip or a flop. Assume for simplicity that both spaces are smooth. Then there exists a fully-faithful functor $\mathrm{D}^b\mathrm{coh}(X^+) \to \mathrm{D}^b\mathrm{coh}(X)$, which is an equivalence in the flop case.

The smoothness hypotheses are unreasonable from the perspective of the MMP, but there are various workarounds for formulating the conjecture with certain classes of singularities, which we will not discuss here, but see for example the article of Abramovich and Chen for some such methods [1]. We also remark that Kawamata [13] proposed a more general conjecture where the notion of a flop is generalized to a **K-equivalence**, which is roughly any birational map that preserves the respective canonical divisors, and the notion of flip is generalized to a K-dominance.

A subtle aspect of the conjecture is that it does not actually state what the hypothetical fully-faithful functor should be. Unlike the case with, say, blow-ups, flips and flops are not birational morphisms, i.e. are not defined everywhere, and so a naive functor like derived pull-back does not make sense. Elsewhere in the superschool, we've see the general Yoga of producing functors between derived categories via Fourier–Mukai kernels. That is, if we can find an interesting object

$$\mathcal{K} \in \mathrm{D}^b\mathrm{coh}(X \times X^+)$$

we get a corresponding Fourier–Mukai functor

$$\Phi_{\mathcal{K}} : \mathrm{D}^b\mathrm{coh}(X^+) \to \mathrm{D}^b\mathrm{coh}(X).$$

One can hope that clever choices of $\mathcal{K}$ may produce the conjectured fully-faithful functors.

Perhaps the most obvious candidate for an interesting Fourier–Mukai kernel would be the structure sheaf of a resolution of the birational map $f : X \dashrightarrow X^+$, i.e. a space Z with two birational morphisms $g : Z \to X$ and $h : Z \to X^+$ and such that $f = g^{-1} \circ h$ on $X \setminus \mathrm{Exc}(f)$. Such a Z always exists by resolution of singularities, and the Fourier–Mukai functor is simply the derived functor

$$h_* g^* : \mathrm{D}^b\mathrm{coh}(X^+) \to \mathrm{D}^b\mathrm{coh}(X).$$

In special situations this does indeed produce the desired fully-faithful functor: in Bondal-Orlov's original article they proved this for the Atiyah flips and flop. The proof is not terribly difficult, but perhaps more subtle than one might expect. The main geometric feature that helps in their proof is that in the Atiyah case the resolution Z is by construction built via the explicit blow-ups, so one can utilize the two semiorthogonal decomposition on $D^b\text{coh}(Z)$ to aid in the calculations.

Unfortunately, a simultaneous resolution of $f : X \dashrightarrow X^+$ does not always result in a fully-faithful functor for more general types of flips or flops. The first example is the so-called Mukai flop. This is a modest variation of the Atiyah example where now X is a smooth variety of dimension $2n$ containing a subvariety $Y \cong \mathbb{P}^n$ with $N_{Y|X} \cong \Omega_Y$. Here one can also blow-up and blow-down in an analogous way to obtain a flop. However, Namikawa [22] observed that one can find explicit line bundles L on X such $\text{Ext}^2(L, L) = 0$ but $\text{Ext}^2(h_*g^*L, h_*g^*L) \neq 0$, thus violating fully-faithfulness.

Soon thereafter, though, Namikawa [23] proved the derived equivalence for the Mukai flop using a different kernel. Recall that flips and flops arised from a small contraction $X \to Y$ coming from the MMP where the birational model X^+ also has a small contraction $X^+ \to Y$. Thus one can form the fiber product

$$X \times_Y X^+$$

and take that as a Fourier–Mukai kernel (note that typically such a fiber product is highly singular, and thus is perhaps a more threatening object to work with than the simultaneous resolution). In addition to giving the correct kernel for the Mukai flop, the fiber product is the correct choice of kernel object for a dramatically large class of flops.

Theorem (Bridgeland [6]): Let $f : X \dashrightarrow X^+$ be a flop with X and X^+ both smooth and $\dim(X) = 3$. Then $D^b\text{coh}(X) \cong D^b\text{coh}(X^+)$.

It was observed a posteriori by Chen [9] that Bridgeland's equivalence is actually implemented by taking the fiber product as the kernel object. Abramovich and Chen [1] generalized Bridgeland's proof to include $\mathbb{Q}$-Gorenstein terminal singularities and establish the fully-faithfulness for threefold flips as well.

Remark: As alluded to by Chen's result, Bridgeland did not actually construct the derived equivalence by studying a Fourier–Mukai kernel object directly. Indeed, for threefold flops Bridgeland gave a complete and explicit answer to a problem alluded to near the start of this article: where do point objects go under a flop equivalence? The answer ends up being so-called "perverse point sheaves." In turn, this allows Bridgeland to actually *construct* the flopped space X^+ as a moduli space of perverse point sheaves with respect to the small contraction $X \to Y$. Bridgeland's argument relies highly on the fact that the small contraction $X \to Y$ contracts only a curve in case of dimension three, and thus regretfully has not yet been fully generalized to other types of small contractions which can arise in higher dimensions. We also remark that shortly after Bridgeland's proof, van den Bergh [30] gave a different approach to establishing threefold flop equivalences by reducing to the case where Y is affine and constructing a projective generator for the category of perverse point sheaves, so that both $D^b\mathrm{coh}(X)$ and $D^b\mathrm{coh}(X^+)$ become mutually equivalent to the derived category of modules over the endomorphism algebra of this generator.

Unfortunately, it is known that the fiber product $X \times_Y X^+$ does *not* always induce a derived equivalence. The main (and I believe, only explicitly known) class of examples are the so-called stratified Mukai flops. This is yet again another generalization of the Atiyah flop, and also a generalization of the Mukai flop, where now the space is the cotangent bundle of a Grassmannian. Namikawa [23] observed that neither the common resolution nor the fiber product give an equivalence (see also Kawamata [14] for the $G(2,4)$ case). Nevertheless, an explicit kernel inducing an equivalence was constructed by Cautis, Kamnitzer, and Licata [7, 8]. Their argument is highly adapted to the geometry of the stratified Mukai flop, and does not appear to suggest any general approach to constructing promising integral kernels. In summary we see that the Bondal-Orlov conjecture is still open, despite a complete answer for threefolds and some important classes of explicit higher dimensional flops.

We end these notes with some discussion of some quite recent approaches to answer similar questions, but by using *geometric invariant theory* (GIT) as an intermediate construction to describe birational maps. Since the fundamentals of geometric invariant theory were discussed in other talks at the superschool, let's give an extremely concise summary mostly to fix notation; we refer the reader to the original source [21] for details as well as the beautiful yet unpublished lecture notes of Reid [24].

Given a variety X equipped with the action of a reductive algebraic group G and an ample line bundle L, we say the action is linearized if we choose a lift of the G action to the total space $\mathrm{Tot}(L)$ such that the projection is equivariant and each $g \in G$ gives an isomorphism $L \cong g^*L$. It is well-known that so long as X is normal, then for any G action on X and ample L, at least some power $L^{\otimes n}$ will admit a linearization. We allow for the case where X is only quasi-projective; for example if X is affine then $L = \mathcal{O}_X$ is ample. Given such data, GIT produces an open subset $X^{ss}(L) \subset X$ called the semistable-points of X for the linearized action, which is designed so that the quotient $X^{ss}(L)/G$ is well-behaved. (Although our notation does not reflect as such, this depends not only on L but on the choice of linearization, which is crucial for what follows).

Example: $\mathbb{C}^*$ acting on an affine variety. Let $X = \mathrm{Spec}(R)$ be an affine variety equipped with the trivial line bundle $\mathbb{C} \times X$. It is easy to see that a $\mathbb{C}^*$ action on X equivalent to specifying a $\mathbb{Z}$-grading $R = \bigoplus R_i$. Likewise, a linearized action corresponds to a $\mathbb{Z}$-grading on $R[x] = \bigoplus R[x]_i$ where $R_i \subseteq R[x]_i$ and $x^n \in R[x]_{-n}$ (the n here corresponds to the choice of linearization). We then define the GIT quotient

$$X//_n\mathbb{C}^* := \mathrm{Proj}R[x]^{\mathbb{C}^*} = \mathrm{Proj}R[x]_0 = \mathrm{Proj}\bigoplus_{i\in\mathbb{N}} R_{ni}z^i. \tag{1}$$

Note that the resulting space only depends on whether $n = 0, n > 0$, or $n < 0$. We thus denote these three spaces by $X//0$, $X//+$ and $X//-$ respectively. By comparing with Mumford's definition of the semistable locus, one can easily check that they each correspond to $X^{ss}(\mathcal{O})/\mathbb{C}^*$ for the respective linearizations.

When $X = \mathbb{C}^n$ and so $R = \mathbb{C}[x_1, \dots x_n]$ we can connect this story to the Atiyah flips/flop. Namely, let $\mathbb{C}^*$ act on X so that $x_1, \dots x_k$ have degree 1, and $x_{k+1}, \dots, x_n$ have degree -1. Notationally, it is cleaner to instead write $n = k + l$ and $R = \mathbb{C}[x_1, \dots x_k, y_1, \dots y_l]$ where the y_i have degree -1. Then the invariant ring $R = k[x_1, \dots x_k, y_1, \dots y_l]^{\mathbb{C}^*}$ is generated by monomials $x_i y_j$ for $i = 1, \dots, k$, $j = 1, \dots l$. These monomials are subject to the relations governed by vanishing of the 2×2 minors of the matrix $k \times l$ matrix $(x_i y_j)$. The corresponding affine variety is the cone over $\mathbb{P}^k \times \mathbb{P}^l$ embedded via the Segre embedding. The two GIT quotients $X//+$ and $X//-$ correspond to removing the unstable loci $x_1 = \dots = x_k = 0$ and $y_1 = \cdots y_l = 0$ respectively. In this way we recover the blow-up diagram which produced the Atiyah flips (and when $k = l$, the flop), and the respective GIT quotients are the total spaces $\mathrm{Tot}_{\mathbb{P}^l}\mathcal{O}(-1)^{\oplus k}$ and $\mathrm{Tot}_{\mathbb{P}^k}\mathcal{O}(-1)^{\oplus l}$.

As suggested above, this example can be generalized dramatically, as was done by Thaddeus [28] and Dolgachev–Hu [10]. Briefly, the point is that for any reductive G acting on a normal projective variety X one can identify the space of linearizations of L as the lattice points of a certain convex cone, called the GIT cone. One can then subdivide this cone into convex locally polyhedral subcones (GIT chambers) where characters in the interior of a given subcone result in the same GIT quotient. In the above example, the GIT cone was all of $\mathbb{R} = \mathbb{Z} \otimes_{\mathbb{Z}} \mathbb{R}$, and the three chambers were $\mathbb{R}_{\leq 0}$, $\{0\}$, and $\mathbb{R}_{\geq 0}$. We now supply a laundry list of results concerning this structure. In general, if two GIT adjacent chambers meet along a wall, we refer to the respective GIT quotients also as $X//0$, $X//+$, $X//-$, where $X//0$ corresponds to choosing a linearization in the interior of the wall itself (this is entirely analogous to the case of a $\mathbb{C}^*$ action, but now depends on the choice of a particular wall). At this level of detail, the choice of sign $\pm$ is arbitrary, but item (3) on the list below may justify this ambiguity to you. We now supply a list of various important facts related to this structure; the first three may be found in [28], and the last is due to Włodarczyk [31].

1. There are always only finitely many walls in the GIT cone.
2. By construction there are always proper morphisms $X//+ \to X//0$ and $X//- \to X//0$ (this follows from the observation that the semistable loci for the $\pm$ chambers are subsets of the semistable locus for the wall). So long as $X//+$ or $X//-$ are both non-empty, these morphisms are birational, and we thus have a birational map $X//+ \dashrightarrow X//-$. If this birational map is small, it satisfies the properties of a flip, except that the relative Picard number need not necessarily be one.
3. If G acts on a variety X with ample line bundle L consider a family of linearizations parametrized by a simplex Δ in the GIT cone. Then there exists another variety Z with an action by a torus $T \cong (\mathbb{C}^*)^k$ and ample line bundle L' such that Δ also lies in the GIT cone for the T action on Z, and for any $t \in \Delta$ one has $X//_t G = Z//_t T$. In particular, if Δ is a line segment, $T \cong \mathbb{C}^*$.

4. If $f : X \to Y$ is *any* birational morphism of projective, normal, $\mathbb{Q}-$ factorial varieties, then there exists a space Z equipped with a $\mathbb{C}^*$ action and a line bundle L such that $X = Z//+$ and $Y = Z//-$ and f is induced from a GIT wall-crossing as in the above. If X and Y are smooth, then Z may be chosen to be smooth as well.

Morally, we may summarize these statements as saying that the study of birational morphisms is secretly encoded in GIT wall-crossings, and moreover in the world of only $\mathbb{C}^*$ actions. Frequently, the space Z appearing in (3) is called a **master space** and the space Z appearing in (4) is called a **birational cobordism**. Both have explicit constructions which we omit for brevity's sake.

One may very well wonder if this machinery could be used to prove conjectures about the fully-faithfulness of the functors appearing the Mori program as discussed above. Indeed, there is a recently developed approach to this going under the nom de plume of **grade restriction windows**. Some aspects of this can be seen in the works of Kawamata and van den Bergh mentioned above, although it was actually the physicists Herbst, Hori, and Page who suggested merging the study of equivalences of derived categories with GIT. These physicists' ideas were brought into the mathematical literature by Segal [25], and the theory was formalized in the works of Halpern-Leistner [11] and Ballard, Favero, Katzarkov [2].

Roughly, the idea is the following: instead of considering just derived categories of coherent sheaves $\mathrm{D}^b\mathrm{coh}(X)$, we consider derived categories of equivariant coherent sheaves $\mathrm{D}^b\mathrm{coh}^G(X)$ to incorporate the G action on X. If $X^{ss} \subset X$ denotes the semistable locus coming from a linearized ample line bundle, we have the obvious derived restriction functor

$$r : \mathrm{D}^b\mathrm{coh}^G(X) \to \mathrm{D}^b\mathrm{coh}^G(X^{ss}).$$

One would like to find an adjoint functor, hopefully fully-faithful, to instead describe $\mathrm{D}^b\mathrm{coh}^G(X^{ss})$ as sitting inside $\mathrm{D}^b\mathrm{coh}^G(X)$. With such a functor, one could hope to compare derived categories across a GIT wall-crossing or across a birational cobordism. Note that taking the derived pushforward along the inclusion $X^{ss} \subset X$ will *not* work as coherent sheaves do not push-forward to coherent sheaves along an open immersion, so such an adjoint functor must be constructed in a non-obvious way. The main results of [2, 11] assert that this does indeed work, albeit with some technicalities.

The core idea is to consider subcategories ("windows") in $\mathrm{D}^b\mathrm{coh}^G(X)$ whose objects have prescribed weights with respect to the action. To make sense of this, one should instead consider actions by $\mathbb{C}^*$ or by a one parameter subgroup

$$\lambda : \mathbb{C}^* \to G$$

of the G action. At a fixed point for the action, one can certainly make sense of the weight of the action on the fiber of a locally free sheaf and this number is constant along connected components of the fixed locus. We thus assume that the fixed locus is connected (or else choose a connected component), and will define the possible "weights" of any complex of coherent sheaves by seeing what weights the terms in a quasi-equivalent complex of locally free equivariant vector bundles has. In other words, given a set of integers I, we define the window

$$W_I^\lambda \subset \mathrm{D}^b\mathrm{coh}^G(X)$$

to be those objects which have a representative with weights in I with respect to λ, again with respect to a chosen connected component of the fixed locus.

One then considers the restriction of the restriction of the restriction functor:

$$r_I : W_I^\lambda \to \mathrm{D}^b\mathrm{coh}^G(X^{ss}).$$

The main result of [2, 11] states that for a particular choice of I the functor r_I is actually an equivalence. The details of the proofs are beyond the scope of these notes; we only comment that the fully-faithfulness portion of the result can be reformulated as the vanishing of the local cohomologies of all objects with such weights along the *un*-stable locus, and such a statement had been observed earlier in the underived setting (i.e. working with cohomologies of vector bundles) by Teleman [27]. The remaining part of the result is the essential surjectivity of r_I. In full generality this requires careful calculations to show that any object of $\mathrm{D}^b\mathrm{coh}^G(X^{ss})$ has a quasi-equivalent representative with weights in I. In concrete examples, such as in the Atiyah flop below, one may have an explicit semi-orthogonal decomposition to compare with which makes this process easier to verify. Towards proving essential surjectivity in general, we remark only that a useful technical observation is that one may twist by objects supported on the unstable-locus (which is a trivial object in $\mathrm{D}^b\mathrm{coh}^G(X^{ss})$) to attempt to alter the weights. In lieu of supplying more details to the above arguments, we review the Atiyah flop from this perspective, which was indeed the observation of [25] on which these results built.

Example: Let $\mathbb{C}^*$ act on $\mathbb{C}^{2n} = \mathrm{Spec}\,\mathbb{C}[x_1, \dots, x_n, y_1, \dots y_n]$ with $n \geq 2$ so that the x_i's have weight 1 and the y_i's have weight -1. We have seen above that performing the GIT wall-crossing gives the local Atiyah flop between the space $\mathrm{Tot}_{\mathbb{P}^n}\mathcal{O}(-1)^{\oplus n}$ and itself. The derived category $\mathrm{D}^b\mathrm{coh}^{\mathbb{C}^*}(\mathbb{C}^{2n})$ is generated by the various linearizations of the trivial line bundle. Let's write $\mathcal{O}[k]$ for the trivial line bundle with weight k under the $\mathbb{C}^*$ action, and so more precisely we have

$$\mathrm{D}^b\mathrm{coh}^{\mathbb{C}^*}(\mathbb{C}^{2n}) = \langle \mathcal{O}[k] \mid k \in \mathbb{Z} \rangle.$$

By a famous result of Beilinson, the derived category of $\mathbb{P}^n$ admits a semiorthogonal decomposition

$$\mathrm{D}^b\mathrm{coh}(\mathbb{P}^n) = \langle \mathcal{O}, \mathcal{O}(1), \dots, \mathcal{O}(n+1) \rangle.$$

Pulling back along the projection induces a similar semiorthogonal decomposition of $D^b\mathrm{coh}(\mathrm{Tot}_{\mathbb{P}^n}\mathcal{O}(-1)^{\oplus n})$. We thus take $I = \{0, \ldots, k+1\}$ and consider the window subcategory

$$W_I = \langle \mathcal{O}, \mathcal{O}[1], \ldots, \mathcal{O}[n+1] \rangle.$$

It is easy to see that the derived restriction along either side of the wall gives the expected equivalence $W_I \cong D^b\mathrm{coh}(\mathrm{Tot}_{\mathbb{P}^n}\mathcal{O}(-1)^{\oplus n})$. Note that if we compose these two isomorphisms, the (auto-) equivalence

$$D^b\mathrm{coh}(\mathrm{Tot}_{\mathbb{P}^n}\mathcal{O}(-1)^{\oplus n}) \cong D^b\mathrm{coh}(\mathrm{Tot}_{\mathbb{P}^n}\mathcal{O}(-1)^{\oplus n})$$

is *not* the identity functor. For example, if E is a sheaf on one side, to compute its image under the isomorphism, one must resolve it by the objects $\mathcal{O}(k)$ and lift to a complex of objects built out of the $\mathcal{O}[k]$'s, but then derive restrict along the *other* semistable locus. For example, if $n = 2$ one can compute explicitly that this process sends $\mathcal{O}$ to $\mathcal{O}$, but sends $\mathcal{O}(1)$ to the complex $\mathcal{O}(-1)^{\oplus 2} \to \mathcal{O}$.

At present the machinery of grade restriction windows has been used to reproduce many interesting flop equivalences (although not enough to reproduce for example Bridgeland's result), but it is not clear at present how successful this method will be in general.

References

1. Abramovich, Dan; Chen, Jiun-Cheng. *Flops, flips and perverse point sheaves on threefold stacks*. J. Algebra 290 (2005), no. 2, 372–407.
2. Ballard, Matthew; Favero, David; Katzarkov, Ludmil. *Variation of geometric invariant theory quotients and derived categories*. To appear in Crelle's Journal.
3. Birkar, Caucher; Cascini, Paolo; Hacon, Christopher D.; McKernan, James. *Existence of minimal models for varieties of log general type*. J. Amer. Math. Soc. 23 (2010), no. 2, 405–468.
4. Bondal, Aleksey; Orlov, Dimitry. *Derived categories of coherent sheaves*. Proceedings of the International Congress of Mathematicians, Vol. II (Beijing, 2002), 47–56,
5. Borisov, Lev; Căldăraru, Andrei. *The Pfaffian-Grassmannian derived equivalence*. J. Algebraic Geom. 18 (2009), no. 2, 201–222.
6. Bridgeland, Tom. *Flops and derived categories*. Invent. Math. 147 (2002), no. 3, 613–632.
7. Cautis, Sabin. *Flops and about: a guide*. Derived categories in algebraic geometry, 61–101, EMS Ser. Congr. Rep., Eur. Math. Soc., Zrich, 2012.
8. Cautis, Sabin; Kamnitzer, Joel; Licata, Anthony. *Derived equivalences for cotangent bundles of Grassmannians via categorical sl2 actions*. J. Reine Angew. Math. 675 (2013), 53–99.
9. Chen, Jiun-Cheng. *Flops and equivalences of derived categories for threefolds with only terminal Gorenstein singularities*. J. Differential Geom. 61 (2002), no. 2, 227–261.
10. Dolgachev, Igor V.; Hu, Yi. *Variation of geometric invariant theory quotients*. With an appendix by Nicolas Ressayre. Inst. Hautes Etudes Sci. Publ. Math. No. 87 (1998), 5–56.
11. Halpern-Leistner, Daniel. *The derived category of a GIT quotient*. J. Amer. Math. Soc. 28 (2015), no. 3, 871–912.
12. Huybrechts, Daniel. *Fourier-Mukai transforms in algebraic geometry*. Oxford Mathematical Monographs. The Clarendon Press, Oxford University Press, Oxford, 2006. viii+307 pp.

13. Kawamata, Yujiro. *D-equivalence and K-equivalence*. J. Differential Geom. 61 (2002), no. 1, 147–171.
14. Kawamata, Yujiro. *Derived equivalence for stratified Mukai flop on G(2,4)*. Mirror symmetry. V, 285–294, AMS/IP Stud. Adv. Math., 38, Amer. Math. Soc., Providence, RI, 2006.
15. Kawamata, Yujiro. *Flops connect minimal models*. Publ. Res. Inst. Math. Sci. 44 (2008), no. 2, 419–423.
16. Kollár, János. *The structure of algebraic threefolds: an introduction to Mori's program*. Bull. Amer. Math. Soc. 17 (1987), 211–273.
17. Kollár, János; Mori, Shigefumi. *Birational geometry of algebraic varieties* With the collaboration of C. H. Clemens and A. Corti. Cambridge Tracts in Mathematics, 134. Cambridge University Press, Cambridge, 1998. viii+254 pp.
18. Kuznetsov, Alexander. *Derived categories view on rationality problems*. Rationality problems in algebraic geometry, 67–104, Lecture Notes in Math., 2172, Fond. CIME/CIME Found. Subser., Springer, Cham, 2016.
19. Meinhardt, Sven; Partsch, Holger. *Quotient categories, stability conditions, and birational geometry*. Geom. Dedicata 173 (2014), 365–392.
20. Mori, Shigefumi. *Flip theorem and the existence of minimal models for 3-folds*. J. Amer. Math. Soc. 1 (1988), no. 1, 117–253.
21. Mumford, D.; Fogarty, J.; Kirwan, F. *Geometric invariant theory*. Third edition. Ergebnisse der Mathematik und ihrer Grenzgebiete (2) [Results in Mathematics and Related Areas (2)], 34. Springer-Verlag, Berlin, 1994. xiv+292 pp.
22. Namikawa, Yoshinori. *Mukai flops and derived categories*. J. Reine Angew. Math. 560 (2003), 65–76.
23. Namikawa, Yoshinori. *Mukai flops and derived categories. II*. Algebraic structures and moduli spaces, 149–175, CRM Proc. Lecture Notes, 38, Amer. Math. Soc., Providence, RI, 2004.
24. Reid, Miles. *What is a flip?* Colloquium Talk, University of Utah, December 1993. Available at http://homepages.warwick.ac.uk/staff/G.Brown/reid.pdf.
25. Segal, Ed. *Equivalence between GIT quotients of Landau-Ginzburg B-models*. Comm. Math. Phys. 304 (2011), no. 2, 411–432.
26. Seidel, Paul; Thomas, Richard. *Braid group actions on derived categories of coherent sheaves*. Duke Math. J. 108 (2001), no. 1, 37–108.
27. Teleman, Constantin. *The quantization conjecture revisited*. Ann. of Math. (2) 152 (2000), no. 1, 1–43.
28. Thaddeus, Michael. *Geometric invariant theory and flips*. J. Amer. Math. Soc. 9 (1996), no. 3, 691–723.
29. Usnich, Alexandr. *Action of the Cremona group on a noncommutative ring*. Adv. Math. 228 (2011), no. 4, 1863–1893.
30. Van den Bergh, Michel. *Three-dimensional flops and noncommutative rings*. Duke Math. J. 122 (2004), no. 3, 423–455.
31. Włodarczyk, Jarosław. *Birational cobordisms and factorization of birational maps*. J. Algebraic Geom. 9 (2000), no. 3, 425–449.

Part II
Approaches to Mirror Symmetry

Introduction to Mirror Symmetry

Richard Derryberry

1 A Guide to the References

- The material on complex geometry, Hodge theory and Kähler geometry can be found in [5].
- A derivation of the Hodge diamond symmetries from mirror symmetry can be found in [4].
- An explanation of the numerology for the quintic threefold can be found in the corresponding lecture of [1].
- The homotopy version of the Lefschetz hyperplane theorem is due to [2].
- The book [3] is a good general reference, as well as containing a lot of content overlap with the above.

2 Review of Differential Forms and Complex Geometry

2.1 Differential Topology

We begin by recalling the definition of *de Rham cohomology*: on a smooth n-manifold M we assign the dg-algebra of *differential forms* on M, $(\Omega^\bullet(M), d_{\mathrm{dR}})$. These are sections of the exterior algebra on the bundle T^*M, with grading given by form degree and differential the de Rham differential $d = d_{\mathrm{dR}}$. Concretely, recall that in local coordinates $(x^1, \dots, x^n)$ the de Rham differential is defined on functions by

$$df = \frac{\partial f}{\partial x^i} dx^i,$$

R. Derryberry (✉)
University of Texas, Austin, USA
e-mail: rhughes@math.utexas.edu

M. Ballard et al. (eds.), *Superschool on Derived Categories and D-branes*,
Springer Proceedings in Mathematics & Statistics 240,
https://doi.org/10.1007/978-3-319-91626-2_8

where we have used the Einstein convention of summing over pairs of raised and lowered indices, and is extended to higher degree forms via the Leibniz rule. We define the *de Rham cohomology* of M to be the cohomology of $(\Omega^\bullet(M), d_{\mathrm{dR}})$, i.e.

$$H^k_{\mathrm{dR}}(M) = \frac{\ker(d : \Omega^k \to \Omega^{k+1})}{\mathrm{im}\ (d : \Omega^{k-1} \to \Omega^k)}.$$

We will define the *Betti numbers* of a closed manifold M to be $b_k = \dim H^k_{\mathrm{dR}}(M)$ – this is a nonstandard definition, but for a closed smooth manifold it is equivalent to the standard definition from algebraic topology. There is in this case a symmetry on the Betti numbers $b_k = b_{n-k}$, implied by the stronger theorem of *Poincaré duality*.

2.2 Complex Geometry

We now review some facts and definitions from complex geometry. Let (X, J) be a complex d-manifold.[1] Unless required for clarity we will omit the complex structure operator J, and we will call X a *(complex) d-fold* for short.

We will write TX and T^*X for the *holomorphic* tangent and cotangent bundles of X, obtained by the natural identification of the tangent bundle with the $+i$-eigenspace of J in the complexified tangent bundle

$$TX \otimes \mathbb{C} \cong \underbrace{T^{1,0}X}_{+i} \oplus \underbrace{T^{0,1}X}_{-i}.$$

Dualizing this decomposition similarly decomposes the complexified cotangent bundle into $(1, 0)$ and $(0, 1)$ summands, and we define

$$(T^{p,q}X)^* = \bigwedge^p (T^{1,0}X)^* \otimes \bigwedge^q (T^{0,1}X)^*$$

the sections of which are the differential (p, q)-forms $\Omega^{p,q}(X)$. By composing the de Rham differential with the projection maps $\Omega^1(X; \mathbb{C}) \to \Omega^{p,q}(X)$, $p + q = 1$, we obtain a decomposition

$$d_{\mathrm{dR}} = \partial + \bar{\partial}, \quad \partial : \Omega^{p,q}(X) \to \Omega^{p+1,q}(X), \quad \bar{\partial} : \Omega^{p,q}(X) \to \Omega^{p,q+1}(X).$$

By taking cohomology with respect to the $\bar{\partial}$-operator, we arrive at the *Dolbeault cohomology groups*

$$H^{p,q}(X) = \frac{\ker(\bar{\partial} : \Omega^{p,q} \to \Omega^{p,q+1})}{\mathrm{im}\ (\bar{\partial} : \Omega^{p,q-1} \to \Omega^{p,q})},$$

[1] The dimension d here is the *complex* dimension of the manifold.

and we define the *Hodge numbers* of X to be the dimensions $h^{p,q}(X) = \dim_{\mathbb{C}} H^{p,q}(X)$.

Remark The *Dolbeault theorem* allows us to identify $H^{p,q}(X)$ with the sheaf cohomology groups $H^q(\Omega^p)$, where Ω^p is the sheaf of *holomorphic* p-forms on X.

We may present the Hodge numbers graphically in the *Hodge diamond* of X; e.g. the Hodge diamond of a 3-fold is

$$\begin{array}{ccccccc} & & & h^{3,3} & & & \\ & & h^{3,2} & & h^{2,3} & & \\ & h^{3,1} & & h^{2,2} & & h^{1,3} & \\ h^{3,0} & & h^{2,1} & & h^{1,2} & & h^{0,3} \\ & h^{2,0} & & h^{1,1} & & h^{0,2} & \\ & & h^{1,0} & & h^{0,1} & & \\ & & & h^{0,0} & & & \\ & & \nwarrow \partial & & \bar{\partial} \nearrow & & \end{array}$$

(Hodge diamond for a complex 3-fold)

We remark that although it is not displayed in the above diamond, there is a symmetry between Hodge numbers of the form $h^{p,q} = h^{d-p,d-q}$. This symmetry can in particular be derived via the theorem of *Serre duality*.

3 Kähler Geometry

3.1 Definition and Examples

Recall that a *symplectic form* on a manifold M is a closed and non-degenerate 2-form $\omega \in \Omega^2(M)$.

Definition 1 A *Kähler manifold* is the data of a complex manifold equipped with a symplectic form, (X, J, ω), satisfying the condition that the symmetric 2-tensor g defined by $g(V, W) = \omega(V, JW)$ is a Riemannian metric for which J is orthogonal.

Remark A Riemannian metric g for which J is orthogonal is called a *Hermitian metric*. We note in passing that it would have been equivalent to require the Hermitian metric g be data and $d\omega = 0$ be a condition.

Observe that the above also implies that $(J^*\omega)(V, W) = \omega(JV, JW) = \omega(V, W)$, i.e. that ω is a $(1, 1)$-form.

Example 1 Consider the projective space $\mathbb{CP}^n = (\mathbb{C}^{n+1} - \{0\})/\mathbb{C}^\times$, and let $s : \mathbb{CP}^n \to \mathbb{C}^{n+1} - \{0\}$ be a local section of the projection map (i.e. a set of local coordinates). Define a 2-form locally by

$$\omega_{FS} = -i\bar{\partial}\partial \log |s|.$$

It is an exercise to show that this is independent of the choice of section, and that the global 2-form obtained is a Kähler form. The form ω_{FS} is called the *Fubini-Study form*.

The example of projective space now provides us with a wealth of further examples of Kähler manifolds via the following proposition.

Proposition 3.1 *Any complex submanifold of a Kähler manifold is naturally Kähler.*

Proof This follows immediately from the fact that the restriction of a Riemannian metric to a submanifold is a Riemannian metric, and the invariance of the tangent bundle of a complex submanifold under the complex structure operator.

3.2 Hodge Theory

On a closed Riemannian manifold (M, g) there is a deep relationship between the de Rham cohomology of M and solutions to the Laplace equation–*harmonic forms*– which goes by the name of Hodge theory. Specifically, letting $\mathcal{H}^k(M)$ denote the space of harmonic k-forms, Hodge theory provides an isomorphism

$$\mathcal{H}^k(M) \cong H^k_{\mathrm{dR}}(M).$$

On closed complex manifolds, a Kähler structure allows us to refine this to a statement about Dolbeault cohomology and its relation to de Rham cohomology, leading to the following extra relations between Hodge and Betti numbers:

$$h^{p,q} = h^{q,p}, \qquad b_k = \sum_{p+q=k} h^{p,q}.$$

Introducing these symmetries and the symmetry obtained from Serre duality, we can refine the Hodge diamond for a *Kähler* 3-fold to the following (we include the Betti numbers also):

$$\begin{array}{ccccccc|l}
 & & & h^{0,0} & & & & b_0 = h^{0,0} \\
 & & h^{1,0} & & h^{1,0} & & & b_1 = 2h^{1,0} \\
 & h^{2,0} & & h^{1,1} & & h^{2,0} & & b_2 = 2h^{2,0} + h^{1,1} \\
h^{3,0} & & h^{2,1} & & h^{2,1} & & h^{3,0} & b_3 = 2h^{3,0} + 2h^{2,1} \\
 & h^{2,0} & & h^{1,1} & & h^{2,0} & & b_2 = 2h^{2,0} + h^{1,1} \\
 & & h^{1,0} & & h^{1,0} & & & b_1 = 2h^{1,0} \\
 & & & h^{0,0} & & & & b_0 = h^{0,0}
\end{array} \qquad \textit{(Hodge diamond for a Kähler 3-fold)}$$

4 Mirror Symmetry and Calabi–Yau Manifolds

4.1 Statement of Mirror Symmetry

The version of mirror symmetry that we will discuss has the following, rough principal at its core:

Two manifolds X and $X^\vee$ are mirror dual if there is a correspondence between the parameters deforming the Kähler structures of one manifold and the parameters deforming the complex structures of the other manifold.

Note that in the above formulation there are many structures that X and $X^\vee$ must have that we have failed to make explicit.

4.2 Calabi–Yau Manifolds

We will partially remedy the omission of any necessary structures on X and $X^\vee$ now.

Definition 2 The *canonical bundle* of a complex d-fold X is $K_X := \bigwedge^d T^*X$, the bundle of holomorphic d-forms on X.

Definition 3 A compact Kähler manifold X is called *Calabi–Yau* if it has trivial canonical bundle.

From here on out we will assume that all manifolds are **connected and simply connected Calabi–Yau**.

Remark The simple connectedness assumption implies that the Calabi–Yau condition is equivalent to the vanishing of the first Chern class $c_1(X)$, as there are then no topologically trivial but holomorphically nontrivial line bundles.

Our assumptions allow us to further reduce the number of parameters in the Hodge diamond for a 3-fold as follows:

(1) Triviality of the canonical bundle implies that $h^{3,0} = 1$ (in words: up to scaling there is a unique holomorphic top form).
(2) Connectedness implies $b_0 = 1$.
(3) Simply connected implies $b_1 = 0$, hence $h^{0,1} = 0$.
(4) Serre duality *together with* triviality of the canonical bundle implies that $h^{0,2} = h^{0,1} = 0$.

This leaves two parameters remaining in the Hodge diamond:

$$\begin{array}{ccccccc|c}
 & & & 1 & & & & 1 \\
 & & 0 & & 0 & & & 0 \\
 & 0 & & h^{1,1} & & 0 & & h^{1,1} \\
1 & & h^{2,1} & & h^{2,1} & & 1 & 2(h^{2,1}+1) \\
 & 0 & & h^{1,1} & & 0 & & h^{1,1} \\
 & & 0 & & 0 & & & 0 \\
 & & & 1 & & & & 1
\end{array}$$

(Hodge diamond for a connected, simply connected Calabi–Yau 3-fold)

4.3 Mirror Symmetry for Simply Connected Calabi–Yau 3-Folds

We may interpret the parameters $h^{1,1}$ and $h^{2,1}$ as follows.

First, recall that a Kähler form ω on X is a closed (1,1)-form. The converse is not true for two reasons: an arbitrary (1,1)-form need not come from a real 2-form, and even if it does it may not satisfy the required positivity condition. Under certain assumptions however–including our case of a simply-connected Calabi–Yau 3-fold–the space of admissible Kähler forms is an *open* cone inside of $H^2_{\text{dR}}(X)$, and so the space of (1,1)-forms whose real part is Kähler is open inside of $H^{1,1}(X)$. We therefore say that the number of *Kähler parameters* is given by $h^{1,1}(X)$.

Second, recall that infinitesimal deformations of the complex structure of a manifold X are parametrized by $H^1(TX)$. Triviality of K_X implies triviality of its dual $\bigwedge^3 TX$, and so the wedge pairing

$$\wedge : TX \otimes \bigwedge^2 TX \to \bigwedge^3 TX$$

induces an identification $TX \cong \bigwedge^2 T^*X$. Hence $H^1(TX) = H^1(\bigwedge^2 T^*X) = H^{2,1}(X)$, and so $h^{2,1}(X)$ is the number of *complex structure parameters* for X.

The rough principal given above now leads us to make the following prediction:

If two simply connected Calabi-Yau 3-folds X and $X^\vee$ are mirror dual, then

$$h^{1,1}(X) = h^{2,1}(X^\vee) \quad \textit{and} \quad h^{1,1}(X^\vee) = h^{2,1}(X).$$

Remark This prediction may be refined to $h^{p,q}(X) = h^{d-p,q}(X^\vee)$ on higher dimensional Calabi–Yau manifolds with $H^2(\mathcal{O}) = 0$.

5 Canonical Example: The Quintic Threefold

Consider the zero set $Q \subset \mathbb{CP}^4$ of a degree 5 polynomial p, i.e. p is a section of $\mathcal{O}(5)$. Since Q is cut out of $\mathbb{CP}^4$ by a single equation, it is a 3-fold. Assuming Q is nonsingular, it inherits a Kähler structure from the Fubini-Study metric for $\mathbb{CP}^4$.

Q is simply connected by the Lefschetz hyperplane theorem ($\pi_1(Q) \xrightarrow{\sim} \pi_1(\mathbb{CP}^4) = 0$) and by the adjunction formula

$$c(Q) = \frac{(1 + c_1(H))^5}{1 + 5c_1(H)} = 1 + O(c_1(H)^2)$$

where H is the hyperplane bundle on $\mathbb{CP}^4$, we see that $c_1(Q) = 0$. Hence Q is a Calabi–Yau 3-fold.

There are 126 degree 5 monomials in 5 variables, hence the dimension of the space $H^0(\mathcal{O}(5)|_Q)$ of homogeneous degree 5 polynomials *not* vanishing on Q is 125 (one simply excludes p). Counting (infinitesimal) deformations of $\mathbb{CP}^4$ gives

$$\dim H^0(T\mathbb{CP}^4) = \dim(PGL_5\mathbb{C}) = \dim(GL_5\mathbb{C}) - \dim(\mathbb{C}^\times) = 5^2 - 1 = 24.$$

Recalling that we have $TQ \cong \bigwedge^2 T^*Q$ (since Q is Calabi–Yau),

$$H^0(TQ) = H^0(\Omega^2_Q) = H^{2,0}(Q) = 0$$

by calculations we have already performed for the Hodge diamond of a simply-connected threefold. Taking the long exact sequence associated to the Euler sequence

$$0 \to \mathcal{O}_{\mathbb{CP}^4} \to \mathcal{O}_{\mathbb{CP}^4}(1)^{\oplus(5)} \to T\mathbb{CP}^4 \to 0$$

gives us $H^1(T\mathbb{CP}^4) = 0$; hence the long exact sequence associated to the adjunction short exact sequence

$$0 \to TQ \to T\mathbb{CP}^4 \to \mathcal{O}(5)|_Q \to 0$$

yields

$$H^1(TQ) = H^0(\mathcal{O}(5)|_Q)/H^0(T\mathbb{CP}^4).$$

Via dimension counting we see that $\dim H^1(TQ) = 125 - 24 = 101$. Hence, $h^{2,1}(Q) = 101$. We also have that $h^{1,1} = 1$ (this is another consequence of the Lefschetz hyperplane theorem: $H_2(Q) = H_2(\mathbb{CP}^4) = \mathbb{Z}$), and so the Hodge diamond for the quintic 3-fold is

$$\begin{array}{ccccccc} & & & 1 & & & \\ & & 0 & & 0 & & \\ & 0 & & 1 & & 0 & \\ 1 & & 101 & & 101 & & 1 \\ & 0 & & 1 & & 0 & \\ & & 0 & & 0 & & \\ & & & 1 & & & \end{array}$$

(Hodge diamond for the quintic 3-fold)

We want to construct a mirror to the quintic. Consider the family of quintics

$$Q_\psi = \{[X_0 : \cdots : X_4] \in \mathbb{CP}^4 \mid f_\psi = X_0^5 + \cdots + X_4^5 - 5\psi X_0 X_1 X_2 X_3 X_4 = 0\}.$$

Calculating the derivative of f_ψ, one shows that Q_ψ is smooth provided ψ is not a fifth root of unity.

Let $G = \{(a_0, \ldots, a_4) \in (\mathbb{Z}/5\mathbb{Z})^5 \mid \sum a_i = 0\}/\langle(a, a, a, a, a)\rangle \cong (\mathbb{Z}/5\mathbb{Z})^3$. G acts on Q_ψ via

$$(a_0, a_1, a_2, a_3, a_4) \cdot [X_0 : X_1 : X_2 : X_3 : X_4] = [X_0\xi^{a_0} : X_1\xi^{a_1} : X_2\xi^{a_2} : X_3\xi^{a_3} : X_4\xi^{a_4}]$$

where $\xi = e^{\frac{2\pi i}{5}}$. This action is not free, and the points with nontrivial stabiliser–where at least two of the homogeneous coordinates vanish–produce singularities in Q_ψ/G. We may find a construct a "good" (in this case meaning *crepant*) resolution of the singularities of this quotient using techniques from toric geometry to obtain a nonsingular space $Q_\psi^\vee$ – as a part of this process, the singularities are replaced by new algebraic cycles which introduce 100 new $h^{1,1}$ parameters. Together with the original hyperplane class, we find that $h^{1,1}(Q_\psi^\vee) = 101$.

Furthermore, we see that we have at least a one parameter family of deformations in complex structure given by the coordinate ψ^5. It is possible to show that this is the only family of deformations, hence $h^{2,1}(Q_\psi^\vee) = 1$, and so the Hodge diamond for $Q_\psi^\vee$ is

$$\begin{array}{ccccccc} & & & 1 & & & \\ & & 0 & & 0 & & \\ & 0 & & 101 & & 0 & \\ 1 & & 1 & & 1 & & 1 \\ & 0 & & 101 & & 0 & \\ & & 0 & & 0 & & \\ & & & 1 & & & \end{array} \qquad \textit{(Hodge diamond for } Q_\psi^\vee\textit{)}$$

which is as predicted for the mirror to the quintic.

References

1. D. Auroux, "Lecture notes from the spring 2009 topics in geometry class." [Online]. Available: https://math.berkeley.edu/~auroux/18.969-S09/
2. R. Bott, "On a theorem of Lefschetz," *Michigan Math. J.*, vol. 6, pp. 211–216, 1959.
3. K. Hori, S. Katz, A. Klemm, R. Pandharipande, R. Thomas, C. Vafa, R. Vakil, and E. Zaslow, *Mirror symmetry*, ser. Clay Mathematics Monographs. American Mathematical Society, Providence, RI; Clay Mathematics Institute, Cambridge, MA, 2003, vol. 1, with a preface by Vafa.
4. C. Voisin, *Mirror symmetry*, ser. SMF/AMS Texts and Monographs. American Mathematical Society, Providence, RI; Société Mathématique de France, Paris, 1999, vol. 1, translated from the 1996 French original by Roger Cooke.
5. ——, *Hodge theory and complex algebraic geometry. I*, ser. Cambridge Studies in Advanced Mathematics. Cambridge University Press, Cambridge, 2002, vol. 76, translated from the French original by Leila Schneps.

Batyrev Mirror Symmetry

Mattia Talpo

1 Introduction

This short note is a survey about an explicit construction for mirror families of Calabi–Yau varieties, due to Batyrev and later generalized by Batyrev–Borisov, that uses toric geometry and polar duality for lattice polytopes. The construction is about Calabi–Yau hypersurfaces in a Fano toric variety.

Historically, after the first example of the quintic threefold [1], many other examples of Calabi–Yau threefolds and mirror pairs were constructed using hypersurfaces in weighted projective spaces. For some of these examples though, the mirror was missing. Batyrev's construction [2] put these examples in a more systematic framework and provided the missing mirrors. Moreover it was later generalized to complete intersections in Fano toric varieties by Batyrev–Borisov [3], and brought combinatorics and toric geometry into the picture. It also partly inspired the Gross–Siebert program [4–6].

The material for this contribution is mostly taken from Cox's expository paper "Mirror Symmetry and Polar Duality of Polytopes" [7], and parts of Cox–Katz, "Mirror Symmetry and Algebraic Geometry" [8] (in particular Sects. 4.1 and 4.2).

M. Talpo (✉)
Department of Mathematics, Simon Fraser University, 8888 University Drive, Burnaby, BC
V5A 1S6, Canada
e-mail: mtalpo@sfu.ca

M. Talpo
Pacific Institute for the Mathematical Sciences, 4176-2207 Main Mall, Vancouver, BC
V6T 1Z4, Canada

M. Ballard et al. (eds.), *Superschool on Derived Categories and D-branes*,
Springer Proceedings in Mathematics & Statistics 240,
https://doi.org/10.1007/978-3-319-91626-2_9

2 Polar Duality of Lattice Polytopes

Batyrev's construction relates mirror pairs with a duality for lattice polytopes.

Definition 1 A *polytope* Δ is the convex envelope $\mathrm{Conv}(x_1, \ldots x_m)$ of a finite number of points in $\mathbb{R}^n$.

A *supporting hyperplane* of a polytope Δ is a hyperplane H in $\mathbb{R}^n$ such that $\Delta \cap H \neq \emptyset$, and Δ is completely contained in one of the two closed half-spaces that H determines in $\mathbb{R}^n$. A *face* of a polytope Δ is the intersection $\Delta \cap H$, where H is a supporting hyperplane. This is again a polytope. The *dimension* of a polytope is the dimension of the affine subspace of $\mathbb{R}^n$ spanned by Δ. Every polytope Δ determines a unique minimal set of points $\{v_1, \ldots, v_k\}$, called its *vertices*, such that $\Delta = \mathrm{Conv}(v_1, \ldots, v_k)$. These points also coincide with the faces of Δ of dimension 0.

Recall also that a *lattice* M is a free abelian group of finite rank, i.e. an abelian group isomorphic to $\mathbb{Z}^n$ for some n. Sometimes it is better not to choose a basis (i.e. the subset corresponding to the standard basis of $\mathbb{Z}^n$ via some isomorphism $M \cong \mathbb{Z}^n$), but we will always assume to have chosen one.

Definition 2 A *lattice polytope* is a polytope in some affine space $\mathbb{R}^n$ whose vertices have coordinates in $\mathbb{Z}^n$.

From now on we will assume that our lattice polytopes are full dimensional (i.e. they are not contained in any proper affine hyperplane of the ambient space) and that $0 \in \mathrm{Int}(\Delta)$. Here $\mathrm{Int}(\Delta)$ denotes the topological interior of Δ, which also coincides with the complement of all proper faces.

The dual or polar Δ° of Δ is another polytope, defined by

$$\begin{aligned}\Delta^\circ &= \{a \in \mathbb{R}^n \mid \langle a, b\rangle \geq -1 \text{ for all } b \in \Delta\} \\ &= \{a \in \mathbb{R}^n \mid \langle a, v\rangle \geq -1 \text{ for all vertices } v \text{ of } \Delta\} \quad \text{(by convexity)}\end{aligned}$$

where we denote by $\langle \cdot, \cdot \rangle$ the standard scalar product of $\mathbb{R}^n$. Note that if one does not want to choose a basis of the lattice M, then the same formulas define the dual of a polytope $\Delta \subseteq M_{\mathbb{R}} := M \otimes \mathbb{R}$ as a polytope in the dual vector space $\Delta^\circ \subseteq M_{\mathbb{R}}^\vee = M^\vee \otimes \mathbb{R}$, and in this case $\langle \cdot, \cdot \rangle \colon M_{\mathbb{R}} \times M_{\mathbb{R}}^\vee \to \mathbb{R}$ denotes the natural pairing $(v, f) \mapsto f(v)$.

It is not hard to check that the set Δ° is indeed a polytope (by the second description it follows that it is a finite intersection of half-spaces, so it is enough to show that it is bounded).

Example 1 If Δ is the square $[-1, 1] \times [-1, 1]$ then Δ° is the polygon with vertices $(\pm 1, 0)$, $(0, \pm 1)$, as in the following picture.

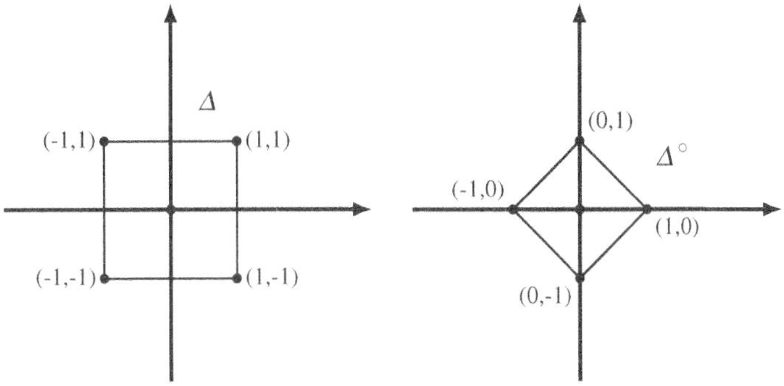

One can also check that $(\Delta^\circ)^\circ = \Delta$, so that this operation is indeed a "duality." Moreover, there is an inclusion-reversing combinatorial correspondence between i-dimensional faces of Δ and $(n-1-i)$-dimensional faces of Δ°.

The polytope Δ° is not always a *lattice* polytope. For example, it is easily verified that $(2\Delta)^\circ = \frac{1}{2}\Delta^\circ$, and the latter might not be a lattice polytope. This applies to the previous example, as $\frac{1}{2}\Delta^\circ = \text{Conv}((\pm\frac{1}{2}, 0), (0, \pm\frac{1}{2}))$ is not a lattice polytope in that case.

Definition 3 A lattice polytope Δ is *reflexive* if ($0 \in \text{Int}(\Delta)$ and) Δ° is a lattice polytope.

There are a few equivalent characterizations of this property. We will mention a couple of these; for details, see for example [9, Chap. 2].

One can prove that every facet (i.e. codimension 1 face) F of a polytope Δ has a unique inward-pointing normal vector u_F such that

$$F = \{a \in \Delta \mid \langle a, u_F \rangle = -1\}.$$

In Example 1, if F is the segment $[-1, 1] \times \{1\}$, then $u_F = (0, -1)$, and for the other facets we get the other vertices of the dual Δ°.

In fact we always have $\Delta^\circ = \text{Conv}(u_F \mid F \text{ a facet of } \Delta)$, so that

Proposition 1 *A lattice polytope Δ is reflexive if and only if every $u_F \in \mathbb{R}^n$ is a lattice point (i.e. is in $\mathbb{Z}^n \subseteq \mathbb{R}^n$).*

Another characterization is the following (which is given as the *definition* of a reflexive polytope in [2]):

Proposition 2 *A lattice polytope Δ is reflexive if and only if for every facet F of Δ there is no lattice point between the affine hyperplane spanned by F and its translate passing through the origin.*

As a consequence, the origin is the *only* lattice point in the interior of a reflexive polytope Δ.

Remark 1 From the last observation, via the results of [10], it follows that in every dimension n there is only a finite number of reflexive lattice polytopes up to integral change of coordinates (i.e. transformation by an element of $\mathrm{GL}(n, \mathbb{Z})$). For $n = 2$ there are 16 equivalence classes, for $n = 3$ they are 4 319 and for $n = 4$ (which is the important case for Mirror Symmetry, since it corresponds to 3-folds) there are 473 800 776 (!) equivalence classes (this was proven in [11]).

The idea for Batyrev Mirror Symmetry is that this duality among lattice polytopes realizes Mirror Symmetry for Calabi–Yau hypersurfaces in Fano toric varieties, as we will explain in the next sections.

3 Varieties from Lattice Polytopes

A lattice polytope in $\mathbb{R}^n$ gives rise to a projective variety. This process is part of a long story, the theory of toric varieties (see [9, 12]).

Definition 4 A *toric variety* is a normal algebraic variety X with an open embedding $T \subseteq X$ of a torus $T = (\mathbb{C}^\times)^n$ and an action $T \times X \to X$ that extends the multiplication action of T on itself.

It turns out that this set of data is completely encoded by a combinatorial polyhedral object in a lattice (the co-character lattice of the torus $\mathrm{Hom}(\mathbb{C}^\times, T)$, usually denoted by N in the literature), called a *fan*: this is a collection of cones, intersecting nicely (i.e. along common faces). The geometry of the toric variety is completely controlled by the combinatorics of this object: geometric properties of the variety can be translated in combinatorial or convex-geometric properties of the fan, and some algebraic invariants (for example sheaf cohomology of divisors) are explicitly computable. Because of this, toric varieties are usually a useful testing ground for new conjectures and theories about varieties in general.

A lattice polytope is an alternative incarnation of the underlying combinatorics of a certain class of toric varieties. Strictly speaking, the polytope also records the information of a torus invariant ample divisor on X, that gives in particular embeddings in projective space.

Here is a quick way to define the toric variety X_Δ associated to a lattice polytope Δ. First note that any lattice point $m = (a_1, \ldots, a_n) \in \mathbb{Z}^n$ gives a "Laurent monomial"

$$t^m = t_1^{a_1} \cdots t_n^{a_n}$$

which is a regular function on the torus $(\mathbb{C}^\times)^n$ (so that negative exponents make perfect sense).

Now we need to assume that Δ has "enough lattice points", or else modify it a bit. This is a technical condition, called normality of the polytope: a lattice polytope $\Delta \subseteq \mathbb{R}^n$ is *normal* if for all $n, m \in \mathbb{N}$ we have

$$(n\Delta) \cap \mathbb{Z}^n + (m\Delta) \cap \mathbb{Z}^n = ((n+m)\Delta) \cap \mathbb{Z}^n.$$

Here $n\Delta$ denotes the dilated polytope $\{a \in \mathbb{R}^n \mid a = nb \text{ for some } b \in \Delta\}$, and $+$ denotes the *Minkowski* sum of polytopes, defined as

$$\Delta + \Delta' = \{a + b \in \mathbb{R}^n \mid a \in \Delta, b \in \Delta'\}$$

for polytopes $\Delta, \Delta' \subseteq \mathbb{R}^n$. For example, one can show that the *standard simplex* Conv $(0, e_1, \ldots, e_n) \subseteq \mathbb{R}^n$ is a normal polytope, while the polytope $\mathrm{Conv}(0, e_1, e_2, e_1 + e_2 + 3e_3) \subseteq \mathbb{R}^3$ is not normal. Here and in what follows, as customary, e_i denote the elements of the standard basis of $\mathbb{R}^n$.

If Δ is not normal, one uses instead the polytope $k\Delta$ (which will be normal for $k \geq n-1$, see [9, Theorem 2.2.12]) in the construction that follows. This is related to ampleness versus very ampleness of the toric divisor encoded by the given polytope Δ. There is also a property of polytopes called *very ampleness*, implied by normality, and relevant for this construction. See [9, Sect. 2.2] for details.

Assuming that Δ is normal, consider $\Delta \cap \mathbb{Z}^n = \{m_0, \ldots, m_k\}$, which is a finite set, and the map

$$(\mathbb{C}^\times)^n \to \mathbb{P}^k \qquad \text{given by} \qquad (t_1, \ldots, t_n) \mapsto [t^{m_0} : \cdots : t^{m_k}]$$

where t^{m_i} is the Laurent monomial described above. This map turns out to be injective, and one defines the toric variety X_Δ as the closure of its image.

Reflexive lattice polytopes give rise, in this manner, to projective Fano toric varieties. Recall that "Fano" means that the anticanonical divisor $-K_{X_\Delta}$ is ample, for a smooth variety. We will allow some singularities and say that a variety X is Fano if it is Gorenstein and the dual of the dualizing sheaf $\omega_X^\vee$ (which is a line bundle) is ample.

Proposition 3 [8, Proposition 3.5.5] *The toric variety X_Δ is Fano if and only if Δ is a reflexive polytope.*

Lattice points on Δ also give interesting hypersurfaces in X_Δ: keeping the notation as before, the equation

$$a_0 t^{m_0} + \cdots + a_k t^{m_k} = 0 \qquad (1)$$

defines a hypersurface in $(\mathbb{C}^\times)^n$ (for any given coefficients $a_0, \ldots, a_k \in \mathbb{C}$), and the closure of this in X_Δ is then a hypersurface $V \subseteq X_\Delta$. Moreover, if Δ is reflexive every such hypersurface is a divisor in the same divisor class, the anticanonical class $|-K_{X_\Delta}|$.

Example 2 The quintic threefold in $\mathbb{P}^4$ can be recovered using this construction. Let Δ_n denote the standard simplex $\mathrm{Conv}(0, e_1, \ldots, e_n)$ in $\mathbb{R}^n$.

Let us take $\Delta \subseteq \mathbb{Z}^4$ to be

$$5\Delta_4 - (1,1,1,1) = \{a \in \mathbb{R}^4 \mid a = 5b - (1,1,1,1) \text{ for some } b \in \Delta_4\}.$$

In other words, Δ is the convex envelope of the vectors

$$(-1,-1,-1,-1),(4,-1,-1,-1),(-1,4,-1,-1),(-1,-1,4,-1),(-1,-1,-1,4)$$

obtained from the vertices $0, 5e_1, 5e_2, 5e_3, 5e_4$ of $5\Delta_4$ by subtracting the vector $(1,1,1,1)$.

This is a reflexive polytope in $\mathbb{R}^4$, and by applying the construction described above, one can check that $X_\Delta = \mathbb{P}^4$, and that (after homogenizing the corresponding Eq. (1)) the hypersurface V is an arbitrary quintic threefold in $\mathbb{P}^4$ (the exponent vectors that show up in the lattice points of Δ give all homogeneous monomials of degree 5 after homogenizing).

4 Batyrev's Construction

We can now talk about Batyrev's construction. Given a reflexive n-dimensional polytope Δ, one can consider the projective toric variety X_Δ (of dimension n), which will be a Fano toric variety, and a general divisor in the anticanonical linear system $V \in |-K_{X_\Delta}|$. For example one can take V to be determined by a Laurent polynomial as in Eq. (1). For the moment let us pretend that everything is smooth (typically this is false).

A (nice) anticanonical hypersurface in a Fano variety is going to have trivial canonical bundle (by the adjunction formula $K_D = (K_X + D)|_D$), so, taking for granted that also the other conditions about vanishing of cohomologies will be satisfied, it is going to be a Calabi–Yau variety, of complex dimension $n-1$. The basic idea is that by considering the dual Δ° and a general divisor in the anticanonical linear system of X_{Δ°, we get a different Calabi–Yau variety V° which should be mirror to V (or rather, the family of hypersurfaces V should be mirror to the family of hypersurfaces V° - we will make this abuse of terminology from now on).

In reality things are more technical, because often X_Δ is too singular, and needs to be resolved via blowups in order for the divisor V to be a "nice" Calabi–Yau variety (i.e. with nice singularities). One also wants the resolution to be *crepant*, i.e. to preserve the canonical bundle, and for $n \geq 3$ the projective toric variety given by an n-dimensional lattice polytope does not need to admit a full crepant resolution (i.e. producing a smooth variety as its outcome), so the best one can do is partially resolve it.

Blowing up along a torus-invariant subvariety is quite convenient using toric language, because it corresponds to combinatorial operations on the fan and polytope associated to the toric variety. We will not go into details here, we will only mention that Batyrev introduces the notion of a "maximal projective crepant partial (MPCP) desingularization" for X_Δ, corresponding to certain triangulations of the polytope Δ. This is a birational map $X' \to X_\Delta$ which partially resolves the singularities of X_Δ and preserves the canonical divisor. By taking a general anticanonical divisor on X' we get a nice Calabi–Yau variety V (see [8, Proposition 4.1.3]). These MPCP desingularizations always exist in this context, and usually there is more than one choice.

By choosing a MPCP for X_Δ and one for X_{Δ°, we get Calabi–Yau varieties V and V° as general anticanonical sections of the partial resolutions, and these should form mirror pairs. In the case of threefolds (so when Δ lives in $\mathbb{R}^4$), V and V° actually turn out to be smooth. Some of the expected consequences of Mirror Symmetry have indeed been proven for Batyrev mirrors V and V°.

Recall that, for a smooth projective complex variety X, the *Hodge number* $h^{p,q}(X)$ is the dimension dim $H^q(X, \Omega_X^p)$ as a complex vector space, where $\Omega_X^p = \Omega_X \wedge \cdots \wedge \Omega_X$ is the wedge product of p copies of the sheaf of Kähler differentials Ω_X of X. The Hodge numbers are usually arranged in a diagram called the *Hodge diamond*, depicted below for dim $X = 3$.

$$
\begin{array}{ccccccc}
 & & & h^{0,0} & & & \\
 & & h^{1,0} & & h^{0,1} & & \\
 & h^{2,0} & & h^{1,1} & & h^{0,2} & \\
h^{3,0} & & h^{2,1} & & h^{1,2} & & h^{0,3} \\
 & h^{3,1} & & h^{2,2} & & h^{1,3} & \\
 & & h^{3,2} & & h^{2,3} & & \\
 & & & h^{3,3} & & &
\end{array}
$$

These numbers have two important symmetries: Hodge theory implies that $h^{p,q} = h^{q,p}$, and Serre duality implies that $h^{n-p,n-q} = h^{p,q}$. If in addition X is a Calabi–Yau threefold, we also have $h^{0,0} = h^{3,0} = 1$ and $h^{1,0} = h^{2,0} = 0$, so that the above diagram can be simplified to the following one

$$
\begin{array}{ccccccc}
 & & & 1 & & & \\
 & & 0 & & 0 & & \\
 & 0 & & h^{1,1} & & 0 & \\
1 & & h^{2,1} & & h^{2,1} & & 1 \\
 & 0 & & h^{1,1} & & 0 & \\
 & & 0 & & 0 & & \\
 & & & 1 & & &
\end{array}
$$

whose only relevant numbers are $h^{1,1}$ and $h^{2,1}$. Recall also that these Hodge numbers $h^{1,1} = \dim\, H^1(X, \Omega_X)$ and $h^{2,1} = \dim\, H^1(X, \Omega_X^2) = \dim\, H^1(X, T_X)$ (where $T_X \cong \Omega_X^\vee$ is the tangent bundle of X, and we used the fact that $\Omega_X^3 \cong \mathcal{O}_X$) give the number of parameters of deformations of a complexified Kähler class on X, and of the complex structure of X respectively. Mirror symmetry predicts that $h^{1,1}(X) = h^{2,1}(X^\vee)$ and $h^{2,1}(X) = h^{1,1}(X^\vee)$, where $X^\vee$ denotes the mirror of X; in other words, the Hodge diamonds of X and $X^\vee$ should be related by a reflection with respect to a diagonal line through the center.

More generally, if X is a Calabi–Yau manifold of dimension bigger than 3, Mirror symmetry predicts (among other facts) that $h^{p,q}(X) = h^{n-p,q}(X^\vee)$ and $h^{n-p,q}(X) = h^{p,q}(X^\vee)$. For Batyrev's construction, indeed this is known to be the case for $p = q = 1$ (see below for some discussion about the general statement).

Theorem 1 ([8, Theorem 4.1.5], [2, Theorem 4.4.3]) *The "Hodge numbers Mirror Symmetry" for $p = q = 1$ holds for Batyrev mirrors, i.e. we have the equality of Hodge numbers $h^{1,1}(V) = h^{\dim V-1,1}(V^\circ)$ and $h^{\dim V-1,1}(V) = h^{1,1}(V^\circ)$.*

If we perform the construction starting from a reflexive lattice polytope $\Delta \subseteq \mathbb{R}^4$, so that $\dim V = \dim V^\circ = 3$, then this is all that is needed to get the full symmetry relation between the Hodge diamonds of V and V°. The proof of the theorem is a computation of the Hodge numbers by using the dictionary of toric geometry to reduce to combinatorics.

There are also other (partial) results about correspondence of complex/Kähler moduli spaces and correlation functions of the A-model and B-model, that we will not get into. See [8, Section 4.1.2] for a thorough discussion.

On the other hand, there are still also some open questions: it is not known

1. whether using this construction with a 4-dimensional reflexive polytope, V and V° give isomorphic SCFTs (this is known for some cases, like the quintic threefold);
2. whether for a reflexive n-dimensional polytope with $n \geq 5$, the relations $h^{p,q}(V) = h^{\dim V-p,q}(V^\circ)$ and $h^{\dim V-p,q}(V) = h^{p,q}(V^\circ)$ hold or not.

Question (2) has been partially answered in later work of Batyrev and Borisov [13]. Namely, they prove that for the *string-theoretic Hodge numbers* $h_{st}^{p,q}$ (defined in [14]), one has the equalities $h_{st}^{p,q}(V) = h_{st}^{\dim V-p,q}(V^\circ)$ and $h_{st}^{\dim V-p,q}(V) = h_{st}^{p,q}(V^\circ)$ where V and V° are Batyrev mirrors. Their result [13, Theorem 4.15] actually also covers the more general case of complete intersections in Fano toric varieties of [3]. Moreover, if V is smooth or $q = 1$, then $h_{st}^{p,q}(V) = h^{p,q}(V)$, so with these assumptions the answer to question (2) is known to be positive.

5 The Quintic Threefold

The original example of Mirror Symmetry for the quintic threefold falls into this general framework. We already saw how to obtain the quintic as a Calabi–Yau hypersurface in the Fano toric variety $\mathbb{P}^4$, using a polytope Δ in Example 2.

The dual of that polytope Δ is

$$\Delta^\circ = \mathrm{Conv}(e_1, e_2, e_3, e_4, (-1, -1, -1, -1)).$$

In fact, Δ has 5 facets F_1, F_2, F_3, F_4, F_5, with supporting hyperplanes with equations $x_i = -1$ for $0 \le i \le 4$ and $x_1 + x_2 + x_3 + x_4 = 1$. The corresponding inner normal vectors (i.e. the vector u_F such that the facet F is described as the intersection of Δ with the hyperplane $\langle a, u_F \rangle = -1$) are then given by e_1, e_2, e_3, e_4 and $(-1, -1, -1, -1)$ respectively. The claim now follows by the description of Δ° as $\Delta^\circ = \mathrm{Conv}(u_F \mid F \text{ a facet of } \Delta)$. Note that both Δ and Δ° are combinatorially standard simplices (in the sense that there is a bijection between their faces and the faces of a standard simplex, compatible with inclusion and intersections), but the way they are positioned in the lattice is important. For example Δ has 125 lattice points, whereas Δ° has only 6.

Using Δ° as lattice polytope, one can check that X_{Δ° can be identified with the quotient $\mathbb{P}^4/G$, where G is the group

$$G = \{(a_1, a_2, a_3, a_4, a_5) \in (\mathbb{Z}/5)^5 \mid a_1 + a_2 + a_3 + a_4 + a_5 = 0\}/(\mathbb{Z}/5).$$

Here the quotient is by the diagonal subgroup, and G acts on $\mathbb{P}^4$ by multiplication by roots of unity in the obvious way.

Indeed, the primitive lattice generators of the rays of the normal fan of Δ° (which is the fan corresponding to the toric variety X_{Δ°) are precisely the vertices

$$(-1, -1, -1, -1), (4, -1, -1, -1), (-1, 4, -1, -1), (-1, -1, 4, -1), (-1, -1, -1, 4)$$

of Δ. if we denote by $M \subseteq \mathbb{Z}^4$ the sublattice generated by these vectors, then by [9, Proposition 3.3.7] there is an isomorphism $X_{\Delta^\circ} \cong X_{\Delta^\circ, M}/(\mathbb{Z}^4/M)$, where $X_{\Delta^\circ, M}$ denotes the toric variety corresponding to the polytope Δ° *with respect to the lattice* M, and the quotient is for the natural action of the finite group $(\mathbb{Z}^4/M)$ on $X_{\Delta^\circ, M}$. The quotient $(\mathbb{Z}^4/M)$ is isomorphic to the group G described above, and $X_{\Delta^\circ, M}$ is isomorphic to $\mathbb{P}^4$, as can be verified by checking that the normal fan of Δ° in M is isomorphic to the fan for $\mathbb{P}^4$. See [9, Example 5.4.10] for more details.

As mentioned above the polytope Δ° has 6 lattice points (the five vertices and the origin), so Eq. (1) in this case becomes

$$c_0 + c_1 t_1 + c_2 t_2 + c_3 t_3 + c_4 t_4 + c_5 t_1^{-1} t_2^{-1} t_3^{-1} t_4^{-1} = 0$$

which by using the coordinates of $\mathbb{P}^4$ and homogenizing (in a "toric" sense - see [9, Sect. 5.4]) becomes

$$c_0 x_0^5 + c_1 x_1^5 + c_2 x_2^5 + c_3 x_3^5 + c_4 x_4^5 + c_5 x_0 x_1 x_2 x_3 x_4 = 0.$$

By rescaling the coordinates one can assume $c_0 = c_1 = c_2 = c_3 = c_4 = 1$. This recovers the equation

$$x_0^5 + x_1^5 + x_2^5 + x_3^5 + x_4^5 + \psi x_0 x_1 x_2 x_3 x_4 = 0$$

that gives the mirror pencil of hypersurfaces (after resolving the singularities).

6 Further Developments

Batyrev–Borisov [3, 15] generalize the above to Calabi–Yau complete intersections in Fano toric varieties. The combinatorics becomes more complicated, but the basic idea is similar.

This time, the starting data is an $(r+d)$-dimensional reflexive polytope Δ, together with a decomposition as a Minkowski sum

$$\Delta = \Delta_1 + \cdots + \Delta_r$$

where Δ_i are lattice polytopes containing the origin (possibly on their boundary). This is called a *nef-partition*. The lattice points of each Δ_i determine a family of hypersurfaces of the Fano toric variety X_Δ, and choosing for each i a generic hypersurface V_i among these, the intersection $V_1 \cap \cdots \cap V_r$ is a a d-dimensional complete intersection Calabi–Yau variety, that needs to be partially resolved, as in the case of hypersurfaces.

To produce the mirror family the idea is to use polar duality again, but with a variation with respect to the hypersurface case, because the origin might not be an interior lattice point of Δ_i. Instead, one defines polytopes ∇_i by the formula

$$\nabla_i = \{a \in \mathbb{R}^d \mid \langle a, b\rangle \geq -1 \text{ for all } b \in \Delta_i \text{ and } \langle a, b\rangle \geq 0 \text{ for all } b \in \Delta_j,\ j \neq i\}.$$

One can prove that ∇_i are lattice polytopes containing the origin, and the Minkowski sum $\nabla = \nabla_1 + \cdots \nabla_r$ is a reflexive polytope of dimension $r+d$. This gives the dual nef-partition, and by applying the same procedure outlined above, one obtains the mirror of the subvariety corresponding to the original nef-partition. See [7, Sect. 6] or the original papers for more details.

The Gross–Siebert program [4–6] mixes SYZ Mirror Symmetry with the Batyrev–Borisov construction. The idea of that is the following: given a Calabi–Yau manifold X, in order to find a mirror $X^\vee$, degenerate it (in a nice way) to a union of toric varieties glued along toric strata (i.e. orbits for the action of the torus on the respective toric variety). Note that this "degenerate" variety will not be smooth.

From the degeneration one can extract combinatorial gadget (which actually has more structure...), called the *dual intersection complex*, that one can dualize via a *discrete Legendre transform*, in a way that is similar to the polar polyhedron construction. From the dual of the dual intersection complex we can construct a central fiber, again union of toric varieties glued along toric strata, and (with a lot of work!) construct a smoothing. The idea is that the smoothing should be mirror to the X that we started with.

In [16] Gross compares this construction to the one of Batyrev–Borisov. He shows that indeed nef-partitions give rise to toric degenerations, and that the algorithm that we crudely outlined above produces the same result as the Batyrev–Borisov construction.

Acknowledgements I am happy to thank the anonymous referee for useful comments and suggestions.

References

1. Philip Candelas, Xenia C. de la Ossa, Paul S. Green, and Linda Parkes. A pair of Calabi-Yau manifolds as an exactly soluble superconformal theory. *Nuclear Phys. B*, 359(1):21–74, 1991.
2. Victor V. Batyrev. Dual polyhedra and mirror symmetry for Calabi-Yau hypersurfaces in toric varieties. *J. Algebraic Geom.*, 3(3):493–535, 1994.
3. Victor V. Batyrev and Lev A. Borisov. On Calabi-Yau complete intersections in toric varieties. In *Higher-dimensional complex varieties (Trento, 1994)*, pages 39–65. de Gruyter, Berlin, 1996.
4. Mark Gross and Bernd Siebert. Affine manifolds, log structures, and mirror symmetry. *Turkish J. Math.*, 27(1):33–60, 2003.
5. Mark Gross and Bernd Siebert. Mirror symmetry via logarithmic degeneration data. I. *J. Differential Geom.*, 72(2):169–338, 2006.
6. Mark Gross and Bernd Siebert. Mirror symmetry via logarithmic degeneration data, II. *J. Algebraic Geom.*, 19(4):679–780, 2010.
7. David A. Cox. Mirror symmetry and polar duality of polytopes. *Symmetry*, 7(3):1633, 2015.
8. D.A. Cox and S. Katz. *Mirror Symmetry and Algebraic Geometry*. Mathematical surveys and monographs. American Mathematical Society, 1999.
9. D.A. Cox, J.B. Little, and H.K. Schenck. *Toric Varieties*. Graduate studies in mathematics. American Mathematical Soc., 2011.
10. Jeffrey C. Lagarias and Günter M. Ziegler. Bounds for lattice polytopes containing a fixed number of interior points in a sublattice. *Canad. J. Math.*, 43(5):1022–1035, 1991.
11. Maximilian Kreuzer and Harald Skarke. Complete classification of reflexive polyhedra in four dimensions. *Adv. Theor. Math. Phys.*, 4(6):1209–1230, 2000.
12. David Cox. What is a toric variety? In *Topics in algebraic geometry and geometric modeling*, volume 334 of *Contemp. Math.*, pages 203–223. Amer. Math. Soc., Providence, RI, 2003.
13. Victor V. Batyrev and Lev A. Borisov. Mirror duality and string-theoretic Hodge numbers. *Invent. Math.*, 126(1):183–203, 1996.
14. Victor V. Batyrev and Dimitrios I. Dais. Strong McKay correspondence, string-theoretic Hodge numbers and mirror symmetry. *Topology*, 35(4):901–929, 1996.
15. Lev Borisov. Towards the Mirror Symmetry for Calabi-Yau Complete intersections in Gorenstein Toric Fano Varieties. Preprint, arXiv:alg-geom/9310001.
16. Mark Gross. Toric degenerations and Batyrev-Borisov duality. *Math. Ann.*, 333(3):645–688, 2005.

Introduction to Differential Graded Categories

Alex A. Takeda

1 Motivation

In past talks at this conference we have seen the definition of a triangulated category and some examples of familiar triangulated categories, including the homotopy category $\mathscr{H}(R)$ and the derived category $\mathscr{D}(R)$ of modules over some ring R. These categories are defined by applying certain constructions to the category $\mathscr{C}(R)$ of complexes of R-modules; the derived category is usually defined by some localization construction. Both these categories get their triangulated structures from the abelian structure of $\mathrm{Mod}(R)$ and the shift operation on complexes.

However, there is a number of ways in which the derived category is insufficient or problematic; one could say that in passing to this localization one forgets too much data. For example, the derived category $\mathscr{D}(R)$ is not abelian: it does not have limits or colimits, and the existence of the kernel or cokernel of a morphism is not guaranteed. In fact one can show the existence of the weaker notion of homotopy limits or colimits, but the derived category with only the triangulated structure does not give a prescription for how to construct them.

Example 1 Here is an example from [14] of how the derived category fails to have kernels. Consider the derived category $\mathscr{D}(\mathbb{Z})$ of $\mathbb{Z}$-modules. For two ordinary $\mathbb{Z}$-modules M, N, seen as objects of the derived category in degree zero, the maps between M and a shift of N in the derived category are calculated by the Ext functors:

$$\mathscr{D}(\mathbb{Z})(M, N[i]) = \mathrm{Ext}^i(M, N) .$$

Let us take $M = N = \mathbb{Z}/2$. There is one nontrivial element in $\mathrm{Ext}^1(\mathbb{Z}/2, \mathbb{Z}/2)$, represented by the $\mathbb{Z}/4$ extension

A. A. Takeda (✉)
Department of Physics, University of California, Berkeley, CA 94720, USA
e-mail: atakeda@berkeley.edu

M. Ballard et al. (eds.), *Superschool on Derived Categories and D-branes*,
Springer Proceedings in Mathematics & Statistics 240,
https://doi.org/10.1007/978-3-319-91626-2_10

$$\mathbb{Z}/2 \xrightarrow{\times 2} \mathbb{Z}/4 \to \mathbb{Z}/2 \,,$$

which represents the nontrivial morphism $f \in \mathrm{Hom}(\mathbb{Z}/2, \mathbb{Z}/2[1])$. Now suppose this map had a kernel K i.e. there was an exact sequence

$$0 \to K \to \mathbb{Z}/2 \to \mathbb{Z}/2[1]$$

in $\mathscr{D}(\mathbb{Z})$. For any i, applying the left exact functor $\mathrm{Hom}^0(\mathbb{Z}[-i], -) = \mathrm{Hom}^i(\mathbb{Z}, -)$ we would get an exact sequence of (ordinary) $\mathbb{Z}$-modules

$$0 \to \mathrm{Hom}^i(\mathbb{Z}, K) \to \mathrm{Hom}^i(\mathbb{Z}, \mathbb{Z}/2) \to \mathrm{Hom}^{i+1}(\mathbb{Z}, \mathbb{Z}/2) \,.$$

Taking $i \neq 0$ we see that $\mathrm{Hom}^i(\mathbb{Z}, K) = 0$ and taking $i = 0$ we get an isomorphism $\mathrm{Hom}^0(\mathbb{Z}, K) \cong \mathrm{Hom}^0(\mathbb{Z}, \mathbb{Z}/2) = \mathbb{Z}/2$. So the morphism $K \to \mathbb{Z}/2$ is an isomorphism in $\mathscr{D}(\mathbb{Z})$ since it induces isomorphisms on all cohomology groups, implying the extension $\mathbb{Z}/2 \to \mathbb{Z}/2[1]$ is trivial, which gives a contradiction.

2 Definitions

2.1 dg Categories

Let k be a commutative ring. A differential graded (dg) module over k is a $\mathbb{Z}$-graded complex of k-modules $V = \oplus_n V^n$ endowed with a differential $d_V : V^n \to V^{n+1}$. A morphism $f : V \to W$ of dg k-modules is a (degree zero) morphism of the chain complexes, i.e. a family of morphisms $f_n : V^n \to W^n$ intertwining the differentials. The category $\mathscr{C}(k)$ of dg k-modules admits a monoidal structure given by the graded tensor product

$$(V \otimes W)^n = \bigoplus_{i+j=n} V^i \otimes W^j \,,$$

whose differential acts on homogeneous objects by a graded version of the Leibniz rule

$$d_{V \otimes W}(a \otimes b) = d_V(a) \otimes b + (-1)^{\deg a} a \otimes d_W(b) \,,$$

and the unit of this monoidal structure is the dg-module given by k in degree zero.

Definition 1 A dg category $\mathscr{A}$ is a category enriched over $\mathscr{C}(k)$, i.e. a category where the morphism spaces $\mathscr{A}(X, Y)$ are dg k-modules and the compositions $\mathscr{A}(X, Y) \otimes \mathscr{A}(Y, Z) \to \mathscr{A}(X, Z)$ are morphisms of dg k-modules.

A dg category with only one object is the same as a differential graded algebra, i.e. a k-algebra with a k-linear differential satisfying $d^2 = 0$ and the graded Leibniz rule. Given any dg category $\mathscr{A}$ we can define the closed category $Z^0(\mathscr{A})$ with the same objects but morphisms spaces given by closed morphisms of degree 0, i.e.

$$Z^0(\mathscr{A})(X, Y) = Z^0(\mathscr{A}(X, Y)) = ker(d^0 : \mathscr{A}(X, Y)^0 \to \mathscr{A}(X, Y)^1) .$$

This forms a category since the composition of two closed morphisms is closed by the Leibniz rule. More importantly, we can form the cohomology category $H^0(\mathscr{A})$ with morphism spaces

$$H^0(\mathscr{A})(X, Y) = H^0(\mathscr{A}(X, Y)) = ker(d^0)/im(d^{-1}) .$$

This also gives a category; one can show that any choices of representatives for two classes in $H^0(\mathscr{A}(X, Y))$ leads to the same class under composition.

Remark 1 The category $\mathscr{C}(k)$ of dg k-modules is not itself a dg category, as the morphism spaces are just usual k-modules without any extra structure. One can enrich this into a dg category as in the next example.

2.2 *The dg Category of R-Modules*

Definition 2 For any k-algebra R, the dg category of right (left) R-modules $\mathscr{C}_{\mathrm{dg}}(R)$ has as objects chain complexes M of right (left) R-modules. The morphisms are first defined as graded k-modules: an element of $\mathscr{C}_{\mathrm{dg}}(R)(M, N)^n$ is a family of morphisms of left (right) R-modules $f^n : M^p \to N^{p+n}$. These graded morphism spaces are then given the structure of dg k-modules by the differential

$$d_f = d_N \circ f - (-1)^n f \circ d_M$$

which endows them with the structure of dg k-modules.

Remark 2 From now one we will use the right module structure by default, noting explicitly when we want left modules.

It is easy to check from the definitions that $Z^0(\mathscr{C}_{\mathrm{dg}}(R))$ is just the category $\mathscr{C}(R)$ of chain complexes of R-modules, with morphisms given by degree zero maps intertwining the differentials. Taking the zeroth cohomology category $H^0(\mathscr{C}_{\mathrm{dg}}(R))$ one gets the homotopy category $\mathscr{H}(R)$, whose morphism spaces given by degree zero maps modulo maps homotopic to zero. We say that $\mathscr{C}_{\mathrm{dg}}(R)$ is a dg enhancement of $\mathscr{H}(R)$; it is in a similar way that we will construct dg enhancements of derived categories.

3 Triangulated Structures and dg Categories

The dg category of modules $\mathscr{C}_{\mathrm{dg}}(R)$ is somewhat special in the sense that its zeroth cohomology category $\mathscr{C}(R)$ is triangulated. Note that in general this is not the case; in a dg category the Hom spaces are graded but not the objects, so it is unclear what taking cones and shifts of objects means. We can enforce this condition for a general dg category by looking at representability of the cone and shift functors acting on the Hom spaces.

Definition 3 A dg category $\mathscr{A}$ is (strongly) pretriangulated if for every object X and integer n, the functor $\mathscr{A}^{op} \to \mathscr{C}(k)$ given by

$$Z \mapsto \mathscr{A}(Z, X)[n]$$

is representable, and for every morphism $f : X \to Y$, the functor $\mathscr{A}^{op} \to \mathscr{C}(k)$ given by

$$Z \to Cone(f : \mathscr{A}(Z, X) \to \mathscr{A}(Z, Y))$$

is representable. We will call the objects representing these functors $X[n]$ and Cf, respectively.

It is easy to see that if a dg category $\mathscr{A}$ is pretriangulated, then its zeroth cohomology category $H^0(\mathscr{A})$ is a triangulated (ordinary) category, with shifts and distinguished triangles inherited from the corresponding representing objects in $\mathscr{A}$.

Example 2 For any k-algebra R, the dg category $\mathscr{C}_{\mathrm{dg}}(R)$ of modules over R is a pretriangulated dg category, with the shift and cone objects naturally just given by the shift and cone of chain complexes. Naturally its zeroth cohomology category is the homotopy category $\mathscr{H}(R)$ with the usual triangulated structure.

In general, every dg category has a *pretriangulated envelope* pretr($\mathscr{A}$) with a fully faithful embedding

$$\mathscr{A} \hookrightarrow \mathrm{pretr}(\mathscr{A})$$

satisfying the universal property that any functor $\mathscr{A} \to \mathscr{B}$ to a pretriangulated category $\mathscr{B}$ factors through it. The pretriangulated envelope can be constructed explicitly with the use of twisted complexes [2], which we will not describe in detail here.

4 Functor Categories and Modules over dg Categories

We have seen that given an (ordinary) k-algebra R one can construct two different dg categories from it: a category with only one object and self-homs given by R in degree zero, or the dg category of R-modules $\mathscr{C}_{\mathrm{dg}}(R)$. We would like to do the same and define a dg category of modules over an arbitrary dg k-algebra $\mathscr{A}$.

However, if we try to naïvely generalize the definition of $\mathscr{C}_{\mathrm{dg}}$ above to a case where $\mathscr{A}$ has nonzero elements in multiple degrees, it would be necessary to keep track of a lot of different degrees by hand, which is very inconvenient. The correct way to do this is to formalize module categories as functor categories, and once we do so it is not any more work to define modules over arbitrary dg categories.

4.1 The Category of dg Categories

A dg functor F between two dg categories $\mathscr{A}$, $\mathscr{B}$ is a functor respecting the dg structure of the morphism spaces, i.e. such that $\mathscr{A}(X, Y) \to \mathscr{B}(FX, FY)$ is a morphism in $\mathscr{C}(k)$ for every pair of objects. This allows us to consider the category of dg categories. For set-theoretic reasons it is wise to restrict to (essentially) small categories, i.e. such that the isomorphism classes of objects form a set.

Definition 4 The category dg-Cat_k of small dg categories over k has as objects small dg categories over k and as morphisms dg functors between them.

Theorem 1 dg-Cat_k *is a symmetric monoidal category with a tensor product* $\otimes$ *and an internal Hom functor* $\mathscr{H}om$, *with an internal adjunctions*

$$\mathscr{H}om(\mathscr{A} \otimes \mathscr{B}, \mathscr{C}) \cong \mathscr{H}om(\mathscr{A}, \mathscr{H}om(\mathscr{B}, \mathscr{C})) .$$

The monoidal structure is given by the following tensor product of categories: $\mathscr{A} \otimes \mathscr{B}$ has objects given by pairs of objects (X_A, X_B) in $\mathscr{A}$, $\mathscr{B}$ and morphism spaces given by tensor of morphism spaces in $\mathscr{C}(k)$:

$$\mathrm{Hom}_{\mathscr{A}\otimes\mathscr{B}}((X_A, X_B), (Y_A, Y_B)) = \mathrm{Hom}_{\mathscr{A}}(X_A, Y_A) \otimes \mathrm{Hom}_{\mathscr{B}}(X_B, Y_B) .$$

The internal hom category $\mathscr{H}om(\mathscr{A}, \mathscr{B})$ has as objects dg functors $\mathscr{A} \to \mathscr{B}$, with the degree n piece $\mathscr{H}om(\mathscr{A}, \mathscr{B})(F, G)^n$ of a morphism space given by a family of degree n morphisms

$$\phi_X \in (\mathscr{B}(FX, GX))^n, (Gf)(\phi_X) = (\phi_Y)(Ff) ,$$

for all $f \in \mathscr{A}(X, Y)$. This graded k-module inherits a differential induced from the differential in $\mathscr{B}(FX, GX)$, giving $\mathscr{H}om(\mathscr{A}, \mathscr{B})$ the structure of a dg category.

4.2 Modules over dg Categories

The internal hom in dg-Cat_k lets us construct new dg categories as categories of functors; consider an arbitrary small dg category $\mathscr{A}$ over k, possibly with multiple objects and hom spaces in many degrees. We can define categories of modules over it as functor categories using $\mathscr{H}om$:

Definition 5 The dg category of right modules over $\mathscr{A}$ is defined by the internal Hom from the opposite category:

$$\mathscr{C}_{\mathrm{dg}}(\mathscr{A}) = \mathscr{H}om(\mathscr{A}^{op}, \mathscr{C}_{\mathrm{dg}}(k)) ,$$

and the category of left modules over $\mathscr{A}$ is analogously defined as $\mathscr{H}om(\mathscr{A}, \mathscr{C}_{\mathrm{dg}}(k))$.

We can get ordinary categories from these by taking the closed and cohomology categories: we define the category of $\mathscr{A}$-modules $\mathscr{C}(\mathscr{A}) = Z^0(\mathscr{C}_{\mathrm{dg}}(\mathscr{A}))$ and the homotopy category of $\mathscr{A}$-modules $\mathscr{H}(\mathscr{A}) = H^0(\mathscr{C}_{\mathrm{dg}}(\mathscr{A}))$. It is easy to check that when $\mathscr{A}$ is the dg category with one object and self-homs given by an ordinary k-algebra R, these notions agree with our previous definitions of $\mathscr{C}_{\mathrm{dg}}(R)$, $\mathscr{C}(R)$ and $\mathscr{H}(R)$.

Lemma 1 *For any dg category $\mathscr{A}$ the dg category of $\mathscr{A}$-modules $\mathscr{C}_{\mathrm{dg}}(\mathscr{A})$ is pretriangulated, with shifts and cones inherited from the target category $\mathscr{C}_{\mathrm{dg}}(k)$.*

4.3 The Yoneda Embedding

For any ring R, there is a distinguished object in $\mathscr{C}_{\mathrm{dg}}(R)$, the unit of the monoidal structure, given by R placed in degree zero, with self-homs given by R itself. Looking at R as a dg category concentrated in degree zero, we see this is just the image of the obvious embedding of dg categories $R \to \mathscr{C}_{\mathrm{dg}}(R)$.

This can be generalized to an arbitrary dg category $\mathscr{A}$ over k in the setting described above: for any object X of $\mathscr{A}$ we define the object $\hat{X}$ in $\mathscr{C}_{\mathrm{dg}}(\mathscr{A})$ given by the functor $\mathrm{Hom}_{\mathscr{A}}(X, -)$. This is the *Yoneda embedding*

$$\mathscr{A} \to \mathscr{C}_{\mathrm{dg}}(\mathscr{A}) ,$$

which one can easily check is a fully faithful dg functor.

As we remark above, the dg category $\mathscr{C}_{\mathrm{dg}}(\mathscr{A})$ is automatically triangulated, even if $\mathscr{A}$ itself is not: for any functor $M : \mathscr{A} \to \mathscr{C}_{\mathrm{dg}}(k)$ one can compose it with the shift $[n]$ in $\mathscr{C}_{\mathrm{dg}}(k)$ to get $M[n]$.

Let us take now the triangulated hull of the collection $\{\hat{X}[n]\}$ of all the shifts of the images of all the objects X under the Yoneda embedding. Remember that the triangulated hull of a collection of objects is the smallest triangulated subcategory containing those objects. In our case we will denote this triangulated hull by $\mathrm{per}_{\mathrm{dg}}(\mathscr{A})$, the dg category of *perfect complexes* over $\mathscr{A}$. From this we can also get the ordinary category of perfect complexes by $\mathrm{per}(\mathscr{A}) = Z^0(\mathrm{per}_{\mathrm{dg}}(\mathscr{A}))$. Note that definition, the Yoneda embedding factors through $\mathrm{per}_{\mathrm{dg}}(\mathscr{A})$, and we will also call this map the Yoneda embedding.

4.4 The dg Derived Category

As we stated in the beginning of the talk, one main objective of defining dg categories is to come up with enhancements of triangulated categories that contain more structure, that is, to find pretriangulated dg categories whose H^0 category recovers some triangulated category we want to study.

It is not clear that we should be able to find a meaningful dg enhancement of an arbitrary triangulated category, but in specific cases, when the triangulated category is given in some algebraic or geometric context, we can often find natural dg enhancements. We have seen an example of this already: for any ring R we defined the dg category $\mathscr{C}_{\mathrm{dg}}(R)$ so that it gives an enhancement of the homotopy category $H^0(\mathscr{C}_{\mathrm{dg}}(R)) = \mathscr{H}(R)$. It is then a natural question to ask whether the derived category $\mathscr{D}(R)$ (and also $\mathscr{D}(\mathscr{A})$ for some general dg category $\mathscr{A}$) also has a similar dg enhancement. More generally, we can ask whether other derived categories of interest, such as derived categories of quasicoherent or coherent sheaves on a scheme X, possess similar dg enhancements.

The answer turns out to be that all these examples do have dg enhancements, and some even have several different dg enhancements: for example the derived category of quasicoherent sheaves $\mathscr{D}(\mathrm{qcoh}X)$ on a separated noetherian scheme X has at least three different constructions of a dg enhancement [13]. One of these constructions involves a familiar construction of quotients of categories.

Proposition 1 ([4, 13]) *For any dg category $\mathscr{A}$ with a full dg subcategory $\mathscr{B}$ there is a dg category denoted $\mathscr{A}/\mathscr{B}$ with an universal morphism (up to quasiequivalence) in* dg-Cat$_k$

$$\mathscr{A} \to \mathscr{A}/\mathscr{B}$$

such that any dg functor $\mathscr{A} \to \mathscr{C}$ with the property that the corresponding map on homotopy categories $H^0(\mathscr{A}) \to H^0(\mathscr{C})$ sends all elements of $\mathscr{B}$ to zero factors through $\mathscr{A} \to \mathscr{A}/\mathscr{B}$.

This quotient dg category can be constructed easily when e.g. the ground ring k is a field; in general the construction is more involved. We can apply this to construct dg enhancements of our familiar derived categories.

Example 3 Consider the dg category $\mathscr{C}_{\mathrm{dg}}(\mathrm{qcoh}X)$ of unbounded complexes of quasicoherent sheaves on a separated noetherian scheme X, and the full dg subcategory $\mathscr{A}c_{\mathrm{dg}}(\mathrm{qcoh}X)$ spanned by all the acyclic complexes. The quotient

$$\mathscr{D}_{\mathrm{dg}}(\mathrm{qcoh}X) = \mathscr{C}_{\mathrm{dg}}(\mathrm{qcoh}X)/\mathscr{A}c_{\mathrm{dg}}(\mathrm{qcoh}X)$$

is an enhancement of the category $\mathscr{D}(\mathrm{qcoh}X)$.

We can apply this same construction to any abelian category C in place of $\mathrm{qcoh}(X)$: applied to $C = \mathrm{Mod}(R)$ we get a dg category $\mathscr{D}_{\mathrm{dg}}(R)$ which is an enhancement of the derived category $\mathscr{D}(R)$. These dg enhancements are referred to in the literature as the *dg derived category* of an abelian category.

When actually computing morphisms in the derived category, it is often more useful to use the formalism of fibrant and cofibrant replacements, which are generalizations of projective and injective resolutions. A more rigorous and thorough treatment of these techniques goes through the discussion of Quillen model structures but we will avoid that and refer to more competent sources [6]. In our specific case we can define the cofibrant and fibrant objects of the category of dg modules over some arbitrary dg category $\mathscr{A}$ as follows:

Definition 6 An object P of $\mathscr{C}(\mathscr{A})$ is cofibrant if for every surjective quasi-equivalence $L \to M$, every $P \to M$ factors through L. An object I of $\mathscr{C}(\mathscr{A})$ is fibrant if for every injective quasi-equivalence $L \to M$, every $L \to I$ extends to M.

Lemma 2 *The category $\mathscr{C}(\mathscr{A})$ admits cofibrant and fibrant replacements; i.e. for any object M there are quasi-isomorphisms $P \to M$ and $M \to I$ where P is cofibrant and I is fibrant. Moreover all objects $\hat{M}$ in the image of the Yoneda embedding $\mathscr{A} \to \mathscr{C}(\mathscr{A})$ are cofibrant.*

So we can also define the derived category and compute its morphisms by using e.g. the fibrant replacement and computing in the homotopy category:

$$\mathscr{D}(\mathscr{A})(X, Y) = \mathscr{H}(\mathscr{A})(P, Y) = H^0(\mathscr{C}_{\mathrm{dg}}(P, Y)) .$$

5 Additive Invariants

5.1 Hochschild Homology of Associative Algebras

Hochschild homology and cohomology were initially defined as invariants of associative algebras, but the definition can be extended to dg algebras and dg categories, and we can use the dg enhancements we constructed above to define invariants of e.g. derived categories of coherent sheaves.

Definition 7 Given an associative k-algebra A and an A-bimodule M, the *Hochschild chain complex* of A with coefficients in the bimodule M is concentrated in non-positive degrees and is defined by [1, 7]

$$C^{-n}(A, M) = M \otimes A^{\otimes n} ,$$

for $n \geq 0$ with a differential $d : C^{-n}(A, M) \to C^{-n+1}(A, M)$ given by

$$d(m \otimes a_1 \otimes \cdots \otimes a_n) = ma_1 \otimes \cdots \otimes a_n + \sum_{i=1}^{n-1} m \otimes \cdots \otimes a_i a_{i+1} \otimes \cdots \otimes a_n + (-1)^n a_n m \otimes a_1 \otimes \cdots \otimes a_{n-1} .$$

The Hochschild homology of A with coefficients in M is defined as the cohomology of this complex: $HH_n(A, M) = H^{-n}(C^*(A, M))$. Hochschild cohomology is defined using a dual complex concentrated in non-negative degrees

$$C^n(A, M) = \mathrm{Hom}(A^{\otimes n}, M) ,$$

with a differential $d : C^n(A, M) \to C^{n+1}(A, M)$ given by

$$df(a_1, \dots, a_{n+1}) = a_1 f(a_2, \dots, a_{n+1}) + \sum_{i=1}^{n}(-1)^i f(a_1, \dots, a_i a_{i+1}, \dots, a_{n+1}) \\ + (-1)^{n+1} f(a_1, \dots, a_n)a_{n+1} .$$

Remark 3 Here we reverse the degrees and direction of the differential in the Hochschild homology complex from the usual conventions just so that it is also cohomologically graded, since it will simplify our notation in the future.

To get an invariant of associative algebras, we can take M to be the diagonal bimodule A_Δ, i.e. just A as a bimodule over itself with the left and right algebra actions, and get the Hochschild (co)homology of A: $HH_n(A) = HH_n(A, A_\Delta)$ and $HH^n(A) = HH^n(A, A_\Delta)$. We can also stop before taking the (co)homology of the complex and define the Hochschild complex as an object of $\mathscr{C}(k)$.

Besides the dg structure Hochschild homology and cohomology of A carry extra structures; for instance $HH_*(A)$ automatically carries an S^1 action which allows us to also define further invariants such as cyclic homology and negative cyclic homology, which we refer to other sources [11]

Example 4 Let A be an associative algebra over k. Then its first two Hochschild homologies are

$$HH_0(A) = A/[A, A], \qquad HH_1(A) = \Omega^1(A) ,$$

where $\Omega^1(A)$ is the vector space of Kähler differentials on A, i.e. spanned over A by symbols da for $a \in A$, modulo the relations

$$dx = 0, \quad d(a + b) = da + db, \quad d(ab) = da\, b + adb$$

for every $x \in k,\ a, b \in A$. Note that if A is the algebra of functions on some manifold then the Kähler differentials is an algebraic version of the space of one-forms. The fact that the first Hochschild homology captures the space of one-forms is our first example of a more general fact we'll get to later, the Hochschild-Kostant-Rosenberg theorem.

The first two Hochschild cohomologies are

$$HH^0(A) = Z(A), \quad HH^1(A) = Der(A)/Inn(A) ,$$

where $Der(A)$ is the space of derivations of A and $Inn(A) \subseteq Der(A)$ are the derivations given by commutators with some element in A. More generally for some A-bimodule M

$$HH_0(A, M) = M/[M, A], \quad HH^0(A, M) = Z_A(M) ,$$

i.e. respectively the coinvariants and the invariants under the adjoint A action.

Hochschild homology can be given an interpretation in terms of derived functors in the category of (A, A)-bimodules, i.e. $A \otimes A^{op}$-modules. Given any bimodule N in $\mathrm{Mod}(A \otimes A^{op})$, there are two functors $\mathrm{Mod}(A \otimes A^{op}) \to \mathrm{Mod}(k)$ given by $N \otimes_{A\otimes A^{op}} -$ and $\mathrm{Hom}_{A\otimes A^{op}}(N, -)$. It is easy to check from the definitions that by taking $N = A_\Delta$, this calculates the zeroth degree Hochschild homology and cohomology

$$HH_0(A, M) = A_\Delta \otimes_{A\otimes A^{op}} M, \qquad HH^0(A, M) = \mathrm{Hom}_{A\otimes A^{op}}(A_\Delta, M)\ .$$

As you might expect the higher Hochschild homologies and cohomologies are the derived functors of the tensor and hom of bimodules: there is a left derived tensor $A_\Delta \otimes^L_{A\otimes A^{op}} -$ and a right derived hom $R\mathrm{Hom}_{A\otimes A^{op}}(A_\Delta, -)$, both functors from the derived category $\mathscr{D}(A \otimes A^{op}) \to \mathscr{D}(k)$, which calculate the Hochschild homology and cohomology

$$HH_\bullet(A, M) = A \otimes^L_{A\otimes A^{op}} M, \qquad HH^\bullet(A, M) = R\mathrm{Hom}_{A\otimes A^{op}}(A, M)\ .$$

The connection between this more abstract definition and the explicit definition above is due to the fact that there is a standard free resolution of the diagonal bimodule $\mathscr{A}_\Delta$ given by the *bar complex* $\bar{C}_n(A)$ given in non-positive degrees by

$$\bar{C}^{-n}(A) = A \otimes A^{\otimes n} \otimes A$$

with the differential $d : \bar{C}^{-n}(A) \to \bar{C}^{-n+1}(A)$ given by

$$d(a_0 \otimes a_1 \otimes \cdots \otimes a_n \otimes a_{n+1}) = \sum_{i=0}^{i=n} (-1)^i a_0 \otimes \cdots \otimes a_i a_{i+1} \otimes \cdots \otimes a_{n+1}\ .$$

and one can check that for any (A, A)-bimodule M there is a quasi-isomorphism of complexes

$$\bar{C}^*(A) \otimes_{A\otimes A^{op}} M \cong C^*(A, M)$$

with the complex we initially used to define Hochschild homology.

Example 5 Consider a finite quiver Q with n vertices and no oriented cycles, and take $A = kQ$ to be its path algebra. Remember that A as a vector space over k is spanned by all the paths in Q, so as an algebra it is generated by the idempotents e_i (one for each vertex) and the elements f_{ij} (one for each edge $i \to j$), subject to the composition rules given by concatenation of paths. Note that since Q is finite, there is an identity element $\mathbf{1} = \sum_i e_i$, and up to scaling it is the only central element, so $Z(A) = k\mathbf{1}$. Note also that every path x with length ≥ 1 is in the commutator ideal; i.e. if x starts at the vertex i, then $x = [e_i, x]$. Thus we have that

$$HH^0(A) \simeq k \qquad HH_0(A) \simeq kQ/\langle f_{ij}\rangle \simeq k^{\otimes n}\ ,$$

where $k^{\otimes n}$ is spanned by the basis $\{e_i\}$.

It can also be shown [3] that in this case where Q has no oriented cycles, there is no higher Hochschild homology: $HH_i(kQ) = 0$ for $i \geq 1$. The Hochschild cohomology groups are more complicated [7]: $HH^i(kQ) = 0$ for $i \geq 2$ but

$$\dim HH^1(kQ) = 1 - n + \sum_{\text{edge } ij} (\text{number of paths } i \to j) .$$

We can check that this is zero if and only if the underlying graph is a tree (even if the quiver has no oriented cycles the underlying graph might have cycles). This examples shows how Hochschild homology and cohomology can have quite different behaviors.

5.2 *Hochschild Homology of dg Algebras and dg Categories*

The definition above can be easily extended to dg algebras and dg categories. For a dg category $\mathscr{A}$ over k, remember that we defined the dg category $\mathscr{C}_{\text{dg}}(\mathscr{A})$ of modules over $\mathscr{A}$. We can also define the dg category $\mathscr{C}_{\text{dg}}(\mathscr{A} \otimes \mathscr{A}^{op})$ of $(\mathscr{A}, \mathscr{A})$-bimodules using the tensor and internal hom structure of dg-Cat_k.

There is also a diagonal bimodule $\mathscr{A}_\Delta$ defined by the Hom functor $\mathscr{A} \otimes \mathscr{A}^{op} \to \mathscr{C}_{\text{dg}}(k)$ of $\mathscr{A}$:

$$\mathscr{A}_\Delta : (Y, X) \mapsto \mathscr{A}(X, Y) .$$

Similarly to what we saw above, for any bimodule $\mathscr{M}$ there is a bimodule tensor functor $\mathscr{M} \otimes_{\mathscr{A} \otimes \mathscr{A}^{op}} -$ and a bimodule Hom functor $\text{Hom}_{\mathscr{A} \otimes \mathscr{A}^{op}}(\mathscr{M}, -)$ both mapping $\mathscr{C}_{\text{dg}}(\mathscr{A} \otimes \mathscr{A}^{op}) \to \mathscr{C}_{\text{dg}}(k)$. They give rise to derived functors $\mathscr{D}_{\text{dg}}(\mathscr{A} \otimes \mathscr{A}^{op}) \to \mathscr{D}_{\text{dg}}(k)$ so just as above we can define the Hochschild homology and cohomology of a dg category by

$$HH_*(\mathscr{A}) = \mathscr{A}_\Delta \otimes^L_{\mathscr{A} \otimes \mathscr{A}^{op}} \mathscr{A}_\Delta, \qquad HH^*(\mathscr{A}) = R\text{Hom}_{\mathscr{A} \otimes \mathscr{A}^{op}}(\mathscr{A}_\Delta, \mathscr{A}_\Delta) .$$

This definition agrees with the earlier definitions when $\mathscr{A}$ is the dg category with one object and with self-homs given by some (dg) algebra.

5.3 *Hochschild Homology of dg Enhancements of Triangulated Categories*

Suppose now that we have a triangulated category $\mathscr{T}$ with a dg enhancement $\mathscr{D}(\mathscr{C})$, i.e. we have some dg algebra $\mathscr{C}$ and a triangulated equivalence $\mathscr{T} \cong \mathscr{D}(\mathscr{C})$. This allows us to define additive invariants for the triangulated category $\mathscr{T}$: in particular we can define the Hochschild (co)homology of $\mathscr{T}$ as the Hochschild (co)homology of the dg algebra $\mathscr{C}$. In general these invariants will depend on the particular dg enhancement we pick, but in some useful contexts the choice of dg enhancement does not matter for any of the additive invariants: e.g. when $\mathscr{T} = \mathscr{D}^b(\mathrm{Coh}X)$ for a smooth proper scheme X [12].

Let X be a smooth projective variety, and consider the derived category $\mathscr{D}^b(X) = \mathscr{D}^b(\mathrm{Coh}\ X)$. Here we follow [10]. It is known that $\mathscr{D}^b(X)$ has a strong generator E, i.e. any object A of $\mathscr{D}^b(X)$ can be generated by E by taking a finite sequence of cones, shifts and finite summands. Consider the dg algebra of endomorphisms $\mathscr{A} = R\mathrm{Hom}(E, E)$. Then we have a triangulated equivalence $\mathscr{D}^b(X) \cong \mathscr{D}^b(\mathscr{A})$, and we can define the Hochschild (co)homology of the derived category of coherent sheaves on X as the Hochschild (co)homology of the dg algebra $\mathscr{A}$:

$$HH^*(X) = HH^*(\mathscr{A}), \qquad HH_*(X) = HH_*(\mathscr{A})\ .$$

Another way of defining Hochschild (co)homology for a scheme X is to just adapt to a geometric setting the notion of Hochschild homology as the self-tensor of the diagonal and Hochschild cohomology as the self-homs of the diagonal, by defining

$$HH_*(X) = \mathbb{H}^*(X \times X, \Delta_*\mathscr{O}_X \otimes^L \Delta_*\mathscr{O}_X), \qquad HH^*(X) = \mathrm{Hom}^*_{X\times X}(\Delta_*\mathscr{O}_X, \Delta_*\mathscr{O}_X)\ .$$

Here $\Delta : X \to X \times X$ is the diagonal embedding, $\mathbb{H}$ calculates the hypercohomology of a complex of sheaves and all the functors are implicitly understood as the corresponding derived functors on the categories of coherent sheaves.

One can prove that these two definitions of Hochschild (co)homology agree regardless of the particular choice of strongly generating object E. Besides being a calculational tool, the definition using the dg enhancement also shows that Hochschild homology satisfies a very nice property under semiorthogonal decomposition. Suppose we have a triangulated category $\mathscr{T}$ with a strong generator E, and a semiorthogonal decomposition

$$\mathscr{T} = \langle \mathscr{T}_1, \dots \mathscr{T}_n \rangle$$

into pieces $\mathscr{T}_i$. Then we can look at the projection E_i of E onto each piece $\mathscr{T}_i$. One can show that this gives strongly generating objects, so using the dg algebras $\mathscr{A}_i = R\mathrm{Hom}(E_i, E_i)$, we get a direct sum decomposition of Hochschild *homology*

$$HH_*(\mathscr{T}) = \bigoplus HH_*(\mathscr{T}_i)\ .$$

Note that we do *not* get a similar decomposition for Hochschild cohomology.

Example 6 Consider the derived category $\mathscr{D}(kQ)$ of representations of an acyclic quiver Q with n vertices. Here we already have a dg enhancement given by $\mathscr{D}_{\mathrm{dg}}(kQ)$. This corresponds as picking as generating object the algebra kQ itself. Remember that we have a decomposition of (right) kQ-modules

$$kQ = \bigoplus_{\alpha \in Q} P_\alpha$$

into the projective modules P_α given by all the paths starting at the vertex α.

Let us choose a total ordering of the vertices of Q compatible with the partial ordering given by the quiver structure, i.e. we require $\alpha < \beta$ if there is a non-zero path going from α to β. Then we see that we have a semiorthogonal orthogonal decomposition of the category $\mathscr{D}(kQ)$ with n pieces $\mathscr{T}_\alpha$, each one with a single object given by the projective P_α. Each piece is equivalent to the category $\mathscr{D}(k)$ and has Hochschild homology given by $HH_0(\mathscr{T}_i) = \mathrm{Hom}(P_\alpha, P_\alpha) = k$ and zero in higher degrees. So from the direct sum decomposition we recover the result

$$HH_0(\mathscr{D}(kQ)) \simeq k^n, \qquad HH_i(\mathscr{D}(kQ)) = 0, i \geq 1 .$$

5.4 The Hochschild-Kostant-Rosenberg Theorem

The classical statement of the Hochschild-Kostant-Rosenberg theorem [8] is a generalization of the fact that for a commutative k-algebra R, the first Hochschild homology gives the space of Kähler differentials.

Theorem 2 *Let R be a finitely presented k-algebra, where k has characteristic zero. Suppose also that R is smooth i.e. the space of Kähler differentials Ω^1_R is a projective R-module. Then we have an isomorphism*

$$HH_n(R) \cong \Omega^n_R = \wedge^n \Omega^1_R .$$

There is also a version of the HKR theorem for the category of coherent sheaves on a smooth projective variety X. Let us again denote the diagonal inclusion by $\Delta : X \to X \times X$ and define two complexes of sheaves in $\mathscr{D}^b(\mathrm{Coh}\, X)$:

$$\mathscr{H}\mathscr{H}_\bullet = \Delta^* \Delta_* \mathscr{O}_X, \qquad \mathscr{H}\mathscr{H}^\bullet = \Delta^! \Delta_* \mathscr{O}_X .$$

These are sheafy versions of the Hochschild homology and cohomology: taking global sections one can show that

$$HH_*(X) = \mathbb{H}^*(X, \mathscr{H}\mathscr{H}_\bullet), \qquad HH^*(X) = \mathbb{H}^*(X, \mathscr{H}\mathscr{H}^\bullet) .$$

The HKR theorem then also holds at the sheaf level:

Theorem 3 *Let X be a smooth projective variety of dimension n. Then there are quasi-isomorphisms*

$$\mathscr{H}\mathscr{H}_\bullet \simeq \bigoplus_{p=0}^{n} \Omega^p_X[p], \qquad \mathscr{H}\mathscr{H}^\bullet \simeq \bigoplus_{p=0}^{n} T^p_X[p] ,$$

where $\Omega_X^p = \wedge^p \Omega_X^1$ *is the sheaf of p-forms on X and* $T_X^p = \wedge^p T_X^1$ *where* T_X^1 *is the tangent sheaf of X.*

So in the case of a smooth projective variety X over $\mathbb{C}$, taking global sections of this sheaf calculates the Hochschild homology of X in terms of the Hodge groups of X

$$HH_k(X) \cong \bigoplus_{p-q=k} H^p(X, \Omega_X^q) = \bigoplus_{p-q=k} H^{p,q}(X) .$$

In particular, since derived equivalences preserves the Hochschild homology we conclude that any derived equivalence $\mathscr{D}(X) \cong \mathscr{D}(X')$ preserves the sum $\sum_{p-q=k} h^{p,q}$, i.e. the column sums of the Hodge diamond.

References

1. D. Aranha, *On Hochschild Cohomology, Koszul Duality and DG Categories*, www.math.uni-bonn.de/ag/stroppel/MA_DhyanAranha.pdf
2. A.I. Bondal, M.M. Kapranov, Enhanced Triangulated Categories, Mat. Sb. **181:5**, 669–683 (1990).
3. C. Cibils, *Hochschild homology of an algebra whose quiver has no oriented cycles*, in *Representation Theory I Finite Dimensional Algebras*, ed. by V. Dlab, P. Gabriel, G. Michler. Lecture Notes in Mathematics **1177** (Springer 1986), 55–59.
4. V. Drinfeld, Dg quotients of dg categories, J. Algebra **272:2**, 643–691 (2004).
5. D. Dugger, B. Shipley, *A curious example of two model categories and some associated differential graded algebras*, arXiv:0710.3070
6. W. Dwyer, L. Spaliński, *Homotopy theories and model categories*, in *Handbook of Algebraic Topology*, ed. by I.M. James. (Elsevier, 1995) 73–126
7. D. Happel, *Hochschild cohomology of finite-dimensional algebras*, Lecture Notes in Mathematics **1404** (Springer, 1989)
8. G. Hochschild, B. Kostant, A. Rosenberg, Differential forms on regular affine algebras, Transactions AMS **102** No.3, 383–408 (1962).
9. B. Keller, *On differential graded categories*, Proceedings ICM **2** (2006)
10. A. Kuznetsov, *Hochschild homology and semiorthogonal decompositions*, arXiv:0904.4330
11. J.-L. Loday, *Cyclic homology*, Grundlehren der mathematischen Wissenschaften **301**, (1998)
12. V. Lunts, D. Orlov, Uniqueness of enhancement for triangulated categories, J. Amer. Math. Soc. **23** 853–908 (2010).
13. D. Orlov, Smooth and proper noncommutative schemes and gluing of dg categories, Advances in Mathematics **302**, 59–105 (2016).
14. B. Toën, *Lectures on DG-categories*, in *Topics in Algebraic and Topological K-Theory*, Lecture Notes in Mathematics **2008** (Springer 2011) 243–301

Introduction to Symplectic Geometry and Fukaya Category

Alex Zhongyi Zhang

1 Symplectic Geometry

Definition 1 (*symplectic form*) Given a vector space V, a **symplectic form** ω is a non-degenerate, anti-symmetric bilinear form. namely, $\forall\, v \in V, \omega(v, w) = 0, \forall\, w \in V \iff v = 0$ and $\omega(v, w) = -\omega(w, v)$. Such a vector space V is called a symplectic vector space.

We know from linear algebra that all symplectic vector spaces must have even dimensions. Let $W \subseteq V$, $W^{\omega} := \{v \in V, \omega(v, w) = 0\ \forall w \in W\}$,

Definition 2 Given a symplectic vector space (V, ω), a subspace $W \subseteq V$ is called isotropic if $W \subseteq W^{\omega}$, *i.e.*$\omega|_W = 0$;
W is coisotropic if $W \supseteq W^{\omega}$,
W is symplectic if $\omega|_W$ is also a symplectic form on W.
W is Lagrangian if it is isotropic and $\dim W = \frac{1}{2}\dim V$.

We have $\forall\ W \subset V$, $\dim W +$ $\dim W^{\omega} = \dim V$, therefore, $W^{\omega\omega} = W$. The Euclidean space $\mathbb{R}^{2n}$ is a symplectic vector space equipped with the standard symplectic form $\omega_0 = \sum_{i=1}^{n} x_i \wedge y_i$. Also, for any symplectic vector space, we have s symplectic basis $u_1, \dots u_n; v_1, \dots, v_n$ such that $\omega(u_j, u_k) = \omega(v_j, v_k) = 0, \omega(u_{,} v_k) = \delta_{kj}$. Namely, we have a map $\Phi : \mathbb{R}^{2n} \to V$ such that $\Phi^*\omega = \omega_0$.

Definition 3 (*symplectomorphism*) $\mathrm{Sp}(V, \omega) = \{\Phi \in Gl(V) \big| \Phi^*\omega = \omega\}$, the linear isomorphisms that preserves the symplectic structure are called symplectomorphisms. Since we know that $V \simeq \mathbb{R}^{2n}$ by the paragraph above, we can identify

This lecture note is written for a talk in Superschool on derived categories and D-branes in July, 2016 at University of Alberta.

A. Z. Zhang (✉)
Columbia College of Columbia University in New York City, New York, NY, USA
e-mail: zz2375@columbia.edu

M. Ballard et al. (eds.), *Superschool on Derived Categories and D-branes*,
Springer Proceedings in Mathematics & Statistics 240,
https://doi.org/10.1007/978-3-319-91626-2_11

$Sp(V, w)$ as the maps $\{\Phi | \Phi^*\omega_0 = \omega_0\} = \{A | A^t J_0 A = J_0\}$. If we identify $\mathbb{R}^{2n}$ with $\mathbb{C}^n$, then J_0 acts as i.

Lemma 4 *$Sp(2n) \cap O(2n) = Sp(2n) \cap Gl(n, \mathbb{C}) = O(2n) \cap Gl(n, \mathbb{C}) \simeq U(n)$, and $U(n)$ is a maximal compact subgroup of the symplectic group and $Sp(2n)$ is homotopy equivalent to $U(n)$.*

Sketch of proof. The first equation is a matter of writing down explicitly the definitions and calculate. We have a polar decomposition $\forall\, \Phi \in Sp(2n)$, $\Phi = UP$ where $U := \Phi \cdot (\Phi^t \Phi)^{-\frac{1}{2}} \in U(n)$. $P = (\Phi^t \Phi)^{\frac{1}{2}}$ is symplectic symmetric and positive definite. Let $U_t := (\Phi^t \Phi)^{-\frac{1}{2}t} \in Sp(2n)$ for $t \in [0, 1]$, this gives a deformation retract from $Sp(2n)$ to $U(n)$. (Further details may be found in Chap. 2 of [1].) □

Corollary 5 *$\pi_1(Sp(2n)) = \pi_1(U(n)) = \pi_1(S^1) \simeq \mathbb{Z}$, where the second equality is induced by the complex determinant function.*

Now we try to associate an integer μ to any loop in the Lagrangian Grassmannian $\Lambda : \mathbb{R}/\mathbb{Z} \to LGr(n)$ such that $\mu(\Lambda_1) = \mu(\Lambda_2)$ if and only if Λ_1 and Λ_2 are homotopic. It should also satisfy $\mu(\Lambda \oplus \Lambda') = \mu(\Lambda) + \mu(\Lambda')$, and $\lambda_0(t) = e^{\pi i t}$ has the number 1 associated to it. This integer is the **Maslov index** of the loop. Actually $LGr(n) \simeq U(n)/O(n)$, therefore, $\pi_1(LG(n)) = \pi_1(U(n)/O(n)) \simeq \mathbb{Z}$, which is induced by μ.

More generally, we have the **Maslov index** for any 2nd relative homotopy group:

$$\mu : \pi_2(M, L) \to \mathbb{Z}$$

defined as follows: if a map $u : (\mathbb{D}^2, S^1) \to (M, L)$ represent a class $[u] \in \pi_2(M, L)$, we trivialize the pullback of the tangent bundle u^*TM on $\mathbb{D}^2$ and get the trivial rank $2n$ bundle. Take the tangent bundle TL restricted to S^1 along this trivialization which gives a loop in $LGr(n)$. Then we define the Maslov index of $[u]$ as the Maslov index for as above.

The **minimal Maslov number** N_L is defined as the smallest positive integer that the image of the map μ hits in $\mathbb{Z}$. We set $N_L = \infty$ if the Maslov index μ vanishes.

Definition 6 (*symplectic manifold*) Now a symplectic structure on a smooth manifold M is a non degenerate closed 2-form ω, namely $(T_q M, \omega_p)$ is a symplectic vector space $\forall p \in M$. Non-degeneracy implies that $\omega^n = \omega \wedge \omega \wedge \ldots \omega$ doesn't vanish, which implies that M is oriented.

A symplectomorphism of $(M, \omega) = Symp(M, \omega) := \{\phi \in Diff(M) | \phi^*\omega = \omega\}$.

There is a systematic way to construct a symplectomorphism from a function $H : M \to \mathbb{R}$. First define a vector field $X_H \in \mathcal{X}(M, \omega)$ by $i_{X_H}\omega = dH$, the non-degeneracy of ω implies the existence of such a vector field. Note since $d(i_{X_H}\omega) = ddH = 0$, we have $\mathcal{L}_{X_H}\omega = (di_X + i_X d)\omega = 0$. Let ψ_t be the local flow generated by X_H, namely $\frac{d\psi_t}{dt} = X_H(\psi_t)$, then ψ_t is a symplectomorphism. In this case, we call ψ_t the **Hamiltonian flow** generated by H.

Now $dH(X_h) = i_{X_H}\omega X_H = 0$, thus X_H is tangent to the level sets of H. For example, if we were to have the height function on the sphere, then with the standard symplectic form on the sphere induced volume form on $\mathbb{R}^3$, we have $\omega = d\theta \wedge dz$, then $X_H = \frac{\partial}{\partial\theta}$, the flow ϕ^t is rotation of S^2 at constant speed.

Basic examples:

(1) $(\mathbb{R}^{2n}, \omega_0)$
(2) any oriented Riemann surface with area form;
(3) $\mathbb{T}^{2n} = \mathbb{R}^{2n}/\mathbb{Z}^{2n}$ with the standard form ω_0 on the quotient space.
(4) Cotangent bundle of any manifold. T^*M with canonical 1-form $\lambda_{can} \in \Omega^1(T^*M), \omega = -d\lambda_{can}$, where $\lambda_{can} = \sum_1^n y_i dx_i$. Here the y_i are the coordinates for dx_i, namely we have coordinate charts $T^*U \to \mathbb{R}^n \times \mathbb{R}^n$, $(q, v^*) \mapsto (x(q), y(q, v^*))$, and $T_{(q,0)}(T^*M) \simeq TM \oplus T_q^*M$.

Proposition 7 *λ_{can} is characterized by the property that $\sigma^*\lambda_{can} = \sigma, \forall \sigma \in \Omega^1(M)$.*

This is because if we write out $\sigma = \sum a_j(x)dx_j$, then as a map in local charts, we should get $(x_1, \ldots, x_n, a_1(x), \ldots a_n(x))$, and $\sigma^*(\sum y_j dx_j) = \sigma$.

Proposition 8 *The image of a 1-form σ is Lagrangian in $T^*M \iff \sigma$ is closed.*

Proof $d\sigma = d\sigma^*\lambda_{can} = \sigma^*(d\lambda_{can}) = d\lambda_{can}\big|\Gamma_\sigma$. □

(5) $\mathbb{CP}^n$ and Fubini-Study form: Consider the function ρ on $\mathbb{C}^n$: $z \mapsto \log(|z|^2 + 1)$. This function is strictly plurisubharmonic, with $\partial\bar{\partial}\rho = \frac{1}{(|z|^2+1)^2}$; therefore $\omega_{FS} := \frac{1}{2}\partial\bar{\partial}\rho$ is Kähler.
Now on a chart $U_0 = (z_1, \ldots, z_n) \subseteq \mathbb{C}P^n$, the transition function on $\mathcal{U} = U_0 \cap U_1$ looks like $\varphi_{0,1}(z_1, \ldots, z_n) = (\frac{1}{z_1}, \cdots \frac{z_n}{z_1})$, this map maps (U) biholomorphically onto itself with $\varphi^*(\log(|z|^2 + 1)) = \log(|z|^2 + 1) + \log(|z_1|^{-2})$. Thus, $\partial\bar{\partial}\varphi^*(\log(|z|^2 + 1)) = \partial\bar{\partial}\varphi^*(\log(|z|^2 + 1)) + \partial\bar{\partial}\log(|z_1|^{-2}) = \partial\bar{\partial}\varphi^*(\log(|z|^2 + 1))$. So we can "glue" $\varphi_i^*\omega_{FS}$ together to give a Kähler structure on $\mathbb{CP}^n$.

Now we introduce a very important property of symplectic manifold, which claims that locally, all symplectic manifolds look the same; however, the global structure would be different.

Theorem 9 (Darboux) *Given a symplectic manifold (M, ω), $\forall\ p \in M$, there exists a neighborhood $U_p \subseteq M$ such that ω restricted to U_p is symplectomorphic to the standard ω_0 in $\mathbb{R}^{2n}$, where dim $M = 2n$.*

The proof of Darboux's theorem uses the so call Moser's trick, details can be found in Chap. 2 of [1].

2 Lagrangian Floer Colomology

Definition 10 (*Lagrangian*) Now let (M, ω) be a symplectic manifold, $N \subset M$ is isotopic if $\omega|_N = 0$.This implies that $\dim N \leq \frac{1}{2}\dim M$ as ω is non-degenerate.If L is isotopic and $\dim L = \frac{1}{2}\dim M$, then we say L is **Lagrangian**.

Now suppose we have compact lagrangians $L_0, L_1 \subset (M, \omega)$, $L_0 \pitchfork L_1 \Rightarrow L_0 \cap L_1$ is a finite set of points.

Definition 11 (*Monotone*) We say a Lagrangian submanifold $L \subseteq M$ is **Monotone** if $\forall\ A \in \pi_2(M, L)$ we have a fixed $\lambda \in \mathbb{R}^+$ such that:

$$\int_A \omega = \lambda \cdot \mu_L(A)$$

From now on we work over monotone Lagrangians with minimal Maslov number N_L at least 2.

Definition 12 The Floer complex

$$CF^*(L_0, L_1) := \Lambda < L_0 \cap L_1 >$$

Which is a Λ- vector space with basis $= L_0 \cap L_1$ $\Lambda := \{\Sigma a_i T^{\lambda_i} | a_i \in \mathbb{K}, \lim_{i \to \infty} \lambda_i = +\infty\}$ is the **Novikov field** with coefficient in $\mathbb{K}$.

If we have $2c_1(TM) = 0$ and the maslov class μ_L vanishes, then we can make $CF^*(L_0, L_1)$ a $\mathbb{Z}-$graded complex, else it is Z_2-graded.

Definition 13 Now given $p, q \in L_0 \cap L_1$, define $\widehat{\mathcal{M}}(p, q, J) = \{u : \mathbb{R} \times [0, 1] \to M \big| D_u \circ j = J \circ D_u, u(s, 0) \in L_0, u(s, 1) \in L_1, \lim_{s \to \infty} u(s, t) = p, \lim_{s \to -\infty} u(s, t) = q\}$.

Then we have an $\mathbb{R}$ action on $\widehat{M}(p, q, J)$ by $r \cdot u(s, t) = u(s + r, t)$, the moduli space $\mathcal{M}(p, q, J) := \widehat{\mathcal{M}}(p, q, J)/\mathbb{R}$.

Remark 14 The equation $D_u \circ j = J \circ D_u$ is just saying $\overline{\partial_J} u = 0$.

Now we define differential on the complex:

Definition 15 $\forall p \in CF^*(L_0 \cap L_1)$,

$$\partial p := \sum_{q \in L_0 \cap L_1, ind(\beta)=1} (\#\mathcal{M}(p, q, \beta, J)) T^{\omega(\beta)} \cdot q, \tag{1}$$

Where $\omega(\beta)$ is the **energy** of the J-holomorphic map u which is represented by β in $\pi_2(M, L_0 \cup L_1)$, it is defined as

$$\omega(u) := \int_{\mathbb{R} \times [0,1]} u^* \omega = \int \int |\frac{\partial u}{\partial s}|^2 ds dt \geq 0$$

Remark 16 If the linearized operator $D_{\bar{\partial}_{J,u}}$ is surjective at $\forall u \in \widehat{\mathcal{M}}(p, q, J)$, then we have $\widehat{\mathcal{M}}(p, q, J)$ is a manifold of dimension $\mu_{L_0 \cup L_1}(u)$ (the Maslov index of u, note that $\pi_1((LGr)) = \pi_1(U(n)/O(n)) \simeq \mathbb{Z}$.)

Remark 17 Gromov's compactness claims that given any positive upper bound E_0 on energy, there are only finitely many homotopy class $\beta = [u]$ such that $\omega(u) \leq E_0$, therefore, we know that the RHS of Eq. 1 is well defined. Namely, for any fixed energy E, $\#\mathcal{M}(p, q, \beta, J))$ is finite.

Proposition 18 *Assume* $[\omega] \cdot \pi_2(M, L_i) = 0$ *for* $i = 0, 1$ *and* L_0, L_1 *are oriented, compact Lagrangians equipped with spin structure, then* ∂ *is well defined and satisfies* $\partial^2 = 0$*, and the* ***Lagrangian Floer cohomology*** $HF^*(L_0, L_1) := H^*(CF(L_0, L_1; \partial))$ *is independent of the almost complex structure* J *and invariant under Hamiltonian isotopes of* L_0 *or* L_1.

The idea of the proof of $\partial^2 = 0$ is to look at a J-holomorphic map u with $\mu(u) = 2$, then Gromov's compactness say that for a sequence of $J-$holomorphic maps with bounded energy, there exists subsequence that converges to nodal configurations. In the case when $\mu(u) = 2$, we have three possible configurations.

(1) Sphere bubbles, a J-holomorphic sphere is connected to the J-holomorphic strip at an interior point of the strip. This is the case when some energy concentrates at the interior point.
(2) Disc bubble: a J-holomorphic disc connected with the J-holomorphic strip at a point on L_0 or L_1, this is the case when some energy concentrates at a point on the boundary.
(3) Broken strip, there are energy concentrates at $\pm\infty$.

Proposition 19 $\omega \cdot \pi_2(M, L_i) = 0$ *implies there are no disc bubbles or sphere bubbles.*

Proof The idea is the energy of the bubbles have to be zero, which implies that they are constant. Look at the long exact sequence of homology groups

$$\cdots \to \pi_2(L) \to \pi_2(M) \to \pi_2(M, L) \xrightarrow{\partial} \pi_1(L) \to \cdots$$

Note that $\omega \cdot \pi_2(M, L_i) = 0$ automatically implies that $\forall\ \beta \in \pi_2(M, L)$, $\int_\beta \omega = 0$, thus no disc bubbles with boundary on L_0 or L_1. Since $\omega|_L = 0$ by definition of Lagrangian manifolds, we have $\forall\ \eta \in \pi_2(L)$, we have $\int_\eta \omega = 0$. By the exactness at $\pi_2(M)$, $\forall\ \alpha \in \pi_2(M)$, we have $\int_\alpha \omega = 0$. Thus no sphere bubbles. □

Gromov compactness claims that after adding the 3 possible configurations, $\overline{\mathcal{M}}(p, q, J)$ is compact. However, since $\omega \cdot \pi_2(M, L_i) = 0$, we are only allowed to have broken strips.

However, the signed count of the number of boundary points of a 1-dimensional manifold is zero. A gluing theorem states that any broken strip is locally the limit of a sequence of index 2 J-holomorphic strips. And

$$\partial\overline{\mathcal{M}}(p, q, [u], J) = \coprod_{\substack{r \in L_0 \cap L_1 \\ [u'] + [u''] = [u] \\ \text{ind}([u']) = \text{ind}([u'']) = 1}} \Big(\mathcal{M}(p, r; [u'], J) \times \mathcal{M}(r, q; [u''], J)\Big) \quad (2)$$

$HF^*(L, L)$ is defined as $HF^*(L_0, \varphi_H(L))$ where $\varphi_H(L)$ is Hamiltonian isotopic to L because HF^* is invariant under Hamiltonian perturbation. namely the original J-holomorphic equation is replaced with

$$\frac{\partial u}{\partial s} + J(u, t)\left(\frac{\partial u}{\partial t} - X_H(t, u)\right) = 0 \quad (3)$$

Example 20 Consider $L \subseteq T^*L$ as the zero section of the cotangent bundle, suppose $f : L \to \mathbb{R}$ is a Morse function, let $H = \pi^* f$, then $\varphi_H(f) = \Gamma_{df} \subseteq T^*L$. Thus $L \cap \varphi_H(L) =$ critical points of f, and $CF^*(L, \varphi(L)) \simeq CM^*(f)$ (the Morse complex) as vector space. The Moduli space of J-holomorphic strips from p to q corresponds $1-1$ to the Moduli space of Morse flow liness from p to q. So we have an iso of chain complex $(CF^*(L, \varphi(L), \partial) \simeq (CM^*(f), d_M)$

The main idea is that under "good" conditions, we have Lagrangian Floer homology is isomorphic to the Morse homology which is isomorphic to the singular homology.

Theorem 21 (Albers, 2007) *For a $2n$-dimensional, closed, symplectic manifold M and a closed, monotone, Lagrangian sub manifold $L \subset M$ of minimal Maslov number $N_L \geq 2$, there exist homomorphisms*

$$\varphi_k : HF_k(L, \phi_H(L)) \to H^{n-k}(L; \mathbb{Z}/2) \textit{ for } k \leq N_L - 2$$

Where $H : S^1 \times M \to \mathbb{R}$ is a Hamiltonian function and ϕ_H the corresponding Hamiltonian diffeomorphism. For $n - N_L + 2 \leq k \leq N_L - 2$, φ_k is an isomorphism.

See [2] for more details.

Remark 22 This morphism above is not always an isomorphism, a counterexample can be found in [3] where a construction by Audin and Polterovich provides Lagrangian embeddings of spheres S^k into $\mathbb{R}^{2n}$.

Remark 23 We might imagine that every Lagrangian can be embedded locally in T^*L in a neighborhood by Weinstein's Lagrangian neighborhood theorem below and use the idea of zero section in Example 20 to think of Lagrangian Floer homology as Morse homology; however, Weinstein's Lagrangian neighborhood theorem is a local result, so we don't always have a rigorous isomorphism globally.

Theorem 24 (Weinstein's Lagrangian neighborhood theorem) *$\forall L \subseteq M$ a Lagrangian sub manifold, there exists a neighborhood U that is symplectic to a neighborhood of $L \subseteq T^*L$.*

Details of the proof can be found in Chap. 3 of [1]

3 Product Structure and Fukaya Category

Definition 25 We define $\mu^1 : CF^*(L_0, L_1) \to CF^*(L_0, L_1)[1]$ as the differential ∂. We can also define

$$\mu^2 : CF^*(L_0, L_1) \otimes_\Lambda CF^*(L_1, L_2) \to CF^*(L_0, L_2)$$

by the following equation:

$$\mu^2(p, q) := \sum_{\substack{q \in L_0 \cap L_2 \\ [u]:\text{ind}([u])=0}} (\#\mathcal{M}(p, q, r; [u], J)) T^{\omega([u])} r. \tag{4}$$

Where $\mathcal{M}(p, p, r; [u], J))$ denotes, for a disc with three given points z_0, z_1, z_2 on the boundary, a J-holomorphic map from $\mathbb{D}$ to M that represents $[u]$ in $\pi_2(M)$ and extends continuously to the closed disc, mapping the boundary arcs from z_0 to z_1, z_1 to z_2, z_2 to z_3 to L_0, L_1, L_2 respectively, while the z_0, z_1, z_2 are mapped to p, q, r respectively.

Proposition 26 *If $\omega \cdot \pi_2(M, L_i) = 0, \forall i \in \{0, 1, 2\}$, then μ^2 satisfies the Leibniz rule with proper signs with respect to ∂; in particular,*

$$\partial(\mu^2(p, q)) = \pm\mu^2(\partial p, q) \pm \mu^2(p, \partial q) \tag{5}$$

The idea of the proof is similar to that of $\partial^2 = 0$, we look at the index 1 moduli spaces of J-holomorphic discs and their compactification. Still assuming transversality, $\mathcal{M}(p, q, r; [u], J)$ is a smooth 1-dimensional manifold and admits a compactification $\overline{\mathcal{M}}(p, q, r; [u], J)$ by adding nodal trees (there is no disc or sphere bubble by the assumption that the symplectic form vanishes on relative homotopy classes). and there can be strip breaking happening at any of the three points p, q, r. If it breaks at p, it represents $\mu^2(\partial p, q)$; at q then represents $\mu^2(p, \partial q)$; if at r, then represents $\partial\mu^2(p, q)$. Since the signed count of the boundary of a 1-dimensional manifold is 0, we have Eq. 5.

Therefore, μ^2 defines a product in Floer cohomology as well, namely

$$[\mu^2] : HF^*(L_0, L_1) \otimes HF^*(L_1, L_2) \to HF^*(L_0, L_2).$$

If $L_0 = L_1 = L_2$, then $[\mu^2]$ is the cup product on $HF^*(L)$.

Proposition 27 (Associativity of μ^2) *We have*

$$\mu^2(p, \mu^2(q, r)) \pm \mu^2(\mu^2(p, q), r) = \pm\mu^3(\partial p, q, r) \pm \mu^3(p, \partial q, r) \pm \mu^3(p, q, \partial r) \pm \partial\mu^3(p, q, r) \tag{6}$$

This is because $\mu^3(p, q, r)$ is defined similar as the sum of the number of J-holomorphic maps of a disc (with four points z_0, z_1, z_2, z_3 on its boundary to M, with the map converges to the points $p, q, r, s \in M$ near the four points and the arcs in between each adjacent pair of z_i to L_i), weighted with the symplectic energy. Then by Gromov compactness, the boundary of 1-dimensional moduli spaces are of two kinds:

(1) Those with a broken strip on the boundary of $\mathbb{D}$ at a nodal point of $\mathbb{D}$ while the other three marked points remain on $\partial\mathbb{D}$, there are four of these, corresponding to the four summands on the RHS of Eq. 6.

(2) Those that corresponds to a degeneration of the domain to the boundary of $\bar{\mathcal{M}}_{0,4}$, namely to a pair of discs, each of whose boundary carries two marked points, and the disc connects to the J-holomorphic strip with a nodal point, there are two marked points left on the disc. There are two of these, corresponding to the two summands on the LHS of Eq. 6.

Thus the singed count of the number of boundary points of a 1-dimensional manifold with boundary give Eq. 6.

More generally, consider $L_0, \dots L_k \subseteq M$, compact, oriented Lagrangians with spin structure. $p_i \in L_{i-1} \cap L_i$, we define

$\mu^k : CF(L_{k-1}, L_k) \otimes \cdots \otimes CF(L_1, L_2) \otimes CF(L_0, L_1) \longrightarrow CF(L_0, L_k)$

$$\mu^k(p_k, \dots, p_1) = \sum_{\substack{q \in L_0 \cap L_k) \\ [u]:\mathrm{ind}([u])=2-k}} (\#\mathcal{M}(p_1, \dots, p_k, q; [u], J))T^{\omega([u])} q, \tag{7}$$

where the dimension of the moduli spaces are

$$\dim \mathcal{M}(p_1, \dots, p_k, q; [u], J) = k - 2 + \mathrm{ind}([u]) = k - 2 + \deg(q) - \sum_{i=1}^{k} \deg(p_i). \tag{8}$$

The special case is when $k = 1$. We had

$$\mu^1 = \partial : CF^*(L_0, L_1) \to CF^*(L_0, L_1),$$

$$\partial p = \sum_{\substack{q \in L_0 \cap L_1) \\ [u]:\mathrm{ind}([u])=1}} (\#\mathcal{M}(p, q; [u], J))T^{\omega([u])} q$$

Proposition 28 *If $\omega \cdot \pi_2(M, L_i) = 0$, $\forall i$, then the operations μ^k satisfy the A_∞-relations*

$$\sum_{\ell=1}^{k} \sum_{j=0}^{k-\ell} (-1)^* \mu^{k+1-\ell}(p_k, \dots, p_{j+\ell+1}, \mu^\ell(p_{j+\ell}, \dots, p_{j+1}), p_j, \dots, p_1) = 0, \tag{9}$$

where $* = j + \deg(p_1) + \cdots + \deg(p_j)$.

Example 29 (1) $k = 1$, Eq. 9 is the same as $\mu^2 = 0$,
(2) $k = 2$, Eq. 9 is the Leibniz' rule
(3) $k = 3$, Eq. 9 is the associativity law of $[\mu^2]$ in HF^*.

For higher k, this gives an explicit homotopy for certain compatibility property among the preceding ones.

The proof is similar to that of the associativity law, we study dimension-1 moduli spaces of J-holomorphic discs and their compactification, fix $p_1, \ldots p_k$ and q, and $[u]$ such that $\text{ind}[u] = 3 - k$, assume J is chosen generically so we have transversality and then $\mathcal{M}(p_1, \ldots p_k, q; [u], J)$ compactifies to a 1-dimensional manifold with boundary, and the boundary points are either of an index 1 J-holomorphic strip breaking off at one of the $(k+1)$ points or a pair of discs each contain at least two marked points. Those consists of the summands of the Eq. 9.

Definition 30 (*Fukaya Category*) Given a symplectic manifold (M, ω) such that $2c_1(TM) = 0$, consider the category consisting of the following data:

(1) Objects: compact, oriented Lagrangians L_i equipped with spin structure, such that $[\omega] \cdot \pi_2(M, L_i) = 0$ with vanishing Maslov index, together with a spin structure.
(2) hom-spaces: $hom^*_{\mathcal{F}(M)}(L_0, L_1) := CF^*(L_0, L_1)$, with differential μ^1 and composition μ^2
(3) higher operations and A_∞ relations (9) for μ^s.

See [4, 5] for more details.

Remark 31 In our previous definition, we may allow $c_1(TM)$, μ_L to be nonzero if we only need a $\mathbb{Z}/2$-grading; and we may also drop the spin structure if we are content to work over characteristic 2.

Acknowledgements I am grateful to David Favero, Matthew Ballard and Eric Sharpe for organizing this super school and giving me the chance to present this material. I am also grateful to my advisor Mohammed Abouzaid for giving me so much guidance along the way.

References

1. Dusa McDuff, Dietmar Salamon, Introduction to Symplectic Topology, Oxford Science Publications, 2nd edition.
2. Peter Albers, A Lagrangian Piunikhin-Salamon-Schwarz morphism and two comparison homomorphisms in Floer homology, arXiv:math/0512037
3. Leonid Polterovich, Monotome Lagrange submanifolds of linear spaces and the Maslov class in cotangent bundles. Mathematische Zeitschrift, vol 207, 1991, No.2, 217–222.
4. Denis Auroux, A beginner's introduction to Fukaya categories, arXiv:1301.7056
5. Paul Seidel, Fakaya categories and Picard-Lefschetz Theory. European Mathematical Society, 2008.

Introduction to Homological Mirror Symmetry

Andrew Harder

1 Introduction

1.1 Background

Mirror symmetry states that to every Calabi-Yau manifold X with complex structure and symplectic symplectic structure there is another dual manifold $X^\vee$, so that the properties of X associated to the complex structure (e.g. periods, bounded derived category of coherent sheaves) reproduce properties of $X^\vee$ associated to its symplectic structure (e.g. counts of pseudo holomorphic curves and discs).

This idea originated in physics, specifically string theory, where the relevant statement is that the A twisted TQFT obtained from X is equivalent to the B twisted TQFT obtained from $X^\vee$ and vice versa. Using this, Candelas and de la Ossa and collaborators [1] in the late 80s and early 90s were able to make predictions regarding counts of rational curves on Calabi-Yau varieties. Subsequently, there was a flurry of activity towards making rigorous these predictions. Early successes include Givental's theorems [2] on the quantum cohomology of complete intersections in toric varieties and periods of their duals, and theorems of Batyrev and Borisov [3, 4] which show that, if X and $X^\vee$ are a mirror pair of dimension d which are constructed combinatorially as complete intersections in toric varieties, then

$$h^{p,q}(X) = h^{d-q,p}(X^\vee).$$

Homological mirror symmetry [5] takes these observations and puts them into a categorical context. In this context, the B-side invariants (B branes) take the form of complexes of coherent sheaves on X, and the A-side invariants (A branes) take the

A. Harder (✉)
Department of Mathematics, University of Miami, Coral Gables, FL 33146, USA
e-mail: a.harder@math.miami.edu

M. Ballard et al. (eds.), *Superschool on Derived Categories and D-branes*,
Springer Proceedings in Mathematics & Statistics 240,
https://doi.org/10.1007/978-3-319-91626-2_12

form of Lagrangian submanifolds, and equivalence between these sets of invariants is interpreted as an equivalence between the bounded derived category of coherent sheaves on X and the Fukaya category of $X^\vee$. There are many problems with this vague statement, since, at first blush, the Fukaya category and the category of coherent sheaves on a variety are very different categories. The Fukaya category is naturally an A_∞ category, without any sort of triangulated structure in general, so explaining what we mean when we say that it is equivalent to a $\mathbb{C}$-linear triangulated category takes some explaining.

1.2 Outline

We will proceed as follows. Section 2 will be devoted to developing some categorical background necessary for stating homological mirror symmetry. We will explain how one can get a k-linear triangulated, Karoubi complete category out of an A_∞ category. We will also explain how to get dg and A_∞ extensions of $\mathrm{D}^b(\mathrm{coh}(X))$, and we will state the expected equivalence for mirror pairs of Calabi-Yau varieties.

In Sect. 3 we will discuss what is known about homological mirror symmetry, and we will explain extensions of the original mirror symmetry conjecture to the case of Fano varieties and varieties of general type.

1.3 Other Sources

Besides the foundational articles cited in the introduction, there are several overviews of homological mirror symmetry in the literature. Ballard's [6] article provides a compact overview of the subject, with lengthy discussions on homological mirror symmetry for both $\mathbb{P}^1$ and the elliptic curve. The reader could do much worse than reading [6], then returning to the current manuscript for some comments on how the subject has changed in the past nine years.

A large amount of background material, along with detailed proofs of many results can also be found in [7, Chap. 8]. Finally, both Kontsevich [5] and Seidel [8] have given ICM talks regarding homological mirror symmetry. Both of these articles are very good starting points - indeed, [5] is *the* starting point for homological mirror symmetry.

2 The Categorical Setup

In this section we will develop the categorical framework necessary to state homological mirror symmetry in enough generality to discuss at least state what is known. The main conjecture of Kontsevich is that if X and $X^\vee$ are mirror manifolds (equipped with appropriate complexified Kähler structures $\omega, \omega^\vee$ and complex structures, $I, I^\vee$), then there is an embedding of triangulated $\mathbb{C}$-linear categories

$$\mathrm{H}^0\mathrm{Tw}\mathcal{F}(X, \omega) \hookrightarrow \mathrm{D}^b(\mathrm{coh}(X^\vee, I^\vee))$$

and vice versa. We will explain what all of this notation means in this section. As well, the more modern formulation involves not just triangulated $\mathbb{C}$-linear categories but an equivalence of A_∞ categories. A proper formulation of mirror symmetry will also involve Karoubi completion.

We assume that the reader is acquainted with notions of Fukaya categories. Basic references on this topic have proliferated in recent years, [9–11]. Equally, we will assume that the reader has some knowledge of algebraic geometry and specifically of derived categories of coherent sheaves on smooth projective varieties [12].

2.1 Dg Categories

On the B-side of mirror symmetry (complex structure), the most natural class of categories are called dg categories. A category is a differential graded (dg) category if for each $a, b \in \mathrm{Ob}(\mathcal{C})$ there is a vector space $\mathrm{hom}_\mathcal{C}(a, b)$ over a field k which satisfies the following conditions;

(1) It is a $\mathbb{Z}$-graded vector space, the graded piece of weight i being denoted $\mathrm{hom}^i_\mathcal{C}(a, b)$.
(2) It has a chosen differential

$$d_\mathcal{C} : \mathrm{hom}_\mathcal{C}(a, b) \longrightarrow \mathrm{hom}_\mathcal{C}(a, b)$$

which increases degree by 1.
(3) If f, g are in $\mathrm{hom}^i_\mathcal{C}(a, b)$, $\mathrm{hom}^j_\mathcal{C}(b, c)$ respectively, then

$$d_\mathcal{C}(g \cdot f) = (d_\mathcal{C} g) \cdot f + (-1)^{i+j} g \cdot (d_\mathcal{C} f)$$

(4) For each $a \in \mathrm{Ob}(\mathcal{C})$, there is some $i_a \in \mathrm{hom}^0_\mathcal{A}(a, a)$ so that $i_a \cdot f = f$ for any $f \in \mathrm{hom}^j_\mathcal{C}(b, a)$ and $g \cdot i_a = g = i_b \cdot g$ for any $g \in \mathrm{hom}^j_\mathcal{C}(a, b)$.

In Sect. 4.4, we will use differential $\mathbb{Z}/2$-graded categories, denoted d($\mathbb{Z}/2$)g categories. These are categories where the grading in (1) is by $\mathbb{Z}/2$ instead of $\mathbb{Z}$, though all other axioms are unchanged.

Example 2.1 The most basic example of a dg category which will be useful later is that of chain complexes over an abelian category $\mathbf{A}$. Let's define $\mathcal{K}(\mathbf{A})$ to be the category whose objects are chain complexes $(a^\bullet, d_a)$ of elements in $\mathbf{A}$, and whose homomorphisms are,

$$\hom^\ell((a^\bullet, d_a), (b^\bullet, d_b)) = \prod_i \hom_{\mathbf{A}}(a^i, b^{\ell+i})$$

equipped with the differential d given by the map

$$df = d_b \cdot f + (-1)^\ell f \cdot d_a$$

where $f \in \hom^\ell((a^\bullet, d_a), (b^\bullet, d_b))$. □

Example 2.2 A dg algebra A gives rise to a dg category with a single object e so that $\hom_{\mathcal{C}}(e, e) = A$. □

To any dg category $\mathcal{C}$, one has its homotopy category $\mathrm{H}^0\mathcal{C}$. We let $\mathrm{Ob}(\mathrm{H}^0\mathcal{C}) = \mathrm{Ob}(\mathcal{C})$, and if we denote by $[a]$ the object in $\mathrm{H}^0\mathcal{C}$ corresponding to $a \in \mathrm{Ob}(\mathcal{C})$, then $\hom_{\mathrm{H}^0\mathcal{C}}([a], [b]) = \mathrm{H}^0(\hom_{\mathcal{C}}(a, b), d)$. This is a k-linear category.

If we have a category $\mathbf{C}$ and a dg category $\mathcal{C}$ so that $\mathrm{H}^0\mathcal{C}$ is equivalent to $\mathbf{C}$, then we say that $\mathcal{C}$ is a *dg enhancement* of $\mathbf{C}$.

2.2 A_∞ Categories

On the A-side of homological mirror symmetry, the most important homological objects are A_∞ categories. We begin with the standard caveat that A_∞ categories are not categories in the classical sense, since composition of morphisms need not be associative, however the point is that we will allow associativity to fail in a controlled manner.

Definition 2.3 An A_∞ category $\mathcal{A}$ is a collection of objects $\mathrm{Ob}(\mathcal{A})$ along with $\mathbb{Z}$-graded vector spaces $\hom_{\mathcal{A}}(a, b)$ for any pair of $a, b \in \mathrm{Ob}(\mathcal{C})$ so that the following conditions hold.

(1) For every $n > 0$ and every set of objects $a_0, \ldots, a_n \in \mathrm{Ob}(\mathcal{A})$, there are maps

$$m_n^{\mathcal{A}}(a_0, \ldots, a_n) : \hom_{\mathcal{A}}(a_{n-1}, a_n) \otimes \cdots \otimes \hom_{\mathcal{A}}(a_0, a_1) \to \hom_{\mathcal{A}}(a_0, a_n)[2 - i]$$

(2) These maps satisfy the quadratic A_∞ associativity relations,

$$\sum_{m,n} (-1)^{\tau_n} m_{d-m+1}^{\mathcal{A}}(f_d, \ldots, f_{m+n+1}, m_m^{\mathcal{A}}(f_{n+m}, \ldots, f_{n-1}), f_n, \ldots, f_1) = 0$$

for $f_i \in \hom_{\mathcal{A}}(a_{i-1}, a_i)$ and $\tau_n = -n + \sum_i |a_i|$

□

For instance, the A_∞ relations imply that

$$m_1^{\mathcal{A}}(m_1^{\mathcal{A}}(f)) = 0,$$

$$m_2^{\mathcal{A}}(f_1, m_2^{\mathcal{A}}(f_2, f_3)) - m_2^{\mathcal{A}}(m_2^{\mathcal{A}}(f_1, f_2), f_3) = m_1^{\mathcal{A}}(m_3^{\mathcal{A}}(f_1, f_2, f_3))$$
$$+m_3^{\mathcal{A}}(m_1^{\mathcal{A}}(f_1), f_2, f_3) + m_3^{\mathcal{A}}(f_1, m_1^{\mathcal{A}}(f_2), f_3) + m_3^{\mathcal{A}}(f_1, f_2, m_1^{\mathcal{A}}(f_3))$$

So $m_1^{\mathcal{A}}$ can be thought of as a differential and if we think about $m_2^{\mathcal{A}}$ as composition of morphisms, then the second relation says that composition is associative up to some factor involving m_3. As in the case of dg categories, we may construct a homotopy category $\mathrm{H}^0\mathcal{A}$ whose objects are those of $\mathcal{A}$ and whose homomorphisms are the 0th cohomology of the morphism complexes of $\mathcal{A}$ with respect to $m_1^{\mathcal{A}}$. We will not assume that our A_∞ categories have units (though we did for dg categories). Instead, we assume that their homotopy categories have units. An A_∞ category whose homotopy category has a unit is called cohomologically unital or c-unital [10].

Remark 2.4 We may also consider $\mathbb{Z}/2$-graded A_∞ categories by insisting that the grading on $\hom_{\mathcal{A}}$ be by $\mathbb{Z}/2\mathbb{Z}$ instead of $\mathbb{Z}$.

Remark 2.5 One can modify this construction to allow a nontrivial m_0 map, by which we mean a map from the underlying field k to $\hom_{\mathcal{A}}(a, a)$ for all $a \in \mathrm{Ob}(\mathcal{A})$. Such categories are called *curved* or *obstructed* A_∞ categories, and they play a role in mirror symmetry for Fano manifolds [13]. However, curved A_∞ categories do not admit homotopy categories, which complicates their homotopy theory. □

In the case where all m_is vanish if $i > 2$, the A_∞ relations say that $\mathcal{A}$ is a dg category, possibly without units. Therefore, the category of dg categories embeds into the category of A_∞ categories.

An A_∞ category $\mathcal{C}$ for which m_1 vanishes is called a *minimal* A_∞ category. In this case, if a, b are two objects in $\mathcal{C}$, then

$$\hom_{\mathrm{H}^0\mathcal{C}}(a, b) = \hom^0_{\mathcal{C}}(a, b).$$

An A_∞ functor between two A_∞ categories $\mathcal{A}$ and $\mathcal{B}$ is a map on objects and homomorphisms in the usual way which satisfies additional conditions with respect to the A_∞ structure maps. We will not reproduce these formulas here but direct the reader to [14] for more details.

An A_∞ functor $f : \mathcal{A} \to \mathcal{B}$ is called a quasi equivalence if the functor that it induces $\mathrm{H}^0\mathcal{A} \to \mathrm{H}^0\mathcal{B}$ is an equivalence of categories.

2.3 *Triangulated Karoubi Closures*

Given a dg or A_∞ category $\mathcal{C}$, one would like to find a triangulated category which contains $\mathcal{C}$. Broadly, a triangulated category is a category which admits a shift functor [1] and in which one can take (perhaps non-canonically), mapping cones of morphisms. The goal now is to formally add these features to an A_∞ category.

One way of doing this is to take the category of *twisted complexes over* $\mathcal{C}$, which we will denote Tw$\mathcal{C}$. Twisted complexes were first defined by Bondal and Kapranov [15] for dg categories, the definition for A_∞ categories that we give here comes from a number of sources, specifically work of Keller [14], Lefèvre-Hasegawa [16] or Seidel [10, 17]. We begin by taking the category $\mathbb{Z}\mathcal{C}$ which formally incorporates shifts into $\mathcal{C}$.

Definition 2.6 Let $\mathbb{Z}\mathcal{C}$ be the category whose objects are formal pairs (a, n) with $a \in \mathrm{Ob}(\mathcal{C})$ and $n \in \mathbb{Z}$. We define

$$\hom_{\mathbb{Z}\mathcal{C}}((a, n), (b, m)) = \hom_{\mathcal{C}}(a, b)[m - n].$$

The A_∞ structure is as in $\mathcal{C}$.

The category $\mathbb{Z}\mathcal{C}$ admits formal shifts of objects sending $\bigoplus_i a_i[i]$ to $\bigoplus_i a_i[i+1]$, and $\mathcal{C}$ is equivalent to the full subcategory of objects in the form $a[0]$. We now need to add formal mapping cones in order to get a triangulated A_∞ category. This is done as follows.

Definition 2.7 Let us take the category Tw$\mathcal{C}$ so that Ob(Tw$\mathcal{C}$) is made up of pairs (b, δ) for $A = (a_1, \ldots, a_n) \in \mathrm{Ob}(\mathbb{Z}\mathcal{C})^n$ and δ a matrix of morphisms, $\delta_{i,j}$ $\hom^1_{\mathbb{Z}\mathcal{C}}$ (a_j, a_i) a *strictly upper triangular matrix*, i.e. $\delta_{i,j} = 0$ if $i \geq j$. We also require that the Maurer-Cartan equation,

$$\sum_{i=1}^{\infty} m_i^{\mathbb{Z}\mathcal{C}}(\delta, \ldots, \delta) = 0$$

is satisfied. Here, we implicitly extend the definition of $m_i^{\mathbb{Z}\mathcal{C}}$ to matrices of homomorphisms in a straightforward way. This is a finite sum by the fact that we have chosen δ to be strictly upper triangular. The space of morphisms between (A, δ) and (B, τ) is $\bigoplus_{i,j} \hom_{\Sigma\mathcal{C}}(a_i, b_j)$ equipped with a twisted set of composition maps. If we take (A_i, δ_i) for $i = 1, \ldots, n$ and $f_i \in \hom_{\mathrm{Tw}\mathcal{C}}((A_{i-1}, \delta_{i-1}), (A_i, \delta_i))$ then we define

$$m_d^{\mathrm{Tw}\mathcal{C}}(f_d, \ldots, f_1) = \sum_{j_0, \ldots, j_d \geq 0} m_i^{\mathbb{Z}\mathcal{C}}(\underbrace{\delta_d, \ldots, \delta_d}_{j_d}, a_d, \underbrace{\delta_{d-1}, \ldots, \delta_{d-1}}_{j_{d-1}}, a_{d-1}, \ldots).$$

□

If $f \in \hom^0_{\mathrm{Tw}\mathcal{C}}((a, \delta_a), (b, \delta_b))$ and $m_1^{\mathrm{Tw}\mathcal{C}}(f) = 0$, then we may define the mapping cone of f,

$$\mathrm{cone}(f) = \left(a[1] \oplus b, \begin{pmatrix} \delta_a & 0 \\ f & \delta_b \end{pmatrix}\right)$$

The category Tw$\mathcal{C}$ should be thought of as the smallest triangulated A_∞ category containing $\mathcal{C}$ as a full subcategory. This is analogous to taking the category of bounded complexes over an abelian category.

Definition 2.8 An A_∞ category is called *triangulated* if the natural embedding

$$\mathcal{C} \hookrightarrow \mathrm{Tw}\mathcal{C}$$

is a quasi equivalence of A_∞ categories. In the dg case, Bondal and Kapranov [15] call such categories *pretriangulated.* □

If an A_∞ category is triangulated, then its homotopy category is a k-linear triangulated category.

One might expect at this point that there is an equivalence between $\mathrm{Tw}\mathcal{F}(X)$ and an A_∞ category whose homotopy category is equivalent to $\mathrm{D}^b(\mathrm{coh}(X))$. However, the category $\mathrm{D}^b(\mathrm{coh}(X))$ is *Karoubi complete*, whereas $\mathrm{Tw}\mathcal{F}(X)$ does not necessarily have this property.

Definition 2.9 A category $\mathcal{T}$ is called *Karoubi complete* (or split closed), if for every $p \in \hom_{\mathcal{T}}(a, a)$ so that $p^2 = p$, there is a pair of morphisms $s : a \to b$ and $i : b \to a$ so that $s \cdot i = \mathrm{id}_b$ and $i \cdot s = p$. We will say that an A_∞ category is Karoubi complete if its homotopy category is Karoubi complete.

The object b in this definition is called the *direct image* of f. □

Given a k-linear triangulated category, $\mathcal{T}$, we say that a functor $F : \mathcal{T} \to \mathcal{S}$ is a Karoubi completion of $\mathcal{T}$ if F is full and faithful and for every $c \in \mathcal{S}$ there is some $p \in \hom_{\mathcal{T}}(a, a)$ so that c is isomorphic to a direct image of $F(p)$. A functor $F : \mathcal{A} \to \mathcal{B}$ of triangulated A_∞ categories is a a Karoubi completion of $\mathcal{A}$ if the functor induced on homotopy categories is a Karoubi completion.

Proposition 2.10 (Seidel [10]) *Every triangulated A_∞ category $\mathcal{C}$ admits a Karoubi completion, and any pair of Karoubi completions of $\mathcal{C}$ are quasi equivalent.*

In fact, Seidel produces such a completion explicitly, which we will not describe in detail, but we will call $\Pi\mathcal{C}$.

2.4 *Enhancements of* $\mathbf{D}^b(\mathbf{coh}(X))$

The goal of this section is to show that there are dg categories whose homotopy categories are equivalent to the bounded derived category of coherent sheaves on a variety X.

Definition 2.11 A dg (resp. A_∞) enhancement of a triangulated category $\mathcal{T}$ is a dg (resp. A_∞) category $\mathcal{C}$ whose homotopy category is equivalent to $\mathcal{T}$. □

In mirror symmetry, categories of B-branes are usually described in terms of categories of coherent sheaves. Let X be a smooth projective variety over a field k. Then associated to X is the category $\mathrm{qcoh}(X)$ of quasi coherent sheaves. One can then take the category of complexes of quasicoherent sheaves, $\mathcal{K}(\mathrm{qcoh}(X))$. This is naturally a dg category, where homomorphism complexes are given as in Example 2.1. A complex $I^\bullet$ is called *h-injective* if for every complex $J^\bullet$ isomorphic to 0 in $\mathrm{D}(\mathrm{qcoh}(X))$,

$$\hom_{\mathcal{K}(\mathrm{qcoh}(X))}(J^\bullet, I^\bullet) \cong 0.$$

Here $\cong$ denotes quasi isomorphism of complexes. The full subcategory $\mathcal{I}(X)$ of h-injective complexes of quasi coherent sheaves has homotopy category which is equivalent to the derived category of quasi coherent sheaves, $\mathrm{D}(\mathrm{qcoh}(X))$. The bounded derived category of coherent sheaves on X can be written as a full subcategory of $\mathrm{D}(\mathrm{qcoh}(X))$ made up of bounded complexes whose cohomology sheaves are coherent. Therefore, there is a full subcategory of $\mathrm{D}(\mathrm{qcoh}(X))$ which is equivalent to $\mathrm{D}^b(\mathrm{coh}(X))$. Since $\mathcal{I}(X)$ has, up to equivalence in the homotopy category, the same objects as $\mathrm{D}(\mathrm{qcoh}(X))$, we can define $\mathrm{D}^b_{\mathrm{dg}}(\mathrm{coh}(X))$ to be the full subcategory of $\mathcal{I}(X)$ made up of objects which are equivalent to objects in $\mathrm{D}^b(\mathrm{coh}(X)) \subseteq \mathrm{D}(\mathrm{qcoh}(X))$.

The category $\mathrm{D}^b_{\mathrm{dg}}(\mathrm{coh}(X))$ has homotopy category which is equivalent to $\mathrm{D}^b(\mathrm{coh}(X))$, hence it is a dg enhancement of $\mathrm{D}^b(\mathrm{coh}(X))$. There are many dg enhancements of categories of coherent sheaves; several are described in [18, 19], but a beautiful result of Lunts and Orlov shows that any pair of dg enhancements of $\mathrm{D}^b(\mathrm{coh}(X))$ are quasi equivalent.

2.5 Homological Mirror Symmetry for Calabi–Yau Manifolds

We are now equipped to state what homological mirror symmetry means for a pair of Calabi–Yau manifolds.

Definition 2.12 Let (X, I, ω) and $(X^\vee, I^\vee, \omega^\vee)$ be a pair of Calabi–Yau varieties. If there is a quasi equivalence of A_∞ categories,

$$\Pi\mathrm{Tw}\mathcal{F}(X, \omega) \cong \mathrm{D}^b_{\mathrm{dg}}(\mathrm{coh}(X^\vee, I^\vee))$$

and vice versa, then we say that (X, I, ω) and $(X^\vee, I^\vee, \omega^\vee)$ are homologically mirror to one another. □

It is quite difficult to show that mirror symmetry holds precisely as in Definition 2.12. Versions of homological mirror symmetry are known for certain Calabi–Yau varieties of higher dimension, though in these cases, one has that the Fukaya category of (X, ω) is defined over the Novikov field, not $\mathbb{C}$, so no such equivalence can hold. Therefore, the right hand side is replaced by a Calabi–Yau variety over $\Lambda_{\mathbb{Q}}$, which should be thought of as a small deformation of a highly degenerate Calabi–Yau variety. This will be explained in more detail in Sect. 3.2.

3 Mirror Symmetry for Calabi–Yau Manifolds

This section will describe some known results regarding mirror symmetry for Calabi–Yau varieties. We begin by explaining the most accessible case, that of the elliptic curve, then we will explain what is known in higher dimensions.

3.1 The Elliptic Curve

The form of homological mirror symmetry for elliptic curves discussed here was sketched by Kontsevich [5], where it was noticed that the Floer product between certain Lagrangian submanifolds on an elliptic curve can be used to recover classical theta functions. Therefore, there is close resemblance between Floer products and compositions of line bundle homomorphisms. This was made precise a little later by Polishchuk and Zaslow [20]. There are other proofs of homological mirror symmetry for elliptic curves that have appeared in recent years, due to Lekili and Perutz [21], as well as Abouzaid and Smith [22] which are similar to one another in spirit, and are similar to the proofs of homological mirror symmetry in higher dimensions discussed in Sect. 3.2.

In this section, we will review some of the most convincing evidence for homological mirror symmetry in the case of an elliptic curve. The reader can refer to [20, Sect. 4] for details. Let's represent E as $\mathbb{R}^2/\mathbb{Z}^2$ be the square torus. We choose a symplectic form

$$\mathrm{d}s^2 = a\mathrm{d}x \wedge \mathrm{d}y$$

so that E has area a. Then every Lagrangian can be written uniquely up to hamiltonian isotopy as the image in E of a line of rational slope. We will also need to choose a B-field, $b \in \mathrm{H}^2(E, \mathbb{R})$ again represented by a form $b\mathrm{d}x \wedge \mathrm{d}y$. The data $(b + \mathrm{i}a)\mathrm{d}x \wedge \mathrm{d}y$ is called a complexified Kähler form on E. From here on, we will denote by E the torus $\mathbb{R}^2/\mathbb{Z}^2$ equipped with the complexified Kähler form $(b + \mathrm{i}a)\mathrm{d}x \wedge \mathrm{d}y$, and by E', the complex torus $\mathbb{C}/(\mathbb{Z} + \mathbb{Z}\tau)$ for $\tau = b + \mathrm{i}a$.

The elements of the Fukaya category of E are quadruples of Lagrangians with gradings, spin structure and flat unitary bundles.

We wish to show that there is a relationship between the subcategory of $\mathcal{F}(E)$ made up of Lagrangians on E descending from lines of slope $d \in \mathbb{Z}$ equipped with unitary rank *one* local system and the category of line bundles on E' of degree d. We will outline some coarse heuristics first.

If we pick a slope d, then there is an S^1 of Lagrangians of slope d depending on the x-intercept of the corresponding line in $\mathbb{R}^2$. A unitary local system on such a Lagrangian is determined by its holonomy, which in this case, is just a number in S^1. Therefore, for a given slope d, there is an $S^1 \times S^1$s worth of Lagrangians branes with this slope. Topologically, this corresponds to the fact that the moduli space of degree d line bundles on E' is isomorphic to E', in other words it is topologically a torus.

The relationship is obviously deeper than this; let's say we have a pair of line bundles $\mathcal{L}_1$ and $\mathcal{L}_2$ of degrees d_1 and d_2. Then we might ask what the space of homomorphisms looks like between $\mathcal{L}_1$ and $\mathcal{L}_2$. We know that if $d_1 < d_2$,

$$\dim_{\mathbb{C}} \hom^0_{E'}(\mathcal{L}_1, \mathcal{L}_2) = \dim_{\mathbb{C}} \mathrm{H}^0(E', \mathcal{L}_2 \otimes \mathcal{L}_1^{-1}) = d_2 - d_1$$

by the Riemann-Roch theorem. Now let us take two Lagrangians branes $L_1^\#$ and $L_2^\#$ with rank one unitary local systems and so that the Lagrangian submanifolds L_1 and L_2 underlying these branes come from lines in $\mathbb{R}$ of slopes d_1 and d_2 respectively. Then $\hom_{\mathcal{F}(E)}(L_1^\#, L_2^\#)$ is simply a vector space whose basis is given by the intersection points of L_1 and L_2. A priori, the degree of these homomorphisms depends on gradings on $L_1^\#$ and $L_2^\#$, but regardless, $\hom_{\mathcal{F}(E)}(L_1^\#, L_2^\#)$ is concentrated in a single degree n. One can arrange matters so that if $d_1 < d_2$ then $n = 0$. It is easy to see that $L_1 \cap L_2$ is a union of $d_2 - d_1$ points, so

$$\dim_{\mathbb{C}} \hom^0_{\mathcal{F}(E)}(L_1^\#, L_2^\#) = d_2 - d_1.$$

So $\dim_{\mathbb{C}} \hom_{E'}(\mathcal{L}_1, \mathcal{L}_2) = \dim_{\mathbb{C}} \hom_{\mathcal{F}(E)}(L_1^\#, L_2^\#)$. What's more interesting is the result of composing homomorphisms. Recall that for a line bundle on an elliptic curve E', where E' is written as $\mathbb{C}/(\mathbb{Z} + \mathbb{Z}\tau)$ for $\tau \in \mathbb{H}$, then sections of any line bundle $\mathcal{L}$ can be represented as entire functions on $\mathbb{C}$ which satisfy certain functional equations. Let us give a simple example of this which appears in [23]. Define theta functions with rational characteristics a, b as

$$\theta_{a,b}(z; \tau) = \sum_{n \in \mathbb{Z}} \exp(\pi \mathrm{i} (n + a)^2 \tau + 2\pi \mathrm{i} (n + a)(z + b)),$$

The function $s = \theta_{0,0}(z; \tau)$ is a basis of sections for $\mathcal{L} = \mathcal{O}_{E'}((0))$ where (0) is the point in E' mapped to by 0 under the covering map $\mathbb{C} \to \mathbb{C}/(\mathbb{Z} + \mathbb{Z}\tau)$. One may write sections of $\mathcal{L}^2$ as $t_1 = \theta_{0,0}(2\tau, 2z)$ and $t_2 = \theta_{\frac{1}{2},0}(2\tau, 2z)$. A classical result is that the identity

$$\theta_{0,0}(\tau, z)^2 = \theta_{0,0}(2\tau, 0)\theta_{0,0}(2\tau, 2z) + \theta_{\frac{1}{2},0}(2\tau, 0)\theta_{\frac{1}{2},0}(2\tau, 2z).$$

This relation as saying that the map

$$\hom_E(\mathcal{L}, \mathcal{L}^2) \otimes \hom_{E'}(\mathcal{O}_{E'}, \mathcal{L}) \to \hom_{E'}(\mathcal{O}_E, \mathcal{L}^2)$$

can be written in terms of the basis s of $\mathrm{H}^0(E', \mathcal{L})$ and the basis t_1, t_2 of $\mathrm{H}^0(E', \mathcal{L}^2)$ as

$$s^2 = \theta_{0,0}(2\tau, 0)t_1 + \theta_{\frac{1}{2},0}(2\tau, 0)t_2,$$

(here we regard s, t_1, t_2 as homomorphisms between bundles obtained by multiplying by the corresponding section). More generally, if we represent sections of line bundles $\mathcal{L}_1, \mathcal{L}_2$ and $\mathcal{L}_1 \otimes \mathcal{L}_2$ as theta functions on E, then the maps

$$\hom_{E'}(\mathcal{L}_1, \mathcal{L}_1 \otimes \mathcal{L}_2) \otimes \hom_{E'}(\mathcal{O}_{E'}, \mathcal{L}_1) \to \hom_{E'}(\mathcal{O}_{E'}, \mathcal{L}_1 \otimes \mathcal{L}_2)$$

have relations whose coefficients are obtained by evaluating theta functions for some values of τ and z.

On the other hand, homomorphisims in $\mathcal{F}(E)$ come with chosen bases, coming from points of intersection between Lagrangians. In the Fukaya category, one has that if $p \in L_1 \cap L_2, q \in L_2 \cap L_3$ then

$$m_2(p, q) = \sum_{r \in L_1 \cap L_3} \mathsf{C}(p, q, r) r$$

where

$$\mathsf{C}(p, q, r) = \sum_{\phi \in \mathcal{M}(p,q,r)} \pm \exp\left(2\pi \mathrm{i} \int_D \phi^* \omega\right) \mathrm{hol}(\phi(\partial D)).$$

Here $\mathcal{M}(p, q, r)$ is the moduli space of immersed pseudoholomorphic discs $\phi : D \to E$ with boundary along L_1, L_2 and L_3 and whose vertices are p, q and r. The term $\mathrm{hol}(\phi(\partial D))$ is a term measuring the holonomy of the induced local system around the boundary of D. In this case, this reduces to counting immersed triangles in E with edges in L_1, L_2 and L_3 and vertices p, q and r with weighs corresponding to the area of the given triangle and holonomy around its boundary. These counts can be done in such a way that the value of $\mathsf{C}(p, q, r)$ is identified with a translate of a theta function evaluated at $\tau = b + \mathrm{i}a, z = 0$. For instance, if we let L_1 be the Lagrangian of slope 0 passing through the origin, L_2 a Lagrangian of slope 1 passing through the origin and L_3 a Lagrangian of slope 2 passing through the origin, then $L_1 \cap L_2 = e_1 = (0, 0) = L_2 \cap L_3$ and $L_1 \cap L_3 = \{e_1, e_2 = (1/2, 0)\}$. Equip all of these Lagrangians with trivial flat unitary bundles of rank 1. Then we have that

$$m_2(e_1, e_1) = \mathsf{C}(e_1, e_1, e_1)e_2 + \mathsf{C}(e_1, e_1, e_2)e_2.$$

Polishchuk and Zaslow compute that

$$\mathsf{C}(e_1, e_1, e_1) = \theta_{0,0}(2(b + \mathrm{i}a), 0), \qquad \mathsf{C}(e_1, e_1, e_2) = \theta_{\frac{1}{2},0}(2(b + \mathrm{i}a), 0).$$

Therefore, the map sending $L_i \mapsto \mathcal{L}_i$ has the same composition maps on both sides.

More generally we match the canonical bases of $\hom_{E'}(\mathcal{L}_1, \mathcal{L}_2)$ coming from theta functions with the canonical bases of $\hom_{\mathcal{F}(E)}(L_1^\#, L_2^\#)$, then m_2 on $\mathcal{F}(E)$ can be matched with composition of homomorphisms of line bundles. Since one can show that m_1 vanishes, this is enough to show that a partial equivalence holds between $\mathrm{D}^b(\mathrm{coh}(E'))$ and $\mathrm{H}^0\mathcal{F}(E)$. The main theorem of [20] extends this to vector bundles and shows that

Theorem 3.1 *There is an equivalence of graded categories,*

$$\mathrm{D}^b(\mathrm{coh}(E')) \cong \mathrm{H}^0\mathcal{F}(E).$$

Polishchuk [24] has sharpened this result to prove that there is a natural minimal A_∞ enhancement of $\mathrm{D}^b(\mathrm{coh}(E'))$ so that the higher multiplications agree with those in $\mathcal{F}(E)$ [24]. See [6, Sect. 4] for a more detailed discussion.

3.2 Calabi–Yau Varieties in Higher Dimension

Seidel outlined an approach to homological mirror symmetry in his 2002 ICM lecture [8] based on deformation theory of Fukaya categories. The basic idea has its roots in classical mirror symmetry. In its roughest form, mirror symmetry is an isomorphism between the complexified Kähler moduli space of X and the complex moduli space of $X^\vee$. The catch is that this isomorphism might only hold between a neighbourhood of a very bad point in the boundary of the complex moduli space of $X^\vee$ corresponding to a very singular Calabi–Yau variety, and a neighbourhood of the large radius limit in of the moduli space of complexified Kähler forms on X. So we might begin by looking a degenerate Calabi–Yau varieties, and try to relate them to the Fukaya categories of Calabi–Yau varieties with large complexified Kähler forms. If we replace a complexified Kähler form with an ample divisor, and we make that divisor very large, this corresponds to giving more and more importance to the complement of this divisor in X.

The idea is then that the Fukaya category of the complement of an ample hypersurface in a compact Calabi–Yau manifold is expected to be equivalent to the category of perfect complexes on a degenerate mirror Calabi–Yau variety.

This is beautifully exhibited in papers of Lekili and Perutz [21] and Lekili and Polishchuk [25], where it is shown that the exact Fukaya category of a 2-torus $\mathbb{T}$ with n points removed satisfies

$$\mathrm{D}^\pi\mathcal{F}(\mathbb{T}\setminus\{p_1,\ldots,p_n\}) \cong \mathrm{Perf}(G_n)$$

where G_n is a cycle of n rational curves and $\mathrm{Perf}(G_n)$ denotes its category of perfect complexes.

The next step in Seidel's approach is to define the relative Fukaya category of X with respect to D, denoted $\mathcal{F}(X, D)$. This is a category with the same objects as $\mathcal{F}(X\setminus D)$, but whose homomorphisms are deformed over $\Lambda_\mathbb{Z}$ by counting discs intersecting D k times with weight q^k. Recall that elements of $\Lambda_\mathbb{Z}$ are certain infinite series in a variable q with fractional exponents. When we specialize to $q = 0$ we recover the Fukaya category $\mathcal{F}(X\setminus D)$, and when we specialize to the algebraic closure of the fraction field of $\Lambda_\mathbb{Z}$, denoted $\Lambda_\mathbb{Q}$, then we should recover $\mathcal{F}(X)$.

Seidel uses this idea to prove a version of mirror symmetry for quartic K3 surfaces. Consider

$$G = \left\{(a_1, a_2, a_3, a_4) \in (\mathbb{Z}/4)^4 : \sum_i a_i = 0\right\}$$

and let $\mathcal{X}_q$ be a minimal resolution of the K3 surface over $\Lambda_{\mathbb{Q}}$ defined by

$$(qx^4 + y^4 + z^4 + w^4) + xyzw \subseteq \mathbb{P}^3/G,$$

where G is used to act component-wise on $\mathbb{P}^3$. Then there is some A_∞ enhancement $\mathrm{D}^b_\infty(\mathrm{coh}(\mathcal{X}_q))$ of $\mathrm{D}^b(\mathrm{coh}(\mathcal{X}_q))$ so that

Theorem 3.2 *There is an equivalence of A_∞ categories over $\Lambda_{\mathbb{Q}}$,*

$$\psi^*\mathrm{D}^b_\infty(\mathrm{coh}(\mathcal{X}_q)) \cong \mathrm{D}^\pi\mathcal{F}(X),$$

for some continuous automorphism ψ of $\Lambda_{\mathbb{Q}}$.

To prove this, Seidel uses the fact that $X\setminus D$ admits a *Lefschetz fibration*, then the work of [10] allows him to compute $\mathcal{F}(X\setminus D)$. He then uses results on the deformation theory of categories to conclude that his equivalence holds.

Nick Sheridan [13] has proved similar results (by slightly different methods) for degree $(n+1)$ hypersurfaces in $\mathbb{P}^n$.

4 Mirror Symmetry for Fano Manifolds

It is expected that a version of homological mirror symmetry holds for Fano manifolds as well. Recall the definition of a Fano manifold:

Definition 4.1 A smooth projective variety is called Fano if its anticanonical line bundle $\omega_X^{-1} = \bigwedge^{\dim X} \Omega_X$ is ample.

Example 4.2 If X is a smooth hypersurface in $\mathbb{P}^n$ cut out by a homogeneous equation of degree $\leq n$, then X is a Fano variety, since $\omega_X^{-1} = \mathcal{O}_{\mathbb{P}^n}(n - \deg X)$ by adjunction. □

Example 4.3 Choose $0 \leq k \leq 8$ points in general position on $\mathbb{P}^2$. Then denote by S_k the blow up of $\mathbb{P}^2$ in these points. The variety S_k is a Fano variety called a del Pezzo surface. Any Fano surface is either equal to S_k for some k, or $\mathbb{P}^1 \times \mathbb{P}^1$. □

If X is a Fano variety and we assume that we have chosen D a normal crossings anticanonical divisor with σ_D a section of ω_X vanishing along D, then σ_D is a nonvanishing section of $\omega_{X\setminus D}$, hence it provides an isomorphism between $\omega_{X\setminus D}$ and $\mathcal{O}_{X\setminus D}$. Therefore $X\setminus D$ is a Calabi–Yau variety, so $U = X\setminus D$ has a mirror Calabi–Yau manifold V. In his seminal work, [26], Auroux demonstrates that putting D back into U corresponds to choosing a regular function on V. Therefore, the mirror to a Fano manifold is a pair of (V, w) where V is a noncompact Calabi–Yau variety and w is a regular function on V. This pair is called a *Landau-Ginzburg model*.

Remark 4.4 This formulation of mirror symmetry for Fano manifolds goes back further than Auroux, of course. It has been understood for a long time in the physics

Table 1 Categories involved in mirror symmetry for Fano varieties and Landau-Ginzburg models

	B side	A side
Fano	$D^b(\mathrm{coh}(X))$	$\coprod_\lambda \mathcal{F}(X)_\lambda$
LG	$\coprod_\lambda \mathsf{MF}(V, \mathsf{w} - \lambda)$	$\mathcal{FS}(V, \mathsf{w})$

literature [27] and even in the mathematical literature [2] that the mirror of a Fano manifold is a Landau-Ginzburg model. We refer to Auroux's work to emphasize the fact that he gives a mathematical explanation for why mirror symmetry takes this form. □

The next two sections are devoted to explaining how homological mirror symmetry is formulated for Fano manifolds. The challenge that we are confronted with is the fact that the function w is an integral part of the Landau-Ginzburg model, therefore we must construct categories which integrate w in some way. A summary is given in Table 1.

In Table 1, the row "Fano" denotes the A side and B side categories associated to a Fano variety, and the row "LG" denotes the categories associated to a Landau–Ginzburg model. Mirror symmetry is a relation between the Fano A side (resp. B side) category and the Landau–Ginzburg B side (resp. A side) category.

4.1 The Directed Fukaya Category

There is an issue when trying to define the Fukaya category of a noncompact manifold, which is that if on allows noncompact Lagrangian branes, then it can be hard to control their behaviour. The solution is to force our Lagrangian branes to have specified behaviour in a neighbourhood of the boundary of M. Not much is known about Fukaya categories defined this way, as far as I'm aware, but Seidel has constructed a category called the directed Fukaya category or the category of vanishing cycles which should capture the same information. The problem with Seidel's category is that it assumes that the function w has only extremely mild singularities. This assumption is fine in low dimensions, or for the mirrors of simple varieties, but in many interesting examples w is very badly behaved.

Let us take (E, ω, J) to be the data of an symplectic manifold E of dimension $2d$ where $\omega \in \Omega^2_E$ is a symplectic form on E, and J is an almost complex structure which is compatible with ω. We will let π be a symplectic morphism from E to an open subset $S \subseteq \mathbb{R}^2$ which is compatible with this almost complex structure. We will assume that this is *Lefschetz fibration*. In effect, this means that π has a finite number of critical points, and near the critical points, π looks like

$$\sum_{i=1}^{d} x_i^2$$

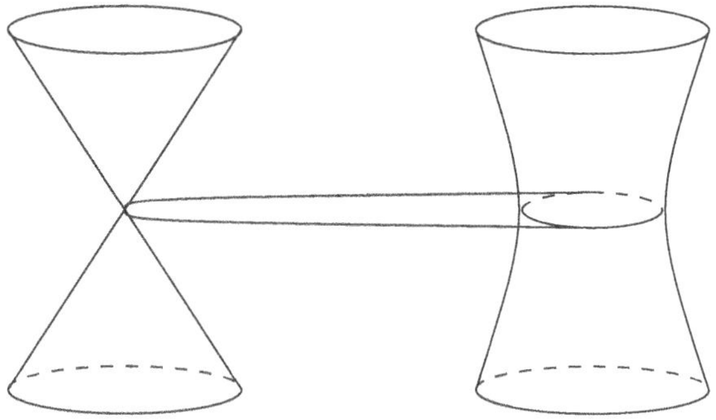

Fig. 1 Vanishing cycle of a node

where x_i are coordinates with respect to the almost complex structure J. We also must assume that π behaves nicely "at infinity" in the fibers. For instance, this means that if one removes fibers of π containing critical points, which we denote $\mathrm{crit}(\pi)$, one obtains a symplectic fiber bundle over $S\backslash\Sigma$. The symplectic structure on E allows us to define a parallel transport between fibers of π along paths in S. If we choose a differentiable map $\gamma : [a, b] \to S$, we get a symplectomorphism $\rho_\gamma : E_{\gamma(a)} \to E_{\gamma(b)}$.

Let us denote by Σ_π the set of critical values of π. Take $p \in \Sigma$ and consider a point s near p and a smooth map $\gamma : [0, 1] \to S$ so that $\gamma(0) = s$ and $\gamma(1) = p$.

If we let x be the critical point of π in E_p then we may define

$$B = \left\{ y \in E_{\gamma(s)} : 0 \leq s \leq 1 \text{ with} \lim_{t\to 1} \rho_{\gamma|[s,t]}(y) = x \right\}$$

which is a ball in E whose boundary in E_s is a Lagrangian sphere. This set is depicted in Fig. 1.

We will choose a base point s in S and a path γ_p from s to every $p \in \Sigma_\pi$ which do not intersect each other except at s. The paths γ_p can be cyclically ordered by choosing one to be γ_1, the one to it's left to be γ_2 and so on. This gives us an ordered set of Lagrangian spheres in E_s, which we denote $L_1, \ldots, L_k$, where $k = |\Sigma_\pi|$. When $d = 1$, we choose the nontrivial spin structure on L_i, otherwise, there is only one possible spin structure on L_i. We can choose gradings arbitrarily, and the resulting structure allows us to promote L_i to Lagrangian branes $L_i^\#$, elements of $\mathcal{F}(E_s)$ where $E_s := \pi^{-1}(s)$.

Let us denote $\Gamma = \{\gamma_p : p \in \Sigma_\pi\}$. Using the notation of Auroux, Katzarkov and Orlov, we define the category $\mathrm{Lag}_{\mathrm{vc}}(\pi, \Gamma)$ to be the category with k objects ℓ_i corresponding to the $L_i^\#$ above, and so that

$$\hom_{\mathrm{Lag}_{\mathrm{vc}}(\pi,\Gamma)}(\ell_i,\ell_j)=\begin{cases}\hom_{\mathcal{F}(E_s)}(L_i^{\#},L_j^{\#}) & \text{if } i<j,\\ k\cdot \mathrm{id}_{\ell_i} & \text{if } i=j,\\ 0 & \text{if } i>j.\end{cases}$$

Composition of homomorphisms is taken inside of $\mathcal{F}(E_s)$ if $i<j$, and composition with id_{ℓ_i} is the identity. This category is called several things in the literature, but we will call it the directed Fukaya category of Lagrangian vanishing cycles. It is clear that the category $\mathrm{Lag}_{\mathrm{vc}}(\pi,\Gamma)$ depends heavily on Γ, but it is a theorem of Seidel [28] that the category of twisted complexes over $\mathrm{Lag}_{\mathrm{vc}}(\pi,\Gamma)$ is invariant under a certain braid group action on the set of all paths, called *mutations*. We will denote $\mathrm{D}^b\mathrm{Lag}_{\mathrm{vc}}(\pi,\Gamma)$ the category $\mathrm{H}^0\mathrm{TwLag}_{\mathrm{vc}}(\pi,\Gamma)$.

The category $\mathrm{D}^b\mathrm{Lag}_{\mathrm{vc}}(\pi,\Gamma)$ has a full exceptional collection of objects given by $\ell_1,\dots,\ell_k$, therefore it is closely related to categories of representations of certain algebras.

4.2 *Mirror Symmetry*

This allows us to state the homological mirror symetry for some Fano varieties.

Conjecture 4.5 *If X is a Fano variety and its mirror Landau–Ginzburg model $((E,\omega,J),\pi)$ is a symplectic Lefschetz fibration then there is an equivalence of categories*

$$\mathrm{D}^b(\mathrm{coh}(X))\cong \mathrm{D}^b\mathrm{Lag}_{\mathrm{vc}}(\pi,\Gamma).$$

This has been proven in a number of cases. The most basic is that of del Pezzo surfaces [29]. In this case, the Landau–Ginzburg mirror of X is given by the following construction.

We take first of all the Landau–Ginzburg mirror of $\mathbb{P}^2$. This is written as the pair

$$(\mathbb{C}^\times)^2, \qquad \pi(x,y)=x+y+\frac{1}{xy}.$$

One can compactify $(\mathbb{C}^\times)^2$ to a surface X_9 which is fibered over $\mathbb{P}^1$ by a function f and so that $\mathsf{f}|_{(\mathbb{C}^\times)^2}$ is equal to π. The fibers over $3\zeta_3^i$ are nodal elliptic curves, the fiber over ∞ is a chain of 9 rational curves and every other fiber is a smooth genus 1 curve. We can deform X_9 smoothly so that for each $0\leq n\leq 8$ there are $3+n$ nodal elliptic curve fibers over points in $\mathbb{A}^1=\mathbb{P}^1\backslash\infty$ and the fiber over ∞ is a chain of $9-n$ curves. Call the resulting surface X_{9-n} for each n, and let f_{9-n} be the morphism from X_{9-n} to $\mathbb{P}^1$. Let

$$Y_{9-n}=X_{9-n}\backslash \mathsf{f}_{9-n}^{-1}(\infty), \qquad \mathsf{w}_{9-n}=\mathsf{f}_{9-n}|_{Y_{9-n}}.$$

Then, for an appropriate choice of symplectic form on Y_{9-n}, Auroux, Katzarkov and Orlov prove that homological mirror symmetry holds.

Theorem 4.6 (Auroux, Katzarkov and Orlov, [23]) *For an appropriate choice of complexified Kähler form $\beta + \mathtt{i}\omega$ on Y_{9-n} and a choice of basis of paths Γ, there is a del Pezzo surface S_n so that*

$$\mathrm{D}^b(\mathrm{coh}(S_n)) \cong \mathrm{D}^b\mathrm{Lag}_{\mathrm{vc}}(\mathsf{w}_{9-n}, \Gamma).$$

In fact, [23] proves much more than this. If one lets $\beta + \mathtt{i}\omega$ be a general symplectic form on Y_{9-n} which does not come from a symplectic form on X_{9-n}, then there is no S_n so that $\mathrm{D}^b(\mathrm{coh}(S_n)) \cong \mathrm{D}^b\mathrm{Lag}_{\mathrm{vc}}(\mathsf{w}_{9-n}, \Gamma)$. Instead, one can interpret $\mathrm{D}^b\mathrm{Lag}_{\mathrm{vc}}(\mathsf{w}_{9-n}, \Gamma)$ as the derived category of coherent sheaves on a *noncomutative* del Pezzo surface. This is said precisely in [23, Sect. 2].

The computations of [23] are similar to those of Polishchuk and Zaslow's, since nontrivial part of $\mathrm{D}^b\mathrm{Lag}_{\mathrm{vc}}(\mathsf{w}_{9-n}, \Gamma)$ occurs in the Fukaya category of a smooth fiber of w_{9-n}, which is just a two dimensional torus. In the case of the Landau–Ginzburg mirror of $\mathbb{P}^2$, if we choose $s = 0$ to be straight line paths to $3\zeta_3^i$, then it not difficult to compute that the vanishing cycles are Hamiltonian isotopic to Lagrangians of slopes 0, 3, 6. Therefore, under Polishchuk and Zaslow's correspondence, there are line bundles of degrees 0, 3, 6 corresponding to these Lagrangians. These three objects should correspond to the restrictions of $\mathcal{O}_{\mathbb{P}^2}, \mathcal{O}_{\mathbb{P}^2}(1), \mathcal{O}_{\mathbb{P}^2}(2)$ to a smooth cubic curve in $\mathbb{P}^2$, which, as we know, should be mirror to the fiber of w_9.

4.3 Other Results

Auroux, Katzarkov and Orlov have proved very similar results for weighted projective planes and Hirzebruch surfaces in [29]. Ueda proved homological mirror symmetry for toric del Pezzo surfaces in [30]. In [31], Futaki and Ueda prove homological mirror symmetry holds for all projective spaces and many weighted projective spaces. In [32], Abouzaid uses a different approach to prove homological mirror symmetry for many toric Fano varieties. A radically different form of homological mirror symmetry for Fano varieties appears in the work of Zaslow and collaborators [33–35], which replaces Fukaya categories with the dg category of perverse sheaves on $(\mathbb{C}^\times)^n$ with microlocal support in a given singular Lagrangian.

4.4 Categories of Matrix Factorizations

We will now discuss the B model category associated to a Landau–Ginzburg model. This category is known as the category of matrix factorizations. Interestingly, the Fukaya category of a Fano variety only has a $\mathbb{Z}/2$-grading, so we will find a d($\mathbb{Z}/2$)g category corresponding to it.

Consider the pair (V, w) where V is an affine variety and $\mathsf{w} \in \mathbb{C}[V]$ is a regular function[1].Define the category $\mathsf{MF}(V, \mathsf{w})$ so that $\mathrm{Ob}(\mathsf{MF}(V, \mathsf{w}))$ is made up of triples

$$P = (P_0, P_1, \delta_1, \delta_2)$$

with $P_0, P_1 \in \mathrm{mod}(\mathbb{C}[V])$, and $\delta_0 \in \hom_{\mathbb{C}[V]}(P_0, P_1)$ and $\delta_1 \in \hom_{\mathbb{C}[V]}(P_1, P_2)$ so that $\delta_0 \cdot \delta_1 = \mathsf{w} \cdot \mathrm{id}_{P_1}$ and $\delta_1 \cdot \delta_0 = \mathsf{w} \cdot \mathrm{id}_{P_0}$. We can also envision objects as pairs,

$$P = \left(P_0 \oplus P_1, \delta = \begin{pmatrix} 0 & \delta_1 \\ \delta_0 & 0 \end{pmatrix} \right).$$

We define homomorphisms as,

$$\begin{aligned} \hom_{\mathsf{MF}(V,\mathsf{w})}(P, Q) = & \hom_{\mathbb{C}[V]}(P_0, Q_0) \oplus \hom_{\mathbb{C}[V]}(P_1, Q_1) \\ & \oplus \hom_{\mathbb{C}[V]}(P_0, Q_1) \oplus \hom_{\mathbb{C}[V]}(P_1, Q_0). \end{aligned}$$

This is given the structure of a $\mathbb{Z}/2$-graded complex where the first two terms are in degree 0 and the second two are in degree 1. We will write homomorphisms then as matrices. There is a differential on this complex given by

$$d\phi = d_Q \phi + (-1)^{\deg \phi} \phi d_P.$$

The category $\mathsf{MF}(V, \mathsf{w})$ is a triangulated d($\mathbb{Z}/2$)g category, and hence its homotopy category is a triangulated category.

Orlov [36] has shown that this category has geometric meaning. There is an equivalence,

$$\mathrm{H}^0\mathsf{MF}(V, \mathsf{w}) \cong \mathrm{D}^b_{\mathrm{sg}}(\mathsf{w}^{-1}(0)).$$

The category $\mathrm{D}^b_{\mathrm{sg}}(\mathsf{w}^{-1}(0))$ is by definition the Verdier quotient of the category $\mathrm{D}^b(\mathrm{coh}(\mathsf{w}^{-1}(0)))$ by its full subcategory of perfect complexes denoted $\mathrm{Perf}(\mathsf{w}^{-1}(0))$. If $\mathsf{w}^{-1}(0)$ is a smooth variety, then $\mathrm{D}^b_{\mathrm{sg}}(\mathsf{w}^{-1}(0))$, so $\mathrm{D}_{\mathrm{sg}}(\mathsf{w}^{-1}(0))$, and by proxy $\mathsf{MF}(V, \mathsf{w})$, measure how singular $\mathsf{w}^{-1}(0)$ is.

4.5 Mirror Symmetry

One expects that if X is a Fano manifold and (V, w) is its Landau–Ginzburg mirror there is an equivalence between $\mathrm{H}^0\mathsf{MF}(V, \mathsf{w})$ and th Fukaya category of a Fano variety. We will now outline how this correspondence is expected to go.

The Fukaya category of a Fano manifold whose symplectic structure is given by the class of ω_X^{-1} is in general obstructed. This means that there is a non trivial m_0

[1]One can forget the condition that X be affine, though this comes at the cost of clarity.

class taking part in the A_∞ relations. For each object $L^\#$, there is a constant λ_L so that if $f \in \hom_{\mathcal{F}(X)}(L_1^\#, L_2^\#)$ then

$$m_1(m_1(f)) = (\lambda_{L_1} - \lambda_{L_2})f.$$

As was mentioned earlier, obstructed A_∞ categories are hard to work with, so our goal will be to break $\mathcal{F}(X)$ into pieces. For each $\lambda \in \mathbb{C}$, the full subcategory of $\mathcal{F}(X)$ made up of A-branes $L^\#$ so that $\lambda_L = \lambda$ can be given the structure of an unobstructed A_∞ category. Therefore, for all $\lambda \in \mathbb{C}$, we get a category $\mathcal{F}(X)_\lambda$. There is only a finite number of values of λ for which $\mathcal{F}(X)_\lambda$ is non trivial.

On the other hand, we have a category of matrix factorizations for each $\lambda \in \mathbb{C}$, obtained as $\mathsf{MF}(X, \mathsf{w} - \lambda)$. Homological mirror symmetry then predicts that

Conjecture 4.7 *If X and (V, w) form a homologically mirror dual pair, then for each $\lambda \in \mathbb{C}$, there is an equivalence of categories,*

$$\mathrm{D}^\pi \mathcal{F}(X)_\lambda \cong \mathrm{H}^0\mathsf{MF}(X, \mathsf{w} - \lambda).$$

This has been proved by Sheridan for degree a Fano hypersurfaces in $\mathbb{P}^{n-1}$. Particularly, Sheridan takes the potential function

$$\mathsf{Z}_a^n := u_1 \dots u_n + \sum_{j=1}^n u_j^a$$

as a polynomial map from $\mathbb{C}^n$ to $\mathbb{C}$. He then lets $\mathsf{W}_a^n = \mathsf{Z}_a^n + w_a^n$ for some constant w_a^n. Then the group

$$(\Gamma_a^n)^* = \{(\zeta_1, \dots, \zeta_n) \in (\mathbb{Z}/a)^n : \zeta_1 + \dots + \zeta_n = 0\}$$

acts naturally on $\mathbb{C}^n$, and so that W_a^n descends to a regular function $\widetilde{\mathsf{W}}_a^n$ on V_a^n. We let $V_a^n - \mathbb{C}^n/(\Gamma_n^a)^*$, then we define the pair

$$(V_a^n, \widetilde{\mathsf{W}}_a^n)$$

to be the Landau–Ginzburg mirror of a smooth degree a hypersurface in $\mathbb{P}^{n+1}$, denoted X_a^n. Then Sheridan shows that for the pair X_a^n and (V_a^n, W_a^n), Conjecture 4.7 holds. Properly, one should first take a smooth crepant resolution of V_a^n, since it is highly singular at the origin, however the resulting category will be equivalent. This is discussed in [28] and [37].

Remark 4.8 We should note that there are expected to be other homological mirrors to a given Fano variety. The construction of mirrors of Fano manifolds of Auroux discussed at the beginning of this section produces a mirror of dimension equal to that of the original Fano manifold. In Sheridan's construction, the mirror has dimension $\dim X_a^n + 2$. In the case of hypersurfaces in projective space, there is another construction of the mirror manifold [2] which should, in theory, produce the same thing as Auroux's construction. There is a construction called Knörrer periodicity [36] which likely relates the categories of matrix factorizations of the LG mirrors constructed by Givental to those of Sheridan.

5 Mirror Symmetry for Varieties of General Type

Recall the following definition.

Definition 5.1 A smooth projective variety is said to be of *general type* if its canonical bundle is ample.

Stereotypical examples of varieties of general type are curves of genus ≥ 2 and smooth hypersurfaces in $\mathbb{P}^n$ of degree $\geq n+2$.

Katzarkov [38] has conjectured that a form of homological mirror symmetry holds for varieties of general type. In this case the mirror is again conjectured to be a Landau–Ginzburg model. Furthermore, if X is a variety of general type and (V, w) is its mirror, then we expect that V is not of the same dimension as X, a phenomenon that was somewhat pathological in Sheridan's proof of homological mirror symmetry.

Conjecture 5.2 *If X and (V, w) are homologically mirror dual, then*

$$\mathsf{MF}(V, \mathsf{w}) \cong \mathcal{F}(X)$$

Note that this differs from homological mirror symmetry for Fano varieties in that we are only considering matrix factorizations with respect to w, not $\mathsf{w} - \lambda$ for all $\lambda \in \mathbb{C}$.

The best understood case of homological mirror symmetry for manifolds of general type, by which I mean the only case in which Conjecture 5.2 has been proven, is that of curves of genus ≥ 2. We will review what is known here.

Consider a compact Riemann surface M_g of genus g, equipped with a natural symplectic form (of which there is, up to equivalence and scaling, only one). We construct its Landau–Ginzburg mirror, starting with the data

$$V = \mathbb{C}^3, \qquad \mathsf{w}_g = -z_1z_2z_3 + z_1^{2g+1} + z_2^{2g+1} + z_3^{2g+1}.$$

Then one takes the quotient of $\mathbb{C}^3$ by a subgroup K_g of $\mathrm{SL}_3(\mathbb{C})$ isomorphic to $\mathbb{Z}/(2g+1)$ generated by the matrix

$$\begin{pmatrix} \zeta & 0 & 0 \\ 0 & \zeta & 0 \\ 0 & 0 & \zeta^{2g-1} \end{pmatrix}.$$

Note that w_g is invariant under the action of this group, thus it descends to a function w'_g on $\mathbb{C}^3/K_g$. The quotient space $\mathbb{C}/K_g$ has a singularity at the image of $(0, 0, 0)$, but this singularity can be crepantly resolved by toric methods. The resulting variety is a smooth variety H_g admits the following morphism

$$H_g \to \mathbb{C}^3/K_g \xrightarrow{\mathsf{w}'_g} \mathbb{C}$$

which we will denote $\widetilde{\mathsf{w}}_g$. The pair $(H_g, \widetilde{\mathsf{w}}_g)$ is the Landau–Ginzburg mirror of M_g. The following theorem is proved by Seidel in the case where $g = 2$ and by Efimov [37] in the case where $g > 2$. One does not know that that category $\mathrm{D}^b_{\mathrm{sg}}(\widetilde{\mathsf{w}}^{-1}(0))$ is Karoubi complete, so it is necessary to add some objects for it to be equivalent to $\mathrm{D}^\pi(\mathcal{F}(M_g))$. Denote by $\overline{\mathrm{D}^b_{\mathrm{sg}}}(\widetilde{\mathsf{w}}^{-1}(0))$ the Karoubi completion of $\mathrm{D}^b_{\mathrm{sg}}(\widetilde{\mathsf{w}}^{-1}(0))$.

Theorem 5.3 (Seidel [28], Efimov [37]) *There is an equivalence of $\mathbb{Z}/2$ graded triangulated categories,*

$$\overline{\mathrm{D}^b_{\mathrm{sg}}}(\widetilde{\mathsf{w}}_g^{-1}(0)) \cong \mathrm{D}^\pi\mathcal{F}(M_g).$$

The proof of this result proceeds as follows. Seidel and Efimov find $2g + 1$ generators of $\mathrm{D}^\pi\mathcal{F}(M_g)$ and compute part of the A_∞ structure on the homomorphisms between them. Then they compute a minimal A_∞ category which is equivalent to a dg extension of $\mathrm{D}^b_{\mathrm{sg}}(\widetilde{\mathsf{w}}_g^{-1}(0))$. Then they compare A_∞ structures to show that these categories are equivalent.

The fact that the Landau–Ginzburg mirror of M_g is obtained as the resolution of a quotient of $\mathbb{C}^3$ plays a key role in the proofs appearing in [28] and [37], however this does not seem to be a general feature. A more general construction of mirrors of varieties of general type which are complete intersections in smooth toric varieties appears in work of Katzarkov, Gross and Ruddat [39]. They do not prove results about homological mirror symmetry, however. Their results are exclusively in terms of the cohomology of X and its mirror.

One would expect that an version of homological mirror symmetry holds between the Fukaya-Seidel category of $(H_g, \widetilde{\mathsf{w}}_g)$ and M_g. Precisely, we should let H_g^0 be the preimage of a small disc in $\mathbb{C}$ containing 0, and let $\widetilde{\mathsf{w}}_g^0$ be the restriction of $\widetilde{\mathsf{w}}_0$ to H_g^0. There is a Fukaya-style category $\mathcal{F}(H_g^0, \widetilde{\mathsf{w}}_g^0)$ associated to the pair $(H_g^0, \widetilde{\mathsf{w}}_g^0)$. We conjecture that;

Conjecture 5.4 *For some choice of symplectic form on H_g and for some choice of complex structure on M_g, there is an equivalence of categories,*

$$\mathrm{D}^b(\mathrm{coh}(M_g)) \cong \mathrm{D}^\pi(\mathcal{F}(H_g^0, \widetilde{\mathsf{w}}_g^0)).$$

The issue with this is similar to the issue we confronted when discussing homological mirror symmetry for more complicated Fano varieties, that is, the category $\mathcal{F}(H_g^0, \widetilde{\mathsf{w}}_g^0)$ is not very well understood. To my knowledge, this version of mirror symmetry is completely open.

References

1. P. Candelas, X. de la Ossa, P. Green, and L. Parkes. An exactly soluble superconformal theory from a mirror pair of Calabi–Yau manifolds. Physics Letters B, 258(1–2):118–126, 1991.
2. A. Givental. A mirror theorem for toric complete intersections. Progress in Mathematics, 160:141–176, 1998.
3. V. Batyrev. Dual polyhedra and mirror symmetry for Calabi–Yau hypersurfaces in toric varieties. In J. Alg. Geom. Citeseer, 1994.
4. V. Batyrev and L. Borisov. Mirror duality and string-theoretic Hodge numbers. Inventiones mathematicae, 126(1):183–203, 1996.
5. M. Kontsevich. Homological algebra of mirror symmetry. In *in Proceedings of the International Congress of Mathematicians*. Citeseer, 1994.
6. M. Ballard. Meet homological mirror symmetry. 2008.
7. P. Aspinwall et al. *Dirichlet branes and mirror symmetry*, volume 4. American Mathematical Soc., 2009.
8. P. Seidel. Fukaya categories and deformations. In *Proceedings of the International Congress of Mathematicians, Vol. II (Beijing, 2002), 351–360, Higher Ed.* Citeseer, 2002.
9. D. Auroux. A beginners introduction to Fukaya categories. In *Contact and symplectic topology*, pages 85–136. Springer, 2014.
10. P. Seidel. *Fukaya categories and Picard-Lefschetz theory*. European Mathematical Society, 2008.
11. I. Smith. A symplectic prolegomenon. Bulletin of the American Mathematical Society, 52(3):415–464, 2015.
12. A. Căldăraru. Derived categories of sheaves: a skimming. Snowbird Lectures in Algebraic Geometry. Contemp. Math, 388:43–75, 2005.
13. N. Sheridan. On the fukaya category of a Fano hypersurface in projective space. Publications mathématiques de l'IHÉS, 124(1):165–317, 2016.
14. B. Keller. A-infinity algebras, modules and functor categories. Contemporary Mathematics, 406:67–94, 2006.
15. A. Bondal and M. Kapranov. Enhanced triangulated categories. Matematicheskii Sbornik, 181(5):669–683, 1990.
16. K. Lefevre-Hasegawa. *Sur les A_∞-catégories*. PhD thesis, Ph. D. thesis, Université Paris 7, UFR de Mathématiques, 2003, math. CT/0310337, 2002.
17. P. Seidel. *Homological mirror symmetry for the quartic surface*, volume 236. American Mathematical Society, 2015.
18. V. Lunts and D. Orlov. Uniqueness of enhancement for triangulated categories. Journal of the American Mathematical Society, 23(3):853–908, 2010.
19. D. Orlov. Smooth and proper noncommutative schemes and gluing of DG categories. Advances in Mathematics, 302:59–105, 2016.
20. A. Polishchuk and E. Zaslow. Categorical mirror symmetry in the elliptic curve. AMS IP Studies in Advanced Mathematics, 23:275–296, 2001.
21. Y. Lekili and T. Perutz. Arithmetic mirror symmetry for the 2-torus. arXiv:1211.4632, 2012.
22. M. Abouzaid and I. Smith. Homological mirror symmetry for the 4-torus. Duke Mathematical Journal, 152(3):373–440, 2010.
23. D. Auroux, L. Katzarkov, and D. Orlov. Mirror symmetry for del Pezzo surfaces: vanishing cycles and coherent sheaves. Inventiones mathematicae, 166(3):537–582, 2006.
24. A. Polishchuk. a_∞-structures on an elliptic curve. Communications in mathematical physics, 247(3):527–551, 2004.
25. Y. Lekili and A. Polishchuk. Arithmetic mirror symmetry for genus 1 curves with n marked points. Selecta Mathematica, 23(3):1851–1907, 2017.
26. D. Auroux. Mirror symmetry and T-duality in the complement of an anticanonical divisor. Journal of Gökova Geometry Topology, 1:51–91, 2007.
27. T. Eguchi, K. Hori, and C.-S. Xiong. Gravitational quantum cohomology. International Journal of Modern Physics A, 12(09):1743–1782, 1997.

28. P. Seidel. Homological mirror symmetry for the genus two curve. Journal of Algebraic Geometry, 20:727–769, 2011.
29. D. Auroux, L. Katzarkov, and D. Orlov. Mirror symmetry for weighted projective planes and their noncommutative deformations. Annals of Mathematics, 167(3):867–943, 2008.
30. K. Ueda. Homological mirror symmetry for toric del Pezzo surfaces. Communications in mathematical physics, 264(1):71–85, 2006.
31. M. Futaki and K. Ueda. Tropical coamoeba and torus-equivariant homological mirror symmetry for the projective space. Communications in Mathematical Physics, 332(1):53–87, 2014.
32. M. Abouzaid. Morse homology, tropical geometry, and homological mirror symmetry for toric varieties. Selecta Mathematica, New Series, 15(2):189–270, 2009.
33. B. Fang, C.-C. M. Liu, D. Treumann, and E Zaslow. A categorification of morellis theorem. Inventiones mathematicae, 186(1):79–114, 2011.
34. B. Fang, C.-C. M. Liu, D. Treumann, and E. Zaslow. The coherent-constructible correspondence for toric Deligne-Mumford stacks. International Mathematics Research Notices, 2014(4):914–954, 2012.
35. D. Nadler and E. Zaslow. Constructible sheaves and the Fukaya category. Journal of the American Mathematical Society, 22(1):233–286, 2009.
36. D. Orlov. Triangulated categories of singularities and D-branes in Landau–Ginzburg models. Trudy Matematicheskogo Instituta imeni VA Steklova, 246:240–262, 2004.
37. A. Efimov. Homological mirror symmetry for curves of higher genus. Advances in Mathematics, 230(2):493–530, 2012.
38. A. Kapustin, L. Katzarkov, D. Orlov, and M. Yotov. Homological mirror symmetry for manifolds of general type. Open Mathematics, 7(4):571–605, 2009.
39. M. Gross, L. Katzarkov, and H. Ruddat. Towards mirror symmetry for varieties of general type. Advances in Mathematics, 308:208–275, 2017.

The SYZ Conjecture via Homological Mirror Symmetry

Dori Bejleri

1 Introduction

These are expanded notes based on a talk given at the Superschool on Derived Categories and D-branes held at the University of Alberta in July of 2016. The goal of these notes is to give a motivated introduction to the Strominger-Yau-Zaslow (SYZ) conjecture from the point of view of homological mirror symmetry.

The SYZ conjecture was proposed in [35] and attempts to give a geometric explanation for the phenomena of mirror symmetry. To date, it is still the best template for constructing mirrors $\check{X}$ to a given Calabi–Yau n-fold X. We aim to give the reader an idea of why one should believe some form of this conjecture and a feeling for the ideas involved without getting into the formidable details. We assume some background on classical mirror symmetry and homological mirror symmetry as covered for example in the relevant articles in this volume.

Should the readers appetite be sufficiently whet, she is encouraged to seek out one of the many more detailed surveys such as [2, 3, 10–12, 18–20] etc.

2 From Homological Mirror Symmetry to Torus Fibrations

Suppose X and $\check{X}$ are mirror dual Kähler Calabi–Yau n-folds. Kontsevich's homological mirror symmetry conjecture [29] posits that there is an equivalence of categories

$$\mathcal{F}uk(X) \cong D^b(\mathrm{Coh}(\check{X}))$$

D. Bejleri
Mathematics Department, Massachusetts Institute of Technology, Cambridge, MA, USA
e-mail: dbejleri@mit.edu

M. Ballard et al. (eds.), *Superschool on Derived Categories and D-branes*,
Springer Proceedings in Mathematics & Statistics 240,
https://doi.org/10.1007/978-3-319-91626-2_13

between[1] the Fukaya category of X and the derived category of $\check{X}$. This should make precise the physical expectation that "the A-model on X is equivalent to the B-model on $\check{X}$." The basic idea of the correspondence is summarized by the following table:

	A-model on X	B-model on $\check{X}$
Objects	Lagrangians with flat $U(m)$-connection (L, ∇)	(complexes) of coherent sheaves $\mathcal{F}$
Morphisms	Floer cohomology groups $HF^*(L, M)$	Ext groups $Ext^*(\mathcal{F}, \mathcal{G})$
Endomorphism algebra	$HF^*(L, L) = H^*(L)$	$Ext^*(\mathcal{F}, \mathcal{F})$

Now we can now try to understand how this correspondence should work in simple cases. The simplest coherent sheaves on $\check{X}$ are structure sheaves of points $\mathcal{O}_p$ and indeed $\check{X}$ is the moduli space for such sheaves:

$$\{\mathcal{O}_p \ : \ p \in \check{X}\} \cong \check{X}.$$

Therefore there must be a family of Lagrangians with flat connections (L_p, ∇_p) parametrized by $p \in \check{X}$ and satisfying

$$H^*(L_p) \cong Ext^*(\mathcal{O}_p, \mathcal{O}_p).$$

Let us compute the right hand side explicitly.

This question is local so we can reduce to an affine neighborhood U of p. Since U is smooth at p, then p is the zero set of a section $\mathcal{O}_U \to V \cong \mathcal{O}_U^{\oplus n}$. Dualizing, we obtain an exact sequence

$$V^* \xrightarrow{s} \mathcal{O}_U \longrightarrow \mathcal{O}_p \longrightarrow 0$$

that we can extend by the *Koszul resolution*

$$0 \longrightarrow \bigwedge^n V^* \xrightarrow{s_n} \bigwedge^{n-1} V^* \xrightarrow{s_{n-1}} \dots \xrightarrow{s_2} V^* \xrightarrow{s} \mathcal{O}_U \longrightarrow \mathcal{O}_p \longrightarrow 0\,.$$

where

$$s_k(v_1 \wedge \dots \wedge v_k) = \sum_{i=1}^{k} s(v_i) v_1 \wedge \dots \wedge \hat{v}_i \wedge \dots \wedge v_k.$$

Truncating and applying $\mathrm{Hom}(-, \mathcal{O}_p)$ gives us

$$0 \longleftarrow \bigwedge^n V_p \longleftarrow \bigwedge^{n-1} V_p \longleftarrow \dots \longleftarrow V_p \longleftarrow k_p \longleftarrow 0$$

[1]One should work with the dg/A_∞ enhancements of these categories but we ignore that here.

where k_p is the skyscraper sheaf at p, V_p is the fiber of V, and all the morphisms are 0 since $s(w)$ vanishes at p for any w. It follows that

$$Ext^*(\mathcal{O}_p, \mathcal{O}_p) = \bigoplus_{k=0}^{n} \bigwedge^k V_p$$

where V_p is an n-dimensional vector space (in fact isomorphic by the section s to T_pU).

Therefore we are looking for Lagrangians L_p in X with

$$H^*(L_p) \cong \bigoplus_{k=0}^{n} \bigwedge^k V_p$$

where V_p is an n-dimensional vector space. If we stare at this for a while, we realize this is exactly the cohomology of an n-torus; $H^*(L_p) \cong H^*(T^n)$. This suggests that points $p \in \check{X}$ might correspond to Lagrangian tori in X with flat connections.

We are led to consider the geometry of Lagrangian tori in the symplectic manifold (X, ω). The first thing to note is that under the isomorphism $TX \cong T^*X$ induced by the symplectic form, the normal bundle of a Lagrangian L is identified with its cotangent bundle:

$$N_L X \cong T^*L.$$

In fact, more is true. There is always a tubular neighborhood of $N_\epsilon(L)$ in X isomorphic to a neighborhood of L in $N_L X$, and under this identification we get that $N_\epsilon(L)$ is *symplectomorphic* to a neighborhood of the zero section in T^*L with the usual symplectic form by the Weinstein neighborhood theorem [38, Corollary 6.2].

On the other hand, if $L \cong T^n$ is an n-torus then $T^*L \cong \mathbb{R}^n \times T^n$ is the trivial bundle. Therefore we can consider the projection

$$\mu : \mathbb{R}^n \times T^n \to \mathbb{R}^n.$$

This is a *Lagrangian torus fibration* over T^*L over an affine space. The restriction of μ to the tubular neighborhood $N_\epsilon(L)$ under the aforementioned identification equips X with the structure of a Lagrangian torus fibration, at least locally around a Lagrangian torus.

The SYZ conjecture predicts that this is true globally: given a Calabi–Yau manifold X for which we expect mirror symmetry to hold, then X should be equipped with a global Lagrangian torus fibration $\mu : X \to B$ which locally around smooth fibers looks like the fibration $T^*T^n \to \mathbb{R}^n$ over a flat base. By the previous discussion, $\check{X}$ should be the moduli space of pairs (L, ∇) where L is a Lagrangian torus fiber of μ and ∇ is a flat unitary connection on the L. However μ can, and often will, have singular Lagrangian fibers (see Remark 2.1.ii) and understanding how these singular fibers affect $\check{X}$ is the greatest source of difficulty in tackling the SYZ conjecture.

Let us momentarily restrict to the open locus $B_0 \subset B$ over which μ has smooth torus fibers and denote the restriction $\mu_0 : X_0 \to B_0$. Then there is an open subset $\check{X}_0 \subset \check{X}$ for which the description as a moduli space of pairs (L, ∇) of a smooth Lagrangian torus fiber of μ_0 equipped with a flat unitary connection makes sense. We can ask what structure does $\check{X}_0$ gain from the existence of $\mu : X \to B$?

Viewing B_0 as the space of smooth fibers of μ, there is a natural map $\check{\mu}_0 : \check{X}_0 \to B_0$ given by $(L, \nabla) \mapsto L$. Now a flat unitary connection ∇ is equivalent to a homomorphism

$$\mathrm{Hom}(\pi_1(L), U(m)).$$

Since $\check{X}_0$ must be $2n$ real dimensional and $\check{\mu}_0$ is a fibration over an n real dimensional base, the fibers must be n real dimensional and so $m = 1$. That is, the fibers of $\check{\mu}_0$ are given by

$$\mathrm{Hom}(\pi_1(L), U(1)) \cong (L)^*$$

the dual torus of L. Ignoring singular Lagrangians, $\check{X}_0 \subset \check{X}$ is equipped with a dual Lagrangian torus fibration $\check{\mu}_0 : \check{X}_0 \to B_0 \subset B$!

Conjecture 1 (Strominger-Yau-Zaslow [35]) *Mirror Calabi–Yau manifolds are equipped with special Lagrangian fibrations*

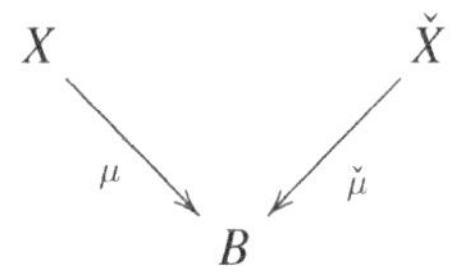

such that μ and $\check{\mu}$ are dual torus fibrations over a dense open locus $B_0 \subset B$ of the base.

Remark 2.1 (i) We will discuss the notion of a *special Lagrangian* and the reason for this condition in 2.1.

(ii) Note that unless $\chi(X) = 0$, then the fibration μ must have singularities. Indeed the only compact CY manifolds with smooth Lagrangian torus fibrations are tori.

(iii) From the point of view of symplectic geometry, Lagrangian torus fibrations are natural to consider. Indeed a theorem of Arnol'd and Liouville states that the smooth fibers of *any* Lagrangian fibration of a symplectic manifold are tori [8, Sect. 49].

This conjecture suggests a recipe for constructing mirror duals to a given Calabi–Yau X. Indeed we pick a $\mu : X \to B$ and look at the restriction $\mu_0 : X_0 \to B_0$ to the smooth locus. Then μ_0 is a Lagrangian torus fibration which we may dualize to obtain $\check{\mu}_0 : \check{X}_0 \to B_0$. Then we compactify X_0 by adding back the boundary $X \setminus X_0 =: D$ and hope that this suggests a way to compactify the dual fibration to obtain a mirror $\check{X}$.

It turns out the story is not so simple and understanding how to compactify $\check{X}_0$ and endow it with a complex structure leads to many difficulties arising from *instanton corrections* and convergence issues for Floer differentials. Furthermore this strategy to construct the dual depends not only on X but also on the chosen fibration μ and indeed we can obtain different mirrors by picking different fibrations, or even from the same fibration by picking a different "compactification" recipe. This leads to mirrors that are *Landau-Ginzburg models* and allows us to extend the statement of mirror symmetry outside of the Calabi–Yau case ([9, 28], etc). Finally, there are major issues in constructing Lagrangian torus fibrations in general. Indeed it is not known if they exist for a general Calabi–Yau, and in fact they are only expected to exist in the *large complex structure limit* (LCSL) [24, 30]. This leads to studying SYZ mirror symmetry in the context LCSL degenerations of CY manifolds as in the *Gross-Siebert program* [20, 21]. We discuss these ideas in more detail in Sect. 5.

2.1 Some Remarks on Special Lagrangians

As stated, the SYZ conjecture is about *special* Lagrangian (sLag) torus fibrations rather than arbitrary torus fibrations. Recall that a Calabi–Yau manifold has a non-vanishing holomorphic volume form $\Omega \in H^0(X, \Omega_X^n)$.

Definition 2.2 A Lagrangian $L \subset X$ is special if there exists a choice of Ω such that

$$\mathrm{Im}(\Omega)|_L = 0.$$

There are several reasons to consider special Lagrangians:

- SLags minimize the volume within their homology class. In physics this corresponds to the fact these are the *BPS branes* (see Sect. 2.2). Mathematically, this corresponds to the existence of a conjectural Bridgeland-Douglas stability condition on the Fukaya category whose stable objects are the special Lagrangians (see for example [27]).
- SLags give canonical representatives within a Hamiltonian isotopy class of Lagrangians. Indeed a theorem of Thomas and Yau [37, Theorem 4.3] states that under some assumptions, there is a unique sLag within each Hamiltonian deformation class.
- The deformation theory of sLag tori is well understood and endows the base B of a sLag fibration with the structures needed to realize mirror symmetry, at least away from the singularities. We will discuss this in more detail in Sect. 4.1.

However, it is much easier to construct torus fibrations than it is to construct sLag torus fibrations and in fact its an open problem whether the latter exist for a general Calabi–Yau. Therefore for many partial results and in many examples, one must get by with ignoring the special condition and considering only Lagrangian torus fibrations.

2.2 A Remark on D-branes and T-Duality

Strominger-Yau-Zaslow's original motivation in [35] differed slightly form the story above. Their argument used the physics of D-branes, that is, boundary conditions for open strings in the A- or B-model.[2]

They gave roughly the following argument for Calabi–Yau threefolds . The moduli space of $D0$[3] B-branes on $\check{X}$ must the moduli space of some BPS A-brane on X. The BPS condition and supersymmetry necessitate that this is a $D3$ brane consisting of a special Lagrangian L equipped with a flat $U(1)$ connection. Topological considerations force $b_1(L) = 3$ and so the space of flat $U(1)$ connections

$$\mathrm{Hom}(\pi_1(L), U(1)) \cong T^3$$

is a 3 torus. Thus $\check{X}$ must fibered by $D3$ A-branes homeomorphic to tori and by running the same argument with the roles of X and $\check{X}$ reversed, we must get a fibration by tori on X as well.

The connection with homological mirror symmetry, which was discovered later, comes from the interpretation of the Fukaya category and the derived category as the categories of topological D-branes for the A- and B-model respectively. The morphisms in the categories correspond to massless open string states between two D-branes.

Now one can consider what happens if we take a $D6$ B-brane given by a line bundle $\mathcal{L}$ on $\check{X}$. By using an argument similar to the one above, or computing

$$Ext^*(\mathcal{L}, \mathcal{O}_p) \cong k[0],$$

we[4] see that there is a one dimensional space of string states between $\mathcal{L}$ and $\mathcal{O}_p$. Therefore the Lagrangian S in X dual to $\mathcal{L}$ must satisfy

$$HF^*(S, L) = k[0].$$

Remembering that the Floer homology groups count intersection points of Lagrangians, this suggests that S must be a section of the fibration μ.

In summary, the SYZ Conjecture states that mirror symmetry interchanges $D0$ B-branes on $\check{X}$ with $D3$ Lagrangian torus A-branes on X and $D6$ B-branes on $\check{X}$ with $D3$ Lagrangian sections on X. On a smooth torus fiber of the fibration, this is interchanging $D0$ and $D3$ branes on dual 3-tori. This duality on each torus is precisely what physicists call T-duality and one of the major insights of [35] is that in the presence of dual sLag fibrations, mirror symmetry is equivalent to fiberwise T-duality.

---

[2] For background on D-branes see for example [2] or the other entries in this volume.

[3] $D0$, $D3$, ...denote 0-dimensional, 3-dimensional, ...D-branes.

[4] That is, k in degree zero and 0 in other degrees.

3 Hodge Symmetries from SYZ

The first computational evidence that led to mirror symmetry was the interchange of Hodge numbers

$$\begin{aligned} h^{1,1}(X) &= h^{1,2}(\check{X}) \\ h^{1,2}(X) &= h^{1,1}(\check{X}) \end{aligned} \tag{1}$$

for compact simply connected mirror Calabi–Yau threefolds X and $\check{X}$. Thus the first check of the SYZ conjecture is if it implies the interchange of Hodge numbers. We will show this under a simplifying assumption on the SYZ fibrations.

Let $f : X \to B$ be a proper fibration and let $i : B_0 \subset B$ be the locus over which f is smooth so that $f_0 : X_0 \to B_0$ is the restriction. Then the higher direct image of the constant sheaf $R^p f_* \mathbb{R}$ is a constructible sheaf with

$$i^* R^p f_* \mathbb{R} \cong R^p (f_0)_* \mathbb{R}$$

for each $p \geq 0$. Furthermore, $R^p(f_0)_* \mathbb{R}$ is the local system on B_0 with fibers the cohomology groups $H^p(X_b, \mathbb{R})$ for $b \in B_0$ since f_0 is a submersion.

Definition 3.1 We say that f is simple if we can recover the constructible sheaf $R^p f_* \mathbb{R}$ by the formula

$$i_* R^p (f_0)_* \mathbb{R} \cong R^p f_* \mathbb{R}$$

for all $p \geq 0$.

Proposition 3.2 *Suppose X and $\check{X}$ are compact simply connected Calabi–Yau threefolds with dual sLag fibrations*

$$\begin{array}{ccc} X & & \check{X} \\ & {\scriptstyle \mu}\searrow \quad \swarrow{\scriptstyle \check{\mu}} & \\ & B & \end{array}$$

such that μ and $\check{\mu}$ are simple. Assume further that μ and $\check{\mu}$ admit sections. Then the Hodge numbers of X and $\check{X}$ are interchanged as in (1).

Before the proof, we will review some facts about tori. If T is an n-torus, there is a canonical identification

$$T \cong H_1(T, \mathbb{R})/\Lambda_T$$

where Λ_T denotes the lattice $H_1(T, \mathbb{Z})/tors \subset H_1(T, \mathbb{R})$. Then the isomorphism $H^1(T, \mathbb{R}) \cong H_1(T, \mathbb{R})^*$ induces an identification

$$T^* \cong H^1(T, \mathbb{R})/\Lambda_T^*$$

where $\Lambda_T^* = H^1(T, \mathbb{Z})/tors \subset H^1(T, \mathbb{R})$. It follows that $H_1(T^*, \mathbb{R}) = H^1(T, \mathbb{R})$ and $\Lambda_{T^*} = \Lambda_T^*$. More generally, denoting $V = H_1(T, \mathbb{R})$, there are isomorphisms

$$H^p(T, \mathbb{R}) \cong \bigwedge^p V^*,$$
$$H^p(T^*, \mathbb{R}) \cong \bigwedge^p V.$$

After fixing an identification $\bigwedge^n V \cong \mathbb{R}$, Poincaré duality gives rise to isomorphisms

$$H^p(T, \mathbb{R}) \cong H^{n-p}(T^*, \mathbb{R})$$

compatible with the identification $\Lambda_T^* = \Lambda_{T^*}$.

Proof of Proposition 3.2 Applying the above discussion fiber by fiber to the smooth torus bundle $\mu_0 : X_0 \to B_0$, we obtain an isomorphism of torus bundles

$$R^1(\mu_0)_*(\mathbb{R}/\mathbb{Z}) := (R^1(\mu_0)_*\mathbb{R})/(R^1(\mu_0)_*\mathbb{Z}/tors) \cong \check{X}_0$$

over B. Similarly $X_0 \cong R^1(\check{\mu}_0)_*(\mathbb{R}/\mathbb{Z})$ and Poincaré duality gives rise to

$$R^p(\mu_0)_*\mathbb{R} \cong R^{3-p}(\check{\mu}_0)_*\mathbb{R}.$$

By the simple assumption on μ and $\check{\mu}$ it follows that

$$R^p\mu_*\mathbb{R} \cong R^{3-p}\check{\mu}_*\mathbb{R}. \tag{2}$$

We want to use this isomorphism combined with the Leray spectral sequence to conclude the relation on Hodge numbers.

Let us analyze the cohomology of X and $\check{X}$. First, $H^1(X, \mathbb{R}) = 0$ by the simply connected assumption and so $H^5(X, \mathbb{R}) = 0$ by Poincaré duality. This implies the Hodge numbers $h^{0,1}(X)$, $h^{1,0}(X)$, $h^{2,3}(X)$ and $h^{3,2}(X)$ are all zero. By Serre duality, $h^{2,0} = h^{0,2}(X) = h^{0,1}(X) = 0$. Furthermore, $h^{1,3} = h^{3,1} = h^1(X, \Omega_X^3) = h^{0,1} = 0$ by the Calabi–Yau condition. Finally, $h^{3,3} = h^{0,0} = 1$ is evident and $h^{0,3} = h^{3,0} = h^0(X, \Omega_X^3) = 1$ again by the Calabi–Yau condition. Putting this together gives us the following relation between Hodge numbers and Betti numbers:

$$h^{1,1}(X) = b_2(X) = b_4(X) = h^{2,2}(X)$$
$$b_3(X) = 2 + h^{1,2}(X) + h^{2,1}(X) = 2(1 + h^{1,2}(X))$$

Of course the same is also true for $\check{X}$. Thus it would suffice to show

$$b_3(\check{X}) = 2 + h^{1,1}(X) + h^{2,2}(X) = 2(1 + h^{1,1}(X)) \tag{3}$$

from which it follows that $h^{1,1}(X) = h^{1,2}(\check{X})$ as well as $h^{1,1}(\check{X}) = h^{1,2}(X)$ by applying the same argument to X.

The sheaves $R^3\mu_*\mathbb{R}$ and $R^0\mu_*\mathbb{R}$ are both isomorphic to the constant sheaf $\mathbb{R}$. As X is simply connected, so is B so we deduce $H^1(B, \mathbb{R}) = 0$ and similarly $H^2(B, \mathbb{R}) = 0$ by Poincaré duality. Thus $H^1(B, R^0\mu_*\mathbb{R}) = H^2(B, R^0\mu_*\mathbb{R}) = H^1(B, R^3\mu_*\mathbb{R}) = H^2(B, R^3\mu_*\mathbb{R}) = 0$ and $H^i(B, R^j\mu_*\mathbb{R}) = \mathbb{R}$ for $i, j = 0, 3$. Next the vanishing $H^1(X, \mathbb{R}) = H^5(X, \mathbb{R})$ imply that $H^0(B, R^1\mu_*\mathbb{R}) = H^3(B, R^2\mu_*\mathbb{R}) = 0$. Applying the same reasoning to $\check{\mu}$ and using the isomorphism (2), we get

$$\begin{aligned} H^0(B, R^2\mu_*\mathbb{R}) &= H^0(B, R^1\check{\mu}_*\mathbb{R}) = 0, \\ H^3(B, R^1\mu_*\mathbb{R}) &= H^3(B, R^2\check{\mu}_*\mathbb{R}) = 0. \end{aligned}$$

Putting this all together, the E_2 page of the Leray spectral sequence for μ becomes

$$\begin{array}{cccc} \mathbb{R} & 0 & 0 & \mathbb{R} \\ 0 & H^1(B, \mathbb{R}^2\mu_*\mathbb{R}) & H^2(B, \mathbb{R}^2\mu_*\mathbb{R}) & 0 \\ 0 & H^1(B, \mathbb{R}^1\mu_*\mathbb{R}) & H^2(B, \mathbb{R}^1\mu_*\mathbb{R}) & 0 \\ \mathbb{R} & 0 & 0 & \mathbb{R} \end{array}$$

(with d_1 from the top-left $\mathbb{R}$ to $H^2(B, \mathbb{R}^2\mu_*\mathbb{R})$, and d_2 from $H^1(B, \mathbb{R}^1\mu_*\mathbb{R})$ to the bottom-right $\mathbb{R}$)

with the only possibly nonzero differentials depicted above. We claim in fact that d_1 and d_2 must also be zero.

Indeed let $S \subset X$ be a section of μ. Then S induces a nonzero section $s \in \mathbb{R} \cong H^0(B, R^3\mu_*\mathbb{R})$ since it intersects each fiber in codimension 3. Furthermore S must represent a nonzero cohomology class on X and so $s \in \ker(d_1)$. This forces d_1 to be the zero map since $H^0(B, R^3\mu_*\mathbb{R})$ is one dimensional. Similarly, the fibers of μ give rise to a nonzero class in $f \in H^3(B, R^0\mu_*\mathbb{R}) \cong \mathbb{R}$. Since the class of a fiber is also nonzero in the cohomology of X as the fibers intersect the section, then f must remain nonzero in $\mathrm{coker}(d_2)$; that is, d_2 must be zero.

This means the Leray spectral sequence for μ degenerates at the E_2 page and similarly for $\check{\mu}$. In particular, we can compute

$$\begin{aligned} h^{1,1}(X) &= b_2(X) = h^1(B, R^1\mu_*\mathbb{R}) = h^1(B, R^2\check{\mu}_*\mathbb{R}), \\ h^{2,2}(X) &= b_4(X) = h^2(B, R^2\mu_*\mathbb{R}) = h^2(B, R^1\check{\mu}_*\mathbb{R}), \end{aligned}$$

where we have again used (2). Therefore we can verify

$$b_3(\check{X}) = 2 + h^1(B, R^2\check{\mu}_*\mathbb{R}) + h^2(B, R^1\check{\mu}_*\mathbb{R}) = 2 + h^{1,1}(X) + h^{2,2}(X)$$

as required. □

Remark 3.3 The argument above (originally appearing in [16]) was generalized by Gross in [17] to obtain a relation between the integral cohomologies of X and $\check{X}$.

The reader may object that there are several assumptions required in the above result. The existence of a section isn't a serious assumption. Indeed all that was required in the proof is the existence of a cohomology class that behaves like a section with respect to cup products. As we already saw in 2.2, mirror symmetry necessitate the existence of such Lagrangians on X dual to line bundles on $\check{X}$ and vice versa. The simplicity assumption, on the other hand, is serious and isn't always satisfied. However, this still gives us a good heuristic check of SYZ mirror symmetry.

4 Semi-flat Mirror Symmetry

In this section we will consider the case where μ and $\check{\mu}$ are smooth sLag fibrations so that $B_0 = B$. This is often called the *semi-flat* case.

In this case we will see that the existence of dual sLag fibrations endows B with the extra structure of an integral affine manifold which results in a toy model of mirror symmetry on B. In fact, we will see that the dual SYZ fibrations can be recovered from this integral affine structure. Finally, we will discuss an approach to realize HMS conjecture in the semi-flat case.

4.1 The Moduli Space of Special Lagrangians

The starting point is the following theorem of McLean:

Theorem 4.1 (McLean [32, Sect. 3]) *Let (X, J, ω, Ω) be a Kähler Calabi–Yau n-fold. Then the moduli space $\mathcal{M}$ of special Lagrangian submanifolds is a smooth manifold. Furthermore, there are natural identifications*

$$H^{n-1}(L, \mathbb{R}) \cong T_L\mathcal{M} \cong H^1(L, \mathbb{R})$$

of the tangent space to any sLag submanifold $L \subset X$.

The idea is that a deformation of L is given by a normal vector field $v \in C^\infty(N_L X, \mathbb{R})$. Then we obtain a 1-form $\alpha \in \Omega^1(L, \mathbb{R})$ and an $n-1$-form $\beta \in \Omega^{n-1}(L, \mathbb{R})$ by contraction with ω and $\mathrm{Im}\Omega$ respectively:

$$\alpha = -i_v\omega,$$
$$\beta = i_v\mathrm{Im}\Omega.$$

It turns out that α and β determine each other and that v induces a sLag deformation of L if and only if α and β are both closed. This gives the above isomorphisms by the maps $v \mapsto [\alpha] \in H^1(L, \mathbb{R})$ and $v \mapsto [\beta] \in H^{n-1}(L, \mathbb{R})$ respectively.

Note in particular that the isomorphism $T_L\mathcal{M} \cong H^1(L, \mathbb{R})$ depends on the symplectic structure ω and the isomorphism $T_L\mathcal{M} \cong H^{n-1}(L, \mathbb{R})$ depends on the complex structure through the holomorphic volume form Ω.

Definition 4.2 An integral affine manifold M is a smooth manifold equipped with transition functions in the affine group $\mathbb{R}^n \rtimes \mathrm{GL}_n(\mathbb{Z})$. Equivalently it is a manifold M equipped with a local system of integral lattices $\Lambda \subset TM$.

The equivalence in Definition 4.2 can be seen by noting that if the transition functions of M are affine transformations, they preserve the integral lattice defined in local coordinates by

$$\Lambda := \mathrm{Span}_{\mathbb{Z}}\left(\frac{\partial}{\partial y_1}, \ldots, \frac{\partial}{\partial y_n}\right) \subset TU. \tag{4}$$

On the other hand, if there exists a local system of integral lattice $\Lambda \subset TM$ with a compatible flat connection ∇ on TM, then on a small enough coordinate patch we can choose coordinates such that Λ is the coordinate lattice and the transition functions must be linear isomorphisms on this lattice.

The vector spaces $H^1(L, \mathbb{R})$ and $H^{n-1}(L, \mathbb{R})$ glue together to form vector bundles on $\mathcal{M}$. Explicitly, if $\mathcal{L} \subset X \times \mathcal{M}$ is the universal family of sLags over $\mathcal{M}$ with projection $\pi : \mathcal{L} \to \mathcal{M}$ then these bundles are $R^1\pi_*\mathbb{R}$ and $R^{n-1}\pi_*\mathbb{R}$ respectively. Similarly, the integral cohomology groups $H^1(L, \mathbb{Z})/tors \subset H^1(L, \mathbb{R})$ and $H^{n-1}(L, \mathbb{Z})/tors \subset H^{n-1}(L, \mathbb{R})$ glue together into local systems of integral lattices $R^1\pi_*\mathbb{Z}/tors \subset R^1\pi_*\mathbb{R}$ and $R^{n-1}\pi_*\mathbb{Z}/tors \subset R^{n-1}\pi_*\mathbb{R}$. Applying Theorem 4.1 fiber by fiber yields two integral affine structures on $\mathcal{M}$:

Corollary 4.3 *There are isomorphisms $R^1\pi_*\mathbb{R} \cong T\mathcal{M} \cong R^{n-1}\pi_*\mathbb{R}$ which endow $\mathcal{M}$ with two integral affine structures given by the integral lattices*

$$R^1\pi_*\mathbb{Z}/tors \subset R^1\pi_*\mathbb{R} \cong T\mathcal{M},$$
$$R^{n-1}\pi_*\mathbb{Z}/tors \subset R^{n-1}\pi_*\mathbb{R} \cong T\mathcal{M}.$$

Poincare duality induces an isomorphism $T\mathcal{M} \cong T^*\mathcal{M}$ exchanging the lattices and their duals.

4.2 Mirror Symmetry for Integral Affine Structures

4.2.1 From SYZ Fibrations to Integral Affine Structures

Now let us return to the case of dual SYZ fibrations

$$\begin{array}{ccc} X & & \check{X} \\ & \searrow^{\mu} \quad {}^{\check{\mu}}\swarrow & \\ & B & \end{array}$$

where both μ and $\check{\mu}$ are smooth. Then dim $B = n =$ dim $H^1(L, \mathbb{R})$ is the dimension of the moduli space of sLag n-tori in X and so B must be an open subset of the moduli space $\mathcal{M}$.

In particular, by Corollary 4.3, the symplectic form ω and the holomorphic volume form Ω on X induces two integral affine structures on B explicitly given by

$$\Lambda_\omega := R^1\mu_*\mathbb{Z}/tors \subset R^1\mu_*\mathbb{R} \cong TB,$$
$$\Lambda_\Omega := R^{n-1}\mu_*\mathbb{Z}/tors \subset R^{n-1}\mu_*\mathbb{R} \cong TB.$$

We call these the Kähler and complex integral affine structures respectively. Similarly the symplectic and holomorphic forms $\check{\omega}$ and $\check{\Omega}$ on $\check{X}$ induce two other integral affine structures

$$\Lambda_{\check{\omega}} := R^1\check{\mu}_*\mathbb{Z}/tors \subset R^1\check{\mu}_*\mathbb{R} \cong TB,$$
$$\Lambda_{\check{\Omega}} := R^{n-1}\check{\mu}_*\mathbb{Z}/tors \subset R^{n-1}\check{\mu}_*\mathbb{R} \cong TB,$$

on B. The fact that these torus fibrations are dual implies natural isomorphisms

$$R^1\mu_*\mathbb{R} \cong R^{n-1}\check{\mu}_*\mathbb{R},$$
$$R^{n-1}\mu_*\mathbb{R} \cong R^1\check{\mu}_*\mathbb{R}.$$

The top isomorphism exchanges Λ_ω and $\Lambda_{\check{\Omega}}$ while the bottom isomorphism exchanges $\Lambda_{\check{\omega}}$ and Λ_Ω. We can summarize this as follows: *SYZ mirror symmetry for smooth sLag torus fibrations interchanges the complex and Kähler integral affine structures on the base B.*

4.2.2 From Integral Affine Structures to SYZ Fibrations

We can go in the other direction and recover the mirror SYZ fibrations μ and $\check{\mu}$ from the integral affine structures on the base B. The key is the following proposition:

Proposition 4.4 *Let $(B, \Lambda \subset TB)$ be an integral affine manifold. Then the torus fibration $TB/\Lambda \to B$ has a natural complex structure and the dual torus fibration $T^*B/\Lambda^* \to B$ has a natural symplectic structure.*

Proof Locally we can find a coordinate chart $U \subset B$ with coordinates $y_1, \ldots, y_n$ such that Λ is a coordinate lattice as in (4). Then the coordinate functions on TU are given by $y_1, \ldots, y_n$ and $x_1 = dy_1, \ldots, x_n = dy_n$ and we can define holomorphic coordinates on TU by $z_j = x_j + \sqrt{-1}y_j$. Since the transition functions on B preserve the lattice, they induce transition functions on TB that are holomorphic with respect to these coordinates giving TB the structure of a complex manifold.

Consider the holomorphic functions defined locally by

$$q_j := e^{2\pi\sqrt{-1}z_j}.$$

These functions are invariant under integral affine transition functions as well as global translations by Λ and so they give a compatible system of holomorphic coordinates for TB/Λ.

Similarly, in local coordinates U where Λ is the coordinate lattice, then $\Lambda^* \subset T^*U$ is generated by $dy_1, \ldots, dy_n$ as a lattice in T^*U. Therefore the standard symplectic structure on T^*B is invariant by Λ^* and descends to T^*B/Λ^*. □

Now suppose B is a smooth manifold equipped with two integral affine structures $\Lambda_0, \Lambda_1 \subset TB$ as well as an isomorphism $TB \cong T^*B$ such that $\Lambda_0 \cong (\Lambda_1)^*$ and $\Lambda_1 \cong (\Lambda_0)^*$. Then we have dual torus fibrations

$$\begin{array}{ccc} X & & \check{X} \\ & {}_{\mu}\searrow \quad \swarrow_{\check{\mu}} & \\ & B & \end{array}$$

where $X := TB/\Lambda_0 \cong T^*B/(\Lambda_1)^*$ and $\check{X} := T^*B/(\Lambda_0)^* \cong TB/(\Lambda_1)$. This construction satisfies the following properties:

(a) if Λ_0 and Λ_1 are the integral affine structures associated to SYZ dual torus fibrations as in Sect. 4.2.1, then this construction recovers the original fibrations;
(b) Λ_0 determines the complex structure of X and the symplectic structure of $\check{X}$;
(c) Λ_1 determines the symplectic structure of X and the complex structure of $\check{X}$.

As a result we recover one of the main predictions of mirror symmetry: *deformations of the complex structure on X are the same as deformations of the symplectic structure on $\check{X}$ and vice versa.*

Remark 4.5 There is an extra piece of structure on B that we haven't discussed. This is a Hessian metric g realizing the identification $TB \cong T^*B$. Recall that a Hessian metric is a Riemannian metric that is locally the Hessian of some smooth potential function K. The two integral affine structures on B endow it with two different sets of local coordinates and the potential functions in these coordinates are related by the Legendre transform. In fact the complex and symplectic structures constructed in Proposition 4.4 can be recovered from the potential function so mirror symmetry in this context is governed by the Legendre transform [25] [2, Sect. 6.1.2].

4.3 The SYZ Transform

To finish off the discussion of semi-flat mirror symmetry, we turn our attention to the HMS conjecture. The goal is to construct a geometric functor

$$\Phi : \mathcal{F}uk(X) \to D^b(\mathrm{Coh}(\check{X}))$$

from the Fukaya category of X to the derived category of coherent sheaves on $\check{X}$ using the geometry of the dual fibrations. The first step is to produce an object of $D^b(\mathrm{Coh}(\check{X}))$ from a Lagrangian $L \subset X$ equipped with a flat unitary connection. We will attempt to do this by exploiting the interpretation of a point $p \in X$ as a flat $U(1)$-connection on the dual fiber.

Let $L \subset X$ be a Lagrangian section of μ corresponding to a map $\sigma : B \to X$, equipped with the trivial connection. By restricting L to each fiber of μ, we obtain a family of flat $U(1)$-connections

$$\{\nabla_{\sigma(b)}\}_{b \in B}$$

on the fibers of $\check{\mu} : \check{X} \to B$. These glue together to give a flat $U(1)$-connection on a complex line bundle $\mathcal{L}$ on $\check{X}$. It turns out this connection gives $\mathcal{L}$ the structure of a holomorphic line bundle on $\check{X}$ (endowed with the complex structure constructed in the last subsection).

This construction was generalized by [7] (see also [31]) as follows. As X is the moduli space of flat $U(1)$-connections on the fibers of $\check{\mu} : \check{X} \to B$, there exists a universal bundle with connection $(\mathcal{P}, \nabla^{\mathcal{P}})$ on $X \times_B \check{X}$. Now given $(L, \mathcal{E}, \nabla)$ where $L \subset X$ is a multisection transverse to the fibers of μ and $(\mathcal{E}, \nabla)$ is a flat unitary vector bundle on L, define the *SYZ transform* by

$$\Phi^{SYZ}(L, \mathcal{E}, \nabla) := (pr_{\check{X}})_*((pr_L)^*\mathcal{E} \otimes (i \times id)^*\mathcal{P})$$

where $pr_L, pr_{\check{X}} : L \times_B \check{X} \to L, \check{X}$ are the projections and $(i \times id) : L \times_B \check{X} \to X \times_B \check{X}$ is the inclusion. Note that $\Phi^{SYZ}(L, \mathcal{E}, \nabla)$ comes equipped with a connection we denote $\nabla_{(L,\mathcal{E},\nabla)}$.

Theorem 4.6 ([7, Theorem 1.1]) *If $L \subset X$ is Lagrangian, then $\nabla_{(L,\mathcal{E},\nabla)}$ endows $\Phi^{SYZ}(L, \mathcal{E}, \nabla)$ with the structure of a holomorphic vector bundle on $\check{X}$. When X and $\check{X}$ are dual elliptic curves fibered over S^1, then every holomorphic vector bundle on $\check{X}$ is obtained this way.*

Viewing holomorphic vector bundles as objects in $D^b(\mathrm{Coh}(\check{X}))$, we hope to extend the SYZ transform to an equivalence $\Phi : \mathcal{F}uk(X) \to D^b(\mathrm{Coh}(\check{X}))$, thus realizing the HMS conjecture. While this hope hasn't been realized in general, it has in some special cases.

When X and $\check{X}$ are dual elliptic curves fibered over S^1, a HMS equivalence Φ is constructed by hand in [34]. One can check that their functor Φ does indeed extend the SYZ transform Φ^{SYZ}. In fact, assuming Theorem 4.6, it is not so hard to construct Φ at least on the level of objects. Each coherent sheaf on the curve X can be decomposed as a direct sum of a torsion sheaf and a vector bundle. Vector bundles are taken care of by Theorem 4.6. Torsion sheaves are successive extensions of skyscrapers at points which correspond to S^1 fibers of $\mu : X \to B$. For more recent work on understanding the SYZ transform see [12] and the references therein.

5 Constructing Mirrors

We now move on to the general problem of constructing mirrors. Given a Kähler Calabi–Yau n-fold (X, J, ω, Ω), the SYZ conjecture suggests the following strategy for constructing a mirror.

5.0.1 Strategy

(i) produce a special Lagrangian fibration $\mu : X \to B$;[5]
(ii) dualize the smooth locus $\mu_0 : X_0 \to B_0$ to obtain a semi-flat mirror $\check{\mu}_0 : \check{X}_0 \to B_0$;
(iii) compactify $\check{X}_0$ to obtain a CY n-fold with a dual SYZ fibration $\check{\mu} : \check{X} \to B$;
(iv) use the geometry of the dual fibrations to construct a HMS equivalence

$$\Phi : \mathcal{F}uk(X) \to D^b(\mathrm{Coh}(\check{X})).$$

5.0.2 Obstacles

There are many obstacles to carrying out 5.0.1 and (ii) is the only step where a totally satisfactory answer is known as we discussed in Sect. 4.

Producing sLag fibrations on a compact Calabi–Yau n-folds is a hard open problem in general. Furthermore, work of Joyce [26] suggests that even when sLag fibrations exist, they might be ill-behaved. The map μ is not necessarily differentiable and may have real codimension one discriminant locus in the base B. In this case B_0 is disconnected and one needs to perform steps (ii) and (iii) on each component and then glue.

Compactifying $\check{X}_0$ to a complex manifold also poses problems. There are obstructions to extending the semi-flat complex structure on $\check{X}_0$ to any compactification. To remedy this, one needs to take a small deformation of $\check{X}_0$ by modifying the complex structure using *instanton corrections*.

[5]This choice is the reason that X may have several mirrors.

Step (iv) has been realized in some special cases (e.g. [1, 3–5, 30] and references therein) but a general theory for producing an equivalence Φ given an SYZ mirror is still elusive.

5.1 Instanton Corrections

The small deformation of the complex structure on the dual $\check{X}_0$ is necessitated by the existence of obstructed Lagrangians. The point is that the Fukaya category of X doesn't contain all pairs (L, ∇) of Lagrangians with flat connection but only those pairs where L is *unobstructed*.

A Lagrangian L is unobstructed if certain counts of holomorphic discs bounded by L cancel out so that the Floer differential satisfies $d^2 = 0$. In particular, if L doesn't bound any nonconstant holomorphic discs, then it is unobstructed. A problem arises if $\mu : X \to B$ has singular fibers because then the smooth torus fibers may bound nontrivial holomorphic discs known as *disc instantons*. For example, any vanishing 1-cycle on a nearby fiber sweeps out such a disc.

To construct the dual $\check{X}$ as a complex moduli space of objects in the Fukaya we need to account for the effect of these instantons on the objects in the Fukaya category. This is done by modifying the semi-flat complex structure using counts of such disc instantons.

In fact, one can explicitly write down the coordinates for the semi-flat complex structure described in Sect. 4 in terms of the symplectic area of cylinders swept out by isotopy of nearby smooth Lagrangian fibers as in Sect. 5.3. Then the discs bounded by obstructed Lagrangians lead to nontrivial monodromy of the semi-flat complex on $\check{X}_0$ which is an obstruction to the complex structure extending to a compactification $\check{X}$. The instanton corrections are given by multiplying these coordinates by the generating series for virtual counts of holomorphic discs bounded by the fibers.

For more details on instanton corrections, see for example [1, 10, 36].

5.2 From Torus Fibrations to Degenerations

Heuristics from physics suggest that X will admit an SYZ fibration in the limit toward a maximally unipotent degeneration.[6] It was independently conjectured in [24, 30] that if $\mathcal{X} \to \mathbb{D}$ is such a degeneration over a disc (where $X = \mathcal{X}_\epsilon$ for for some small $\epsilon \ll 1$) and g_t is a suitably normalized metric on $\mathcal{X}_t$, then the Gromov-Hausdorff limit of the metric spaces $(\mathcal{X}_t, g_t)$ collapses the Lagrangian torus fibers onto the base B of an SYZ fibration. Furthermore, this base should be recovered as the dual complex of the special fiber of $\mathcal{X} \to \mathbb{D}$ endowed with the appropriate singular integral affine

[6] That is, a degeneration with maximally unipotent monodromy. These are sometimes known as large complex structure limits (LCSL).

structure. Then one can hope to reconstruct the instanton corrected SYZ dual directly from data on B.

This allows one to bypass the issue of constructing a sLag fibration by instead constructing a maximally unipotent degeneration. Toric degenerations are particularly well suited for this purpose. This is the point of view taken in the Gross-Siebert program [20, 21] and gives rise to a version of SYZ mirror symmetry purely within algebraic geometry. In this setting the instanton corrections should come from logarithmic Gromov-Witten invariants of the degeneration as constructed in [6, 13, 23] and these invariants can be computed tropically from data on the base B. For more on this see for example [18, 19, 22].

5.3 Beyond the Calabi–Yau Case

The SYZ approach can also be used to understand mirror symmetry beyond the case of Calabi–Yau manifolds. The most natural generalization involves log Calabi–Yau pairs (X, D) where $D \subset X$ is a boundary divisor and the sheaf $\omega_X(D)$ of top forms with logarithmic poles along D is trivial. That is, D is a section of the anticanonical sheaf ω_X^{-1} and $X \setminus D$ is an open Calabi–Yau.

In this case the mirror should consist of a pair (M, W) consisting of a complex manifold M with a holomorphic function $W : M \to \mathbb{C}$. The pair (M, W) is known as a *Landau-Ginzburg model* and the function W is the *superpotential* [28]. Homological mirror symmetry takes the form of an equivalence

$$\Phi : \mathcal{F}uk(X, D) \to MF(M, W)$$

between a version of the Fukaya category for pairs (X, D) and the category of *matrix factorizations* of (M, W). Recall that a matrix factorization is a 2-periodic complex

$$\left(\cdots \longrightarrow P_0 \xrightarrow{\ d\ } P_1 \xrightarrow{\ d\ } P_0 \longrightarrow \cdots \right)$$

of coherent sheaves on M satisfying $d^2 = W$. By a theorem of Orlov [33], the category $MF(M, W)$ is equivalent to the derived category of singularities $D^b_{sing}(\{W = 0\})$.[7]

The SYZ conjecture gives a recipe for constructing the Landau-Ginzburg dual (M, W). Here we give the version as stated in [9]:

Conjecture 2 *Let (X, J, ω) be a compact Kähler manifold and D a section of K_X^{-1}. Suppose $\mu : U = X \setminus D \to B$ is an SYZ fibration where U is equipped with a holomorphic volume form Ω. Then the mirror to (X, D) is the Landau-Ginzburg model $(\check{U}, W)$ where*

[7] Here we've assumed for simplicity that the only critical value of W is at $0 \in \mathbb{C}$.

$$\check{\mu} : \check{U} \to B$$

is the SYZ dual fibration equipped with the instanton corrected complex structure and the superpotential W is computed by counting holomorphic discs in (X, D).

We briefly recall the construction of the superpotential. Let $\mu_0 : U_0 \to B_0$ be the smooth locus of the fibration so that $\check{U}_0$ is the semi-flat dual. Consider a family of relative homology classes $A_L \in H_2(X, L; \mathbb{Z})$ as the Lagrangian torus fiber L varies. Then the function

$$z^A : \check{U}_0 \to \mathbb{C} \quad z^A(L, \nabla) = \exp\left(-\int_{A_L} \omega\right) \mathrm{hol}_\nabla(\partial A_L).$$

is a holomorphic local coordinate on $\check{U}_0$.

Let

$$m_0(L, \nabla) = \sum_{\beta \in H_2(X, L; \mathbb{Z})} n_\beta(L) z^\beta$$

where $n_\beta(L)$ is Gromov-Witten count of holomorphic discs in X bounded by L and intersecting D transversally.[8] This is a holomorphic function on $\check{U}_0$ when it is defined but in general it only becomes well defined after instanton correcting the complex structure. The idea is that the number $n_\beta(L)$ jumps across an obstructed Lagrangian L that bounds disc instantons in $X \setminus D$. Instanton corrections account for this and so m_0 should extend to a holomorphic function W on the instanton corrected dual $\check{U}$.

In fact m_0 is the obstruction to Floer homology constructed in [15]. That is, $d^2 = m_0$ where d is the Floer differential on the Floer complex $CF^*(L, L)$. This explains why the Landau-Ginzburg superpotential W should be given by m_0. If one believes homological mirror symmetry, then obstructed chain complexes in the Fukaya category should lead to matrix factorizations with $W = m_0$ on the mirror.

Example 5.1 Let $X = \mathbb{P}^1$ with anticanonical divisor $\{0, \infty\} = D$. Then $U = \mathbb{C}^*$ admits a sLag fibration $\mu : U \to B$ where B is the open interval $(0, \infty)$ and $\mu^{-1}(r) = \{|z| = r\}$ is a circle. The dual is $\check{U} = \mathbb{C}^*$ is also an algebraic torus and there are no instanton corrections since all the fibers of μ are smooth. Each sLag circle $L \subset U \subset X$ cuts X into two discs D_0 and D_∞ whose classes satisfy $[D_0] + [D_\infty] = [\mathbb{P}^1]$ in $H_2(X, L; \mathbb{Z})$ so that the corresponding coordinate functions z_0 and z_∞ on $\check{U}$ satisfy $z_0 z_\infty = 1$. Furthermore,

$$\exp\left(-\int_{D_0} \omega\right) \exp\left(-\int_{D_\infty} \omega\right) = e^{-A}$$

where $A = \int_{\mathbb{P}^1} \omega$ is the symplectic area. Furthermore, it is easy to see that $n_{[D_0]}(L) = n_{[D_1]}(L) = 1$. Putting it together and rescaling by a factor, we obtain the superpotential

[8]More precisely, the sum is over curve classes β with Maslov index $\mu(\beta) = 2$.

$$W = z_0 + \frac{e^{-A}}{z_0} : \mathbb{C}^* \to \mathbb{C}.$$

A similar argument works for any Fano toric pair (X, D) where μ is the moment map, B is the interior of the moment polytope P, $\check{U} = (\mathbb{C}^*)^n$ is an algebraic torus, and W is given as a sum over facets of P [9, 14].

References

1. Mohammed Abouzaid, Denis Auroux, and Ludmil Katzarkov. Lagrangian fibrations on blowups of toric varieties and mirror symmetry for hypersurfaces. Publ. Math. Inst. Hautes Études Sci., 123:199–282, 2016.
2. Paul S. Aspinwall, Tom Bridgeland, Alastair Craw, Michael R. Douglas, Mark Gross, Anton Kapustin, Gregory W. Moore, Graeme Segal, Balázs Szendr" oi, and P. M. H. Wilson. *Dirichlet branes and mirror symmetry*, volume 4 of *Clay Mathematics Monographs*. American Mathematical Society, Providence, RI; Clay Mathematics Institute, Cambridge, MA, 2009.
3. Mohammed Abouzaid. Family Floer cohomology and mirror symmetry. *ArXiv e-prints*, April 2014.
4. Mohammed Abouzaid. The family Floer functor is faithful. *ArXiv e-prints*, August 2014.
5. Mohammed Abouzaid. Homological mirror symmetry without corrections. *ArXiv e-prints*, March 2017.
6. Dan Abramovich and Qile Chen. Stable logarithmic maps to Deligne-Faltings pairs II. Asian J. Math., 18(3):465–488, 2014.
7. D. Arinkin and A. Polishchuk. Fukaya category and Fourier transform. In *Winter School on Mirror Symmetry, Vector Bundles and Lagrangian Submanifolds (Cambridge, MA, 1999)*, volume 23 of *AMS/IP Stud. Adv. Math.*, pages 261–274. Amer. Math. Soc., Providence, RI, 2001.
8. V. I. Arnol′d. *Mathematical methods of classical mechanics*, volume 60 of *Graduate Texts in Mathematics*. Springer-Verlag, New York, [1989?]. Translated from the 1974 Russian original by K. Vogtmann and A. Weinstein, Corrected reprint of the second (1989) edition.
9. Denis Auroux. Mirror symmetry and T-duality in the complement of an anticanonical divisor. J. Gökova Geom. Topol. GGT, 1:51–91, 2007.
10. Denis Auroux. Special Lagrangian fibrations, wall-crossing, and mirror symmetry. In *Surveys in differential geometry. Vol. XIII. Geometry, analysis, and algebraic geometry: forty years of the Journal of Differential Geometry*, volume 13 of *Surv. Differ. Geom.*, pages 1–47. Int. Press, Somerville, MA, 2009.
11. Matthew Robert Ballard. Meet homological mirror symmetry. In *Modular forms and string duality*, volume 54 of *Fields Inst. Commun.*, pages 191–224. Amer. Math. Soc., Providence, RI, 2008.
12. K. Chan. Homological mirror symmetry for local Calabi-Yau manifolds via SYZ. *ArXiv e-prints*, July 2016.
13. Qile Chen. Stable logarithmic maps to Deligne-Faltings pairs I. *Ann. of Math. (2)*, 180(2):455–521, 2014.
14. Cheol-Hyun Cho and Yong-Geun Oh. Floer cohomology and disc instantons of Lagrangian torus fibers in Fano toric manifolds. Asian J. Math., 10(4):773–814, 2006.
15. Kenji Fukaya, Yong-Geun Oh, Hiroshi Ohta, and Kaoru Ono. *Lagrangian intersection Floer theory: anomaly and obstruction.*, volume 46 of *AMS/IP Studies in Advanced Mathematics*. American Mathematical Society, Providence, RI; International Press, Somerville, MA, 2009.
16. Mark Gross. Special Lagrangian fibrations. I. Topology. In *Integrable systems and algebraic geometry (Kobe/Kyoto, 1997)*, pages 156–193. World Sci. Publ., River Edge, NJ, 1998.

17. Mark Gross. Special Lagrangian fibrations. II. Geometry. A survey of techniques in the study of special Lagrangian fibrations. In *Surveys in differential geometry: differential geometry inspired by string theory*, volume 5 of *Surv. Differ. Geom.*, pages 341–403. Int. Press, Boston, MA, 1999.
18. Mark Gross. *Tropical geometry and mirror symmetry*, volume 114 of *CBMS Regional Conference Series in Mathematics*. Published for the Conference Board of the Mathematical Sciences, Washington, DC; by the American Mathematical Society, Providence, RI, 2011.
19. Mark Gross. Mirror symmetry and the Strominger-Yau-Zaslow conjecture. In *Current developments in mathematics 2012*, pages 133–191. Int. Press, Somerville, MA, 2013.
20. Mark Gross and Bernd Siebert. Affine manifolds, log structures, and mirror symmetry. Turkish J. Math., 27(1):33–60, 2003.
21. Mark Gross and Bernd Siebert. Mirror symmetry via logarithmic degeneration data. I. J. Differential Geom., 72(2):169–338, 2006.
22. Mark Gross and Bernd Siebert. From real affine geometry to complex geometry. *Ann. of Math. (2)*, 174(3):1301–1428, 2011.
23. Mark Gross and Bernd Siebert. Logarithmic Gromov-Witten invariants. J. Amer. Math. Soc., 26(2):451–510, 2013.
24. Mark Gross and P. M. H. Wilson. Large complex structure limits of $K3$ surfaces. J. Differential Geom., 55(3):475–546, 2000.
25. Nigel J. Hitchin. The moduli space of special Lagrangian submanifolds. *Ann. Scuola Norm. Sup. Pisa Cl. Sci. (4)*, 25(3-4):503–515 (1998), 1997. Dedicated to Ennio De Giorgi.
26. Dominic Joyce. Singularities of special Lagrangian fibrations and the SYZ conjecture. Comm. Anal. Geom., 11(5):859–907, 2003.
27. Dominic Joyce. Conjectures on Bridgeland stability for Fukaya categories of Calabi-Yau manifolds, special Lagrangians, and Lagrangian mean curvature flow. EMS Surv. Math. Sci., 2(1):1–62, 2015.
28. Anton Kapustin and Yi Li. Topological correlators in Landau-Ginzburg models with boundaries. Adv. Theor. Math. Phys., 7(4):727–749, 2003.
29. Maxim Kontsevich. Homological algebra of mirror symmetry. In *Proceedings of the International Congress of Mathematicians, Vol. 1, 2 (Zürich, 1994)*, pages 120–139. Birkhäuser, Basel, 1995.
30. Maxim Kontsevich and Yan Soibelman. Homological mirror symmetry and torus fibrations. In *Symplectic geometry and mirror symmetry (Seoul, 2000)*, pages 203–263. World Sci. Publ., River Edge, NJ, 2001.
31. Naichung Conan Leung, Shing-Tung Yau, and Eric Zaslow. From special Lagrangian to Hermitian-Yang-Mills via Fourier-Mukai transform. Adv. Theor. Math. Phys., 4(6):1319–1341, 2000.
32. Robert C. McLean. Deformations of calibrated submanifolds. Comm. Anal. Geom., 6(4):705–747, 1998.
33. D. O. Orlov. Triangulated categories of singularities and D-branes in Landau-Ginzburg models. *Tr. Mat. Inst. Steklova*, 246(Algebr. Geom. Metody, Svyazi i Prilozh.):240–262, 2004.
34. Alexander Polishchuk and Eric Zaslow. Categorical mirror symmetry: the elliptic curve. Adv. Theor. Math. Phys., 2(2):443–470, 1998.
35. Andrew Strominger, Shing-Tung Yau, and Eric Zaslow. Mirror symmetry is T-duality. Nuclear Phys. B, 479(1–2):243–259, 1996.
36. Junwu Tu. On the reconstruction problem in mirror symmetry. Adv. Math., 256:449–478, 2014.
37. R. P. Thomas and S.-T. Yau. Special Lagrangians, stable bundles and mean curvature flow. Comm. Anal. Geom., 10(5):1075–1113, 2002.
38. Alan Weinstein. Symplectic manifolds and their Lagrangian submanifolds. Advances in Math., 6:329–346 (1971), 1971.

Part III
Physical Motivations

The Derived Category of Coherent Sheaves and B-model Topological String Theory

Stephen Pietromonaco

1 Topological Closed String Theories

The starting point for closed string topological string theories is the non-linear sigma model which studies maps $\phi : \Sigma \to X$, where Σ is a compact, oriented Riemann surface called the 'worldsheet' and we take X to be a Calabi–Yau threefold, called the 'target space.' If only closed strings are present, Σ is taken to be without boundary. We can take local complex coordinates $(z, \bar{z})$ on Σ, and $w^i = \phi^i(z, \bar{z})$ on X. We have a Kähler metric $g_{i\bar{j}}$, as well as an anti-symmetric B-field $B_{i\bar{j}}$ on X. Of course, the indices here correspond to tensor components in the complex coordinates w^i.

The theory becomes *topological* after performing one of two possible twists. In what sense is the theory topological? Such a non-linear sigma model is a two-dimensional quantum field theory defined on the fixed Riemann surface Σ. Therefore, to say the twisted theory is topological is to say there exists a subsector of operators such that the correlation functions are independent of the metric *on the worldsheet*. It is crucial to not confuse the metric on the worldsheet with the metric on the target Calabi–Yau. I will review the two topologically twisted models which Witten [4] called the A and B models. The A-model will depend only on the Kähler structure on X while the B-model will depend only on the complex structure. So there will indeed be partial dependence on the target space metric, the exact form of which will depend on the model under consideration. In addition, I will define a BRST operator Q (this operator will be different in the A and B models). The physical observables of the topological subsector will consist of products of local operators, each of which is invariant under the BRST operator Q. By convention, we denote the target space by Y in the A-model and as X in the B-model.

Let T_X be the complexified tangent bundle of X, which can be decomposed as $T_X = T_X^{(1,0)} \oplus T_X^{(0,1)}$. The fermions in the theory require a choice of square-root

---

S. Pietromonaco (✉)
University of Toronto, Toronto, Canada
e-mail: scpietromonaco@gmail.com

M. Ballard et al. (eds.), *Superschool on Derived Categories and D-branes*, Springer Proceedings in Mathematics & Statistics 240,
https://doi.org/10.1007/978-3-319-91626-2_14

bundles $K^{1/2}$ and $\overline{K}^{1/2}$, where K and $\overline{K}$ are the canonical and anti-canonical bundles on Σ, respectively. The non-linear sigma model action is given by: (equation (2.4) in [4])

$$S = \int_{\Sigma} d^2z \left(\frac{1}{2} g_{ij} \partial_z \phi^i \partial_{\bar{z}} \phi^j + \frac{i}{2} B_{ij} \partial_z \phi^i \partial_{\bar{z}} \phi^j + i\psi_-^{\bar{i}} D_z \psi_-^i g_{i\bar{i}} + i\psi_+^{\bar{i}} D_{\bar{z}} \psi_+^i g_{\bar{i}i} + R_{i\bar{i}j\bar{j}} \psi_+^i \psi_+^{\bar{i}} \psi_-^j \psi_-^{\bar{j}} \right), \tag{1}$$

where $R_{i\bar{i}j\bar{j}}$ is the Riemann tensor on X, D_z is the ∂ operator on $\overline{K}^{1/2} \otimes \phi^* T_X^{(1,0)}$, arising by pulling back the holomorphic part of the Levi–Civita connection on T_X. Likewise, $D_{\bar{z}}$ is the $\bar{\partial}$ operator on $K^{1/2} \otimes \phi^* T_X^{(1,0)}$. The fermion fields are sections of the following bundles,

$$\begin{aligned} \psi_+^i &\in \Gamma\big(K^{1/2} \otimes \phi^* T_X^{(1,0)}\big), & \psi_+^{\bar{i}} &\in \Gamma\big(K^{1/2} \otimes \phi^* T_X^{(0,1)}\big), \\ \psi_-^i &\in \Gamma\big(\overline{K}^{1/2} \otimes \phi^* T_X^{(1,0)}\big), & \psi_-^{\bar{i}} &\in \Gamma\big(\overline{K}^{1/2} \otimes \phi^* T_X^{(0,1)}\big). \end{aligned} \tag{2}$$

The sigma model action above is really a worldsheet action; the integral is over two-forms on Σ. Therefore, all of the structures described above need to be pulled back to Σ via ϕ, which implies that the pullback of the metric, the B-field, and the connection will all inherit ϕ dependence. As mentioned, $\psi_\pm^i$, $\psi_\pm^{\bar{j}}$ are the fermionic fields and the bosonic fields are the local coordinates ϕ^i and $\phi^{\bar{j}}$.[1]

The supersymmetry (SUSY) transformations are generated by the four infinitesimal fermionic parameters $\alpha_+, \tilde{\alpha}_+, \alpha_-, \tilde{\alpha}_-$. The first two are anti-holomorphic sections of $\overline{K}^{-1/2}$ and the latter two are holomorphic sections of $K^{-1/2}$. We refer the reader to Eq. (2.5) in [4] for the full form of the supersymmetry transformations. Since we have four SUSY parameters, two of each chirality, we say the resulting theory has "worldsheet $\mathcal{N} = (2, 2)$ supersymmetry."

1.1 Closed String A-model

Let Y be the Calabi–Yau target space in the A-model. We consider here a restricted symmetry such that $\tilde{\alpha}_- = \alpha_+ = 0$ and $\alpha = \alpha_- = \tilde{\alpha}_+$. In other words, we have only one SUSY parameter which we call α. We now perform the first of two possible topological twists to construct the A-model topological string theory. Consider the field $\chi \in \Gamma(\phi^* T_X)$ which projects into $\phi^* T_X^{(1,0)}$ as $\chi^i = \psi_+^i$ and into $\phi^* T_X^{(0,1)}$ as $\chi^{\bar{i}} = \psi_-^{\bar{i}}$. We regard $\psi_+^{\bar{i}}$ as a $(1, 0)$ form on Σ valued in $\phi^* T_X^{(0,1)}$ and following [4], denote it as $\psi_z^{\bar{i}}$. Likewise, ψ_-^i is a $(0, 1)$ form valued in $\phi^* T_X^{(1,0)}$, denoted $\psi_{\bar{z}}^i$. The A-model SUSY transformations are

[1] Having the bosonic fields correspond to the local coordinates on a Riemannian manifold is an idea originating in 'supersymmetric quantum mechanics.'

$$\begin{aligned}\delta\phi^i &= i\alpha\chi^i \\ \delta\phi^{\bar{i}} &= i\alpha\chi^{\bar{i}} \\ \delta\chi^i &= \delta\chi^{\bar{i}} = 0 \\ \delta\psi_z^{\bar{i}} &= -\alpha\partial_z\phi^{\bar{i}} - i\alpha\chi^{\bar{j}}\Gamma^{\bar{i}}_{\bar{j}\bar{m}}\psi_z^{\bar{m}} \\ \delta\psi_{\bar{z}}^i &= -\alpha\partial_{\bar{z}}\phi^i - i\alpha\chi^j\Gamma^i_{jm}\psi_{\bar{z}}^m\end{aligned} \tag{3}$$

where Γ^i_{jm} is the holomorphic part of the Levi–Civita connection on the complexified tangent bundle and $\Gamma^{\bar{i}}_{\bar{j}\bar{m}}$ is the anti-holomorphic part. Corresponding to the single SUSY parameter α, we define the operator Q to be its generator. As such, the variation of any local operator W under a SUSY transformation with parameter α, is given by

$$\delta W = -i\alpha\{Q, W\}. \tag{4}$$

One can show from the action that $Q^2 = 0$, on-shell. This means that though there may be non-zero terms equated to Q^2, they will vanish if the equations of motion are satisfied. Thus, we have a nilpotent operator Q which is commonly referred to as a BRST operator. With this in hand, we can rewrite the sigma model action as,

$$S = \int_\Sigma i\{Q, V\} - 2\pi i \int_\Sigma \phi^*(B + iJ), \tag{5}$$

where $V = 2\pi g_{i\bar{j}}(\psi_z^{\bar{j}}\bar{\partial}\phi^i + \partial\phi^{\bar{j}}\psi_{\bar{z}}^i)$ and $B + iJ \in H^2(Y, \mathbb{C})$ is the complexified Kähler form. Given an operator W, we say W is Q-closed if $\{Q, W\} = 0$ and we say it is Q-exact if $W = \{Q, W'\}$, for some operator W'. We also call a Q-closed operator 'BRST invariant.' We will take it as a fact that a correlation function of a Q-exact operator must vanish

$$\langle\{Q, W_1 W_2 \ldots\}\rangle = 0. \tag{6}$$

Let us assume that $W_2, W_3, \ldots$ are Q-closed operators, and consider the correlation function $\langle\{Q, W_1 W_2 \ldots\}\rangle$ for any operator W_1. By the fact cited above, this correlation function vanishes. Moreover, since Q behaves like a differential, we can apply Leibniz' rule to get

$$0 = \langle\{Q, W_1 W_2 \ldots\}\rangle = \langle W_1\{Q, W_2 W_3 \ldots\}\rangle + \langle\{Q, W_1\}W_2 W_3 \ldots\rangle. \tag{7}$$

Since $W_2, W_3, \ldots$ are Q-closed operators, the term $\langle W_1\{Q, W_2 W_3 \ldots\}\rangle$ will vanish when expanded using Leibniz' rule. All that remains is the correlation function $\langle\{Q, W_1\}W_2 W_3 \ldots\rangle$ involving one Q-exact operator and the rest, Q-closed. Since the original correlation function vanished, clearly this one must too. Therefore, the presence of even one Q-exact operator annihilates the correlation function. We should then restrict attention to Q-closed operators:

In the topological subsector, the physical observables are products of local operators, all of which are Q-closed (i.e. BRST invariant).
We note that a shift in the action by a Q-exact operator $S \to S + \int_\Sigma \{Q, S'\}$ will leave all correlation functions invariant. In the sigma model action, the only place the complex structure of Y appears is in the term V. If we deform the complex structure $V \to V + \delta V$, this leads to a deformation of the action $S \to S + \int_\Sigma \{Q, \delta V\}$, which will leave all physical observables invariant. Thus, it appears that the A-model topological field theory is independent of the complex structure we choose to endow Y with. Clearly, it explicitly depends on the Kähler structure on the target space, through the term $2\pi i \int_\Sigma (B + iJ)$.

By the SUSY transformations (3) we have $\delta\chi^i = \delta\chi^{\bar{i}} = 0$, where χ^i and $\chi^{\bar{i}}$ are the fermionic superpartners of ϕ_i and $\phi^{\bar{i}}$, respectively. This means the operators χ^i and $\chi^{\bar{i}}$ are Q-closed. Thus, we have a basis of local BRST invariant operators on Σ, which we can use to write a general operator as

$$W_a = a_{I_1 \cdots I_p} \chi^{I_1} \cdots \chi^{I_p}, \tag{8}$$

where here the capital I_q denotes unbarred indices, and

$$a = a_{I_1 \cdots I_p} d\phi^{I_1} \cdots d\phi^{I_p}, \tag{9}$$

is a p-form on Y. By computing the variation of the operator W_a, we find that $\{Q, W_a\} = -W_{da}$, with a rather remarkable conclusion:

A local operator W_a is Q-closed (BRST invariant) if and only if $da = 0$. In other words, we can identify the Q-cohomology in the A-model with the de Rham cohomology $H^*(Y, \mathbb{C})$ on the target space. Notice this is consistent with the A-model being independent of the complex structure on Y.
A correlation function in the closed string A-model is given by the following path integral,

$$\langle W_a W_b \cdots \rangle = \int \mathscr{D}\phi \mathscr{D}\psi \mathscr{D}\chi e^{-S} W_a W_b \cdots . \tag{10}$$

Here, we will focus just on the bosonic map $\phi : \Sigma \to Y$. It turns out that in the topological sector, we want to restrict to maps such that the term $\{Q, V\}$ in the action vanishes. Looking at the form of V, we see that we must insist $\bar{\partial}\phi^i = \partial\phi^{\bar{i}} = 0$, i.e. ϕ is a holomorphic map. So instead of performing the path integral over *all* maps, we localize to only the holomorphic ones. In this context, such a holomorphic map is called a *worldsheet instanton*. We can consider the degree-d worldsheet instantons and their moduli space $\mathscr{M}_d$. For example, a degree-0 map simply sends all of Σ to a point in Y, implying $\mathscr{M}_0 = Y$. We get the following reduction of the path integral

$$\int \mathscr{D}\phi \mathscr{D}\psi \mathscr{D}\chi \longrightarrow \sum_d \int_{\mathscr{M}_d} (\mathscr{D}\phi)_d \int \mathscr{D}\psi \mathscr{D}\chi. \tag{11}$$

Since the relevant space of operators in the A-model is identified with the de Rham cohomology $H^*(Y, \mathbb{C})$, there is a natural grading by the degree of the forms. In physics, this is called the "ghost number," meaning if $a \in H^p(Y, \mathbb{C})$, then the operator W_a is said to have ghost number p. One should imagine the worldsheet instantons to be "wrapped" on the two-cycles in Y. Roughly speaking, this explains the dependence of the A-model on the Kähler structure of Y. The theory turns out to be independent of the complex structure.

1.2 Closed String B-model

If we perform the opposite twist we get the closed string B-model where certain fields are simply sections of different bundles over Σ. For purposes of anomaly cancellation, we will take $c_1(X) = 0$, i.e. take the target space to be Calabi–Yau. Define the following combinations of the fermionic fields, $\eta^{\bar{j}} = \psi_+^{\bar{j}} + \psi_-^{\bar{j}}$, and $\theta_j = g_{j\bar{k}}(\psi_+^{\bar{k}} - \psi_-^{\bar{k}})$ where now the fermionic fields are sections of the following bundles

$$\psi_\pm^{\bar{i}} \in \Gamma\left(\phi^* T_X^{(0,1)}\right), \quad \psi_+^i \in \Gamma\left(K \otimes \phi^* T_X^{(1,0)}\right), \quad \psi_-^i \in \Gamma\left(\overline{K} \otimes \phi^* T_X^{(1,0)}\right). \quad (12)$$

Let ρ^i be a one-form on Σ valued in $\phi^* T_X^{(1,0)}$ whose $(1, 0)$ part is ψ_+^i and $(0, 1)$ part is ψ_-^i. The B-model SUSY transformations are,

$$\begin{aligned} \delta\phi^i &= 0 \\ \delta\phi^{\bar{i}} &= i\alpha\eta^{\bar{i}} \\ \delta\eta^{\bar{i}} &= \delta\theta_i = 0 \\ \delta\rho^i &= -\alpha d\phi^i. \end{aligned} \quad (13)$$

The physical local observables are again given by products of BRST invariant fields,

$$W_A = A_{\bar{k}_1 \ldots \bar{k}_q}^{j_1 \ldots j_p} \eta^{\bar{k}_1} \ldots \eta^{\bar{k}_q} \theta_{j_1} \ldots \theta_{j_p}. \quad (14)$$

Clearly, such an object is a $(0, q)$-form, valued in the bundle $\bigwedge^p T_X^{(1,0)}$. Analogously to the A-model, we find that

$$\{Q, W_A\} = -W_{\bar{\partial} A}. \quad (15)$$

In other words, in the B-model the Q-cohomology is the Dolbeault cohomology on the target space $H^{0,q}(X, \bigwedge^p T_X^{(1,0)})$, with forms valued in an exterior power of the holomorphic tangent bundle.

Also, like the A-model, the path integral localizes to only certain maps ϕ, but in this case the condition is that $\bar{\partial}\phi^{\bar{k}} = \partial\phi^{\bar{k}} = 0$. This can only be satisfied if ϕ is a constant map from the worldsheet into X. Clearly, the moduli space of such maps is simply $\mathcal{M}_0 = X$. The upshot of this is that physical observables in the B-model are given simply by *ordinary* integrals over the target space. These are essentially the *period integrals* over the non-vanishing holomorphic $(3,0)$-form Ω. When considering mirror symmetry, people often say something like, "a hard computation on one side can be converted to a trivial computation on the other side." This idea applies here: on the A-side, correlation functions require a sum of integrals over non-trivial moduli spaces, while on the B-side, the computation reduces to simply period integrals. These period integrals are indicative of the dependence of the B-model on the complex structure of X as well as the independence of the Kähler structure.

1.3 Topological Field Theory Versus Topological String Theory?

It is a good time to rectify a common confusion between *topological field theories* and *topological string theories*. Simply put, we take a topological field theory to be such that there exists a subsector where the correlation functions are independent of the metric on the spacetime; in our case, the string worldsheet. The correlation functions are then given by a path integral over the bosonic fields ϕ^i and well as the fermionic fields, described in the previous section. However, we only implicitly mention a fixed metric $h_{\alpha\beta}$ on the string worksheet Σ itself. We certainly do not allow for dynamics of $h_{\alpha\beta}$, as it is not summed over in the path integral. Topological string theory arises from including the worldsheet metric as a dynamical field, which we include in the path integral prescription for correlation functions. We describe this as "coupling a topological field theory to worldsheet gravity." Thus, our correlation functions now involve a sum over the genus g of Σ, as well as an integral over the moduli space of complex structures on Σ. This should come as no surprise, since string theory is a theory of quantum gravity. Indeed, quantum gravity is by definition a quantum field theory where the metric on spacetime (in this case, the worldsheet) is dynamical and included in the path integral. The mathematically rigorous foundation of topological string theory is known as *Gromov-Witten theory*.

2 The Open String B-model

With the closed string theory in hand, we now endeavor to include open strings in the theory. This simply amounts to allowing the worldsheet Σ to have a boundary, denoted $\partial\Sigma$. These worldsheet boundaries have the interpretation of open string endpoints. Under the map $\phi : \Sigma \to X$, the image of $\partial\Sigma$ is required to live on certain special submanifolds of X called *D-branes*. One should interpret the

D-branes as providing boundary conditions on the open string endpoints: the endpoints are forced to lie *on* the D-brane (Dirichlet boundary conditions), while they are allowed to move freely *within* the D-brane itself (Neumann boundary conditions). In more physical language, we say that D-branes are non-perturbative solutions of an effective field theory. Interestingly enough, these non-perturbative solutions were actually expected for quite a long time. However, the true magic of their discovery [5] is that they allow for a two-dimensional analysis, via the open string worldsheet. This was quite exciting and unexpected. In other words, we expected some non-perturbative solutions to exist, but had no idea these objects would support open string endpoints.

We should immediately exorcise any confusions about the distinction between *boundaries* of Σ and *punctures* in Σ. With worldsheets involving only closed strings, the strings themselves are represented by "loops" stretching out to the infinite past or future. Using the conformal invariance of the worldsheet theory, we can map these to simply point-like punctures on the surface of Σ. In the path integral prescription, these punctures are superficially filled in to give a compact Riemann surface, at the expense of inserting a vertex operator at that point, representing the closed string state. Genuine *boundaries* of Σ are different, however. A boundary component of Σ is superficially partitioned by punctures. These punctures represent open string states stretching out to the infinite past or future, while the remaining segments of the boundary component are precisely what we think of as the open string endpoints "moving in time."

Let X be a Calabi–Yau threefold. To the roughest approximation, a *Dp-brane* in the context of topological string theory is a real p-dimensional submanifold of X, i.e. a representative of a class in $H_p(X, \mathbb{Z})$. The p refers to the spatial dimensions of the D-brane and hence, a Dp-brane has a $(p+1)$-dimensional worldvolume. The time direction lies outside the Calabi–Yau and plays essentially no role in a topological string theory.

A D-brane however, is much more than just a submanifold. As introduced above, D-branes support open string endpoints. Hence, these open string endpoints appear as "particle worldlines" in the $(p+1)$-dimensional worldvolume of the Dp-brane. Indeed there are good physical reasons to interpret this as the D-brane giving rise to a quantum field theory or gauge theory on its worldvolume. In the context of topological strings, we ignore the time direction and consider a gauge theory on simply the p-dimensional subspace of X. In a gauge theory on a spacetime Z, the physical fields are connections on, or sections of a vector bundle associated to a principal bundle defined on Z. Since the endpoints of open strings appear as gauge-theoretic particles in the D-brane, we are inclined to consider a D-brane as a submanifold along with a vector bundle supported on it. In the B-model, the objects are holomorphic, so we take the bundles to be holomorphic. Therefore as a first pass, we make the following naïve definition of a D-brane:

Naïve Definition 1 A single Dp-brane in the B-model topological string theory, for $p = 0, 2, 4, 6$ is a complex dimension $p/2$ holomorphic submanifold Z of a Calabi–Yau threefold X along with a holomorphic line bundle $L \to Z$.

It will soon become apparent that a stack of multiple D-branes will correspond to higher rank bundles. We would like to build the category of B-model D-branes such that the objects are defined on the ambient Calabi–Yau X. Under the natural inclusion $Z \hookrightarrow X$ we can pushforward holomorphic vector bundles to sheaves on X. Clearly such a pushforward is not a holomorphic vector bundle on X: vector bundles always have sections on small enough open sets, whereas this pushforward has no sections on any open set outside Z. We must broaden our consideration from merely the geometrical category of holomorphic vector bundles to the algebraic or sheaf-theoretic category of coherent sheaves. As we will see later, we actually must further enlarge our category. We will be compelled to understand B-model D-branes as *complexes* of coherent sheaves, modulo various equivalences. To explain these ideas we introduce now some of the required algebraic geometry.

2.1 Coherent Sheaves and D-branes

For some of the foundational algebraic geometry to follow, I refer the reader to [6, 7]. Let X be a compact, smooth complex manifold, or more generally a scheme, with $\mathcal{O}_X$ its structure sheaf of regular functions. We begin by defining a sheaf-theoretic generalization of the notion of a module over a ring. This is known as an $\mathcal{O}_X$-module, and is the largest category of sheaves we will need to consider. It contains as subcategories the coherent sheaves and locally-free sheaves, which we will introduce shortly.

Definition 1 For $\mathcal{E}$ a sheaf on X, we say $\mathcal{E}$ is an $\mathcal{O}_X$-module, if for all open sets $U \subseteq X$, the sections $\mathcal{E}(U)$ constitute an $\mathcal{O}_X(U)$-module. In addition, the restriction morphisms must be compatible with the module structure, in the following sense: consider nested open sets $V \subseteq U$ and define sections $f \in \mathcal{O}_X(U)$, $s \in \mathcal{E}(U)$. We require that $(f \cdot s)|_V = f|_V \cdot s|_V$, where we denote the restriction morphism as the familiar function restriction.

Notice that $\mathcal{O}_X$-modules are a generalization of modules over a ring. The intrinsic geometry of X gives rise to the structure sheaf $\mathcal{O}_X$ which naturally assigns a ring $\mathcal{O}_X(U)$ to each open set. It is precisely this ring of local functions which provides the multiplication, turning $\mathcal{E}(U)$ into an $\mathcal{O}_X(U)$-module. Hence, an $\mathcal{O}_X$-module is really a sheaf of modules. The $\mathcal{O}_X$-modules constitute an abelian category. This should come as no surprise given that abelian categories are in some sense modeled on the category of modules over a ring.

Trivially, $\mathcal{O}_X$ itself is an $\mathcal{O}_X$-module. More generally, $\mathcal{O}_X^{\oplus N}$ is an $\mathcal{O}_X$-module called 'the free $\mathcal{O}_X$-module of rank N.' A particularly refined subcategory of $\mathcal{O}_X$-modules is those which look locally like $\mathcal{O}_X^{\oplus N}$ for some N. This leads to the following definition, which will allow us to identify certain special $\mathcal{O}_X$-modules with holomorphic vector bundles.

Definition 2 A sheaf $\mathscr{E}$ on X is called locally-free of rank N if there exists an open cover $\{U_\alpha\}$ of X such that $\mathscr{E}|_{U_\alpha} \cong \mathscr{O}_X|_{U_\alpha}^{\oplus N}$.

One can show that locally-free sheaves of rank N form a category. Given that vector bundles trivialize over special open sets, locally-free sheaves seem to correspond exactly to holomorphic vector bundles. The correspondence is made precise by the following Proposition.

Proposition 1 *There exists a one-to-one correspondence between holomorphic vector bundles of rank N on X and locally-free sheaves of rank N on X.*

Proof The proof here is very elementary, and we only sketch it. Given a holomorphic vector bundle E on X, for all open sets U, define $\mathscr{E}(U)$ to be the sections of the vector bundle over U. Since the vector bundle must trivialize, this resulting sheaf will of course be locally-free. Conversely, given a locally-free sheaf $\mathscr{E}$, using the given isomorphism $\mathscr{E}|_{U_\alpha} \cong \mathscr{O}_X|_{U_\alpha}^{\oplus N}$, we can define holomorphic transition functions, which will produce a holomorphic vector bundle E.

Given holomorphic vector bundles E and F, we will usually denote their corresponding locally-free sheaves by $\mathscr{E}$ and $\mathscr{F}$, respectively.

D6-branes and Locally-Free Sheaves

In topological string theory on a Calabi–Yau threefold X, when we talk about "space-filling branes" we mean a D6-brane whose underlying homology class is a multiple of the fundamental class of X. Quite simply, D6-branes are in one-to-one correspondence with locally-free sheaves on X. This provides a translation between a precise mathematical notion and a phrase appearing frequently in the physics literature:

A stack of N D6-branes wrapping a Calabi–Yau threefold X corresponds to a rank N locally-free sheaf on X.

On a D6-brane, we specify purely Neumann boundary conditions, which allow the open string endpoint to move freely within X. This choice corresponds to the constraint

$$\theta_j = g_{j\bar{k}}(\psi_+^{\bar{k}} - \psi_-^{\bar{k}}) = 0. \tag{16}$$

Like we saw in the brief analysis of the closed string B-model, the BRST operator Q is taken to be the Dolbeault operator $\bar{\partial}$, and we only take our local operators on the worldsheet to consist of Q-closed local operators. Recalling the SUSY transformations (13), these are precisely θ_j and $\eta^{\bar{j}}$. But the space-filling condition forces the θ_j to vanish, so our local operators will only depend on $\eta^{\bar{j}}$, and of course ϕ. Thus, since $\bar{j}$ is an anti-holomorphic index, we conclude that our local operators must be $(0, q)$-forms, possibly valued in some bundle.

Let us attempt to construct a well-defined D-brane category, assuming at first that the only objects are D6-branes. By the above correspondence, the objects are simply given by a bundle $E \to X$. To give a pair of objects, is to give a pair of bundles on

X, $E_1 \to X$ and $E_2 \to X$. Since these are bundles over the same base manifold, we can define $\mathrm{Hom}(E_1, E_2)$ to be the bundle morphisms between them. It will be useful to note here that $\mathrm{Hom}(E_1, E_2) \simeq E_1^* \otimes E_2$ is itself a vector bundle with fiber defined as $\mathrm{Hom}(E_1, E_2)(x) = \mathrm{Hom}(E_1(x), E_2(x))$, for all $x \in X$.

We then take our local operators representing an open string state to be W_A, where A is a $(0, q)$-form valued in the bundle $\mathrm{Hom}(E_1, E_2)$. Therefore, it is natural to define the morphisms from $E_1 \to X$ to $E_2 \to X$ (equivalently the open string states stretching from one D6-brane to the other), to be the Dolbeault cohomology group

$$H^{0,q}_{\bar{\partial}}\big(X, \mathrm{Hom}(E_1, E_2)\big).$$

by the familiar Čech-Dolbealt isomorphism, the Dolbeault cohomology group above is isomorphic to Čech cohomology

$$H^{0,q}_{\bar{\partial}}\big(X, \mathrm{Hom}(E_1, E_2)\big) \simeq \check{H}^q\big(X, \mathscr{H}om(\mathscr{E}_1, \mathscr{E}_2)\big), \tag{17}$$

where $\mathscr{E}_1$ and $\mathscr{E}_2$ are the locally-free sheaves corresponding to the vector bundles E_1 and E_2. In the B-model, specifically in the case of space-filling branes, we can unambiguously assign a 'ghost number' q to an open string. We will see that this will be less clean when considering branes of non-zero codimension.

As a simple example, we can compute a three-point correlator of open string states [3]. Consider three D6-branes corresponding to holomorphic vector bundles E_1, E_2, and E_3. Let us call the three local operators W_A, W_B, and W_C, where

$$A \in H^{0,1}_{\bar{\partial}}(X, \mathrm{Hom}(E_1, E_2)), \quad B \in H^{0,1}_{\bar{\partial}}(X, \mathrm{Hom}(E_2, E_3)), \quad C \in H^{0,1}_{\bar{\partial}}(X, \mathrm{Hom}(E_3, E_1)). \tag{18}$$

Recall that in the B-model, since instantons are suppressed, the correlation functions are given simply by integrals over X. Indeed, the path integrals in the topological sector include only contributions from the moduli space $\mathscr{M}_0$ of degree zero harmonic maps into X. But of course, these are simply constant maps, and $\mathscr{M}_0 = X$. This implies,

$$\langle W_A W_B W_C \rangle = \int_X \mathrm{Tr}(A \wedge B \wedge C) \wedge \Omega. \tag{19}$$

The 'integrand' is a (3, 3)-form, which is natural to integrate over a threefold. Of course, when wedging forms valued in the bundle $\mathrm{Hom}(E_i, E_j)$, we implicitly compose the morphisms.

The Mukai Vector and D-Brane Charges

We have seen that when considering only D6-branes on a threefold, it sufficed to model them as objects in the category of locally-free sheaves. The goal of this section is to gently acquaint the reader with some of the more general coherent sheaves needed to formalize D4, D2, and D0-branes. For a rigorous definition of coherent sheaves, see [6]. For my purposes, it will suffice to think very roughly of coherent sheaves as the minimal, full abelian category arising as the "completion" of the category of locally-free sheaves upon adding all kernels and cokernels.

Naïve Definition 2 In the B-model topological string on a Calabi–Yau threefold X, a D-brane corresponds to a coherent sheaf $\mathscr{F}$ on X. The support of the sheaf $\mathrm{supp}(\mathscr{F})$ defines the underlying homology class of the D-brane.

Branes need not be pure dimensional. For example, a coherent sheaf $\mathscr{F}$ can be supported on curves and points. We interpret such an $\mathscr{F}$ as a *bound state* of D0-D2 branes. Such brane configurations occur, for example, in Donaldson–Thomas theory. A helpful device for guiding intuition here is the *Mukai vector* or equivalently, the *D-brane charges* [8] associated to a coherent sheaf.

Definition 3 Let X be a smooth n-dimensional variety and let $\mathscr{F}$ be a coherent sheaf on X. The Mukai vector is defined to be

$$v(\mathscr{F}) = \mathrm{ch}(\mathscr{F})\sqrt{\mathrm{td}(X)} = (v_0, \ldots, v_n) \in H^{2*}(X, \mathbb{Q}). \tag{20}$$

If X is also projective, then the D-brane charge is given simply by the Poincaré dual of the Mukai vector[2]

$$\mathscr{Q}(\mathscr{F}) = \mathrm{PD}\big(\mathrm{ch}(\mathscr{F})\sqrt{\mathrm{td}(X)}\big) \in H_{2*}(X, \mathbb{Q}). \tag{21}$$

By convention, we order the charges as $\mathscr{Q}(\mathscr{F}) = (\mathscr{Q}_n, \ldots, \mathscr{Q}_0)$, where $\mathscr{Q}_i \in H_{2i}(X, \mathbb{Q})$ is called the D$2i$-charge.

Recall that based on the naïve definition, we concluded that the coherent sheaves which most directly correspond to physical D-branes are pushforwards of holomorphic vector bundles along inclusions.[3] Let X be an n-dimensional smooth, projective variety and let $\iota : Z \hookrightarrow X$ be the inclusion of the m-dimensional subvariety Z into X. In addition, let E be a rank N holomorphic vector bundle on Z.

Lemma 1 *Given X, Z, and E as described above, we have*

$$\begin{aligned} ch_k(\iota_* E) &= 0, \quad \textit{for all } k < n - m, \\ PD\big(ch_{n-m}(\iota_* E)\big) &= N[Z] \in H_{2m}(X, \mathbb{Q}). \end{aligned} \tag{22}$$

Proof This is a straightforward computation which can be found, for example, in [10].

This simple result about the Chern character of pushforwards of vector bundles, immediately implies the following corollary about the D-brane charges.

---

[2] In [9], the authors introduce 'gamma classes' which encode corrections to the factor of $\sqrt{\mathrm{td}X}$.

[3] This is not *quite* true. Due to a phenomenon related to the *Freed-Witten anomaly*, one must also tensor by $K_Z^{-1/2}$ where K_Z is the canonical bundle of Z. There is a nice discussion of this in [2, 3].

Corollary 1 *Again, given X, Z, and E as above, the D-brane charges satisfy*

$$\begin{aligned} \mathcal{Q}_k(\iota_* E) &= 0, \quad \text{for all } k > n-m, \\ \mathcal{Q}_m(\iota_* E) &= N[Z] \in H_{2m}(X, \mathbb{Q}). \end{aligned} \tag{23}$$

Proof Using the Lemma, we note that $\left(\mathrm{ch}(\iota_* E)\sqrt{\mathrm{td}(X)}\right)_k = 0$ for all $k < n - m$. Poincaré dualizing, this shows that all entires in the D-brane charge vanish for $k > n - m$, thus proving the first claim. Note that $\left(\sqrt{\mathrm{td}(X)}\right)_0 = 1$, and so

$$\left(\mathrm{ch}(\iota_* E)\sqrt{\mathrm{td}(X)}\right)_{n-m} = \mathrm{ch}_{n-m}(\iota_* E). \tag{24}$$

By Poincaré dualizing and applying the Lemma once more, the second claim follows.

This corollary provides a precise mathematical translation of a phrase, prevalent in the physics literature, generalizing one made earlier about D6-branes and locally-free sheaves:

In physics, one often hears about "a stack of N D-branes wrapping a holomorphic cycle $Z \subseteq X$." Mathematically, this corresponds to a rank N holomorphic vector bundle on Z.

Let us introduce now a few of the familiar coherent sheaves one might encounter on a Calabi–Yau threefold X. It has been previously observed that D6-branes correspond to locally-free sheaves. In non-zero codimension, D4, D2, and D0-branes correspond to *torsion sheaves*. A torsion sheaf is a coherent sheaf $\mathscr{F}$ of *rank zero*, which is encoded into the Mukai vector as $v_0 = 0$, or equivalently into the D-brane charges as $\mathcal{Q}_3 = 0$.

Let Z be a holomorphic subvariety of X. This gives rise to a short exact sequence

$$0 \to \mathscr{I}_Z \to \mathscr{O}_X \to \mathscr{O}_Z \to 0, \tag{25}$$

where $\mathscr{O}_Z$ is the structure sheaf on Z and $\mathscr{I}_Z$ is called an *ideal sheaf*. In algebraic geometry, an ideal sheaf on X is a rank one torsion-free sheaf $\mathscr{I}_Z$ with trivial determinant. There is necessarily an injective sheaf morphism $\mathscr{I}_Z \to \mathscr{O}_X$ and the cokernel defines a subscheme $Z \subseteq X$ along with the short exact sequence above. If Z is a divisor, then $\mathscr{I}_Z$ is actually a line bundle, and $\mathscr{O}_Z$ is an example of a D4-brane. If Z is supported only on curves and points, then $\mathscr{O}_Z$ indeed corresponds to D2 or D0-branes, as expected. However, in that case $\mathscr{I}_Z$ is a rank one torsion-free sheaf which is not locally-free.

Ideal sheaves have no immediate interpretation as D-branes. However, notice that because the Chern character is additive on short exact sequences, applying $\mathcal{Q}$ to (25), the D-brane charges are seen to satisfy

$$\mathcal{Q}(\mathscr{O}_X) = \mathcal{Q}(\mathscr{I}_Z) + \mathcal{Q}(\mathscr{O}_Z), \tag{26}$$

which looks like a manifestation of charge conservation. This is perhaps hinting that an ideal sheaf may have an interpretation as a bound-state of a brane ($\mathcal{O}_X$) and a suitably defined anti-brane ($\mathcal{O}_Z$) coupled via a map $\mathcal{O}_X \to \mathcal{O}_Z$.

2.2 Summary and Outlook

Roughly speaking, one may think of the category of coherent sheaves $\mathrm{Coh}(X)$ as containing all of the locally-free sheaves on X, plus all of the ideal sheaves, structure sheaves, and pushforwards of sheaves arising from holomorphic vector bundles on subvarieties. Thus, if we want to expand beyond the world of vector bundles, considering the coherent sheaves is the most natural first step. We hope to argue that the derived category $D^b\mathrm{Coh}(X)$ will be large enough to account for all B-model D-branes at this point. In the following section we will introduce some of the machinery of homological algebra and sheaf cohomology. There are at least two indications so far that such machinery should be important.

Recall that we have only done one computation in this section: in the case of two D6-branes, we computed the spectrum of open string states stretching between the branes. Here we used that D6-branes correspond to vector bundles $E_1 \to X$ and $E_2 \to X$, and since they share a common base space, the group of morphisms $\mathrm{Hom}(E_1, E_2)$ was well-defined. But in higher codimension, branes need not intersect, and certainly will not be given simply by a locally-free sheaf. For example, we can have a brane supported on a divisor, and another supported on a curve with open strings stretching between. Or we can have a stack of N D0-branes supporting open string endpoints. In this setting, it is natural to expect the *Ext Groups* to encode the open string spectra, as they are a natural generalization of bundle morphisms.

In addition, the ideal sheaf short exact sequence we encountered is perhaps hinting that we should consider *complexes* of coherent sheaves. We saw that the application of the D-brane charge $\mathcal{Q}$ to such a short exact sequence seems to encode a charge conservation. The physical BRST formalism provides a natural grading by the ghost number, so we can consider a D-brane as a direct sum, graded by the ghost number. Turning on VEVs for a tachyon field, will deform this direct sum to a genuine complex. The second motivation to consider complexes, comes from the general philosophy of resolutions. It's often beneficial to replace an arbitrary element of a category by a tower of "pleasant" objects. In other words, you have resolved the object by a complex of nice objects. The coherent sheaves we find to be particularly pleasant are the locally-free sheaves associated to space-filling branes. Given a coherent sheaf which is not locally-free (coming from a D0-, D2-, or D4-brane) we can find a locally-free resolution.

Once in the category of complexes of coherent sheaves, the glaring question is, are there physical reasons to identify complexes up to homotopy and quasi-isomorphism? Remarkably, the answer is conjecturally, yes. Identifying homotopic maps between D-branes will be natural from the BRST formalism. We will interpret quasi-isomorphic complexes to be in the same "universality class" of

Renormalization Group flow on the worldsheet. Moreover, we can realize this flow as brane/anti-brane annihilation via a non-zero tachyon VEV.

3 Sheaf Cohomology, Derived Functors, and Ext Groups

We begin with a few remarks pertaining to the global sections of a sheaf. We assume the reader is familiar with Čech cohomology.

Remark 1 Given a sheaf $\mathscr{F}$ on X, the zeroth Čech cohomology group computes the global sections,

$$\Gamma(X, \mathscr{F}) \simeq \check{H}^0(X, \mathscr{F}).$$

Remark 2 Given an $\mathscr{O}_X$-module $\mathscr{F}$, the global sections of $\mathscr{F}$ correspond to morphisms $\mathscr{O}_X \to \mathscr{F}$,

$$\Gamma(X, \mathscr{F}) \simeq \mathrm{Hom}(\mathscr{O}_X, \mathscr{F}).$$

We should also record the familiar isomorphism between Čech cohomology and Dolbeault cohomology.

Remark 3 Let Ω^p be the sheaf of holomorphic p-forms on X. The Čech-Dolbeault isomorphism states that

$$H^{p,q}_{\bar{\partial}}(X) \simeq \check{H}^q(X, \Omega^p). \tag{27}$$

More generally, we can let E be a holomorphic vector bundle on X, with corresponding locally-free sheaf $\mathscr{E}$. The generalized Čech-Dolbeault isomorphism relates (p, q)-forms valued in E to the sheaf $\mathscr{E} \otimes \Omega^p$

$$H^{p,q}_{\bar{\partial}}(X, E) \simeq \check{H}^q(X, \mathscr{E} \otimes \Omega^p). \tag{28}$$

One important idea will be that of *resolutions*. The general philosophy of resolutions is that given an arbitrary object A in some category, it might be preferable to replace A by a tower of especially pleasant objects in the category. One often speaks of injective, projective, flasque/flabby, or free resolutions. These focus our attention on especially nice, or rigid objects in the category. This provides a way of defining *derived functors* which can be evaluated at such arbitrary objects A. This can be done in some generality in the category of R-modules over a ring R. However, we will focus on the category of $\mathscr{O}_X$-modules. In this category, using resolutions to define derived functors will immediately give a definition of sheaf cohomology. This sheaf cohomology is extremely abstract, so is not terribly helpful in explicit computations, but it agrees with Čech cohomology, and will allow for the definition of the Ext groups.

Given an injective resolution of some object A,

$$0 \longrightarrow A \longrightarrow \mathscr{I}_0 \longrightarrow \mathscr{I}_1 \longrightarrow \dots \tag{29}$$

and a left-exact functor F, we get a complex

$$0 \longrightarrow F(\mathscr{I}_0) \longrightarrow F(\mathscr{I}_1) \longrightarrow F(\mathscr{I}_2) \longrightarrow \dots \tag{30}$$

We define the nth right derived functor of F at A, denoted $\mathbf{R}^n F(A)$, to be the nth cohomology of the above sequence. From here on, we will restrict attention to the category of $\mathscr{O}_X$-modules, with the primary left-exact functor of interest being $\mathrm{Hom}(\mathscr{O}_X, -)$. By an earlier remark, we may also refer to $\mathrm{Hom}(\mathscr{O}_X, -)$ as the *global section functor* since acting on any sheaf results in the group of global sections. We arrive finally at the important definition of sheaf cohomology

Definition 4 We define sheaf cohomology for $\mathscr{O}_X$-modules to be the right derived functor of the left-exact global sections functor $\mathrm{Hom}(\mathscr{O}_X, -)$. Given an $\mathscr{O}_X$-module $\mathscr{F}$, then the nth sheaf cohomology group of $\mathscr{F}$ is

$$H^n(X, \mathscr{F}) = \boldsymbol{R}^n \mathrm{Hom}(\mathscr{O}_X, -)(\mathscr{F}). \tag{31}$$

The most pressing point to be made after a definition using derived functors, is that the result is *independent* of the particular resolution we chose. Resolutions of a given object are generally far from unique, and it would clearly be problematic if we got a different result for sheaf cohomology depending on which resolution we chose; this is not the case.

Remark 4 The 0th sheaf cohomology group computes the global sections of the sheaf. In particular, it agrees with Čech cohomology.

Proof The proof here is very straightforward. In general for a right derived functor, we have $\mathbf{R}^0 F(A) = F(A)$, since the functor F is left-exact. Using this, we compute

$$H^0(X, \mathscr{F}) = \mathbf{R}^0 \mathrm{Hom}(\mathscr{O}_X, -)(\mathscr{F}) = \mathrm{Hom}(\mathscr{O}_X, \mathscr{F}) = \mathscr{F}(X) \tag{32}$$

The definition of sheaf cohomology using derived functors is quite abstract. In fact, it's so abstract, that it's essentially immune to computations. However, this same abstraction makes it incredibly elegant to use in theory building. When needed for actual computations, the best approach is to prove that it is isomorphic to something like Čech cohomology which is far more computable.

Theorem 1 *Given an $\mathscr{O}_X$-module $\mathscr{F}$, the Čech and sheaf cohomologies are isomorphic,*

$$H^n(X, \mathscr{F}) \simeq \check{H}^n(X, \mathscr{F}). \tag{33}$$

Proof I refer the reader to Theorem 4.5 in [6].

We have defined sheaf cohomology for arbitrary $\mathcal{O}_X$-modules, however we can simply restrict attention to the coherent sheaves if we like. The reason being, the coherent sheaves are a *full* subcategory, meaning the morphisms are the same in the subcategory, as they are in the original category. In fact, in applications to D-branes, we will usually regard the locally-free sheaves as the particularly nice objects within the category of coherent sheaves. Thus we'll want to take an arbitrary coherent sheaf, and resolve it using a tower of locally-free sheaves. First, we must introduce the Ext groups.

Definition 5 Let $\mathcal{E}$ be an $\mathcal{O}_X$-module. The functor $\mathrm{Hom}(\mathcal{E}, -)$ is left-exact, so we may consider its right derived functor evaluated at an $\mathcal{O}_X$-module $\mathcal{F}$. This allows for the following definition of the Ext groups,

$$\mathrm{Ext}^n(\mathcal{E}, \mathcal{F}) = \boldsymbol{R}^n\mathrm{Hom}(\mathcal{E}, -)(\mathcal{F}). \tag{34}$$

There are a few simple examples where the Ext groups correspond to familiar quantities:

$$\mathrm{Ext}^0(\mathcal{E}, \mathcal{F}) = \mathbf{R}^0\mathrm{Hom}(\mathcal{E}, -)(\mathcal{F}) = \mathrm{Hom}(\mathcal{E}, \mathcal{F}), \tag{35}$$

and

$$\mathrm{Ext}^n(\mathcal{O}_X, \mathcal{F}) = \mathbf{R}^n\mathrm{Hom}(\mathcal{O}_X, -)(\mathcal{F}) = H^n(X, \mathcal{F}). \tag{36}$$

Thus, we see that Ext groups at least encode abelian groups of sheaf morphisms and sheaf cohomology groups. In addition, we have the following useful result, known as Serre duality.

Theorem 2 *In the case where X is a Calabi–Yau m-fold, for all $n = 0, \ldots, m$*

$$Ext^n(\mathcal{E}, \mathcal{F}) \simeq Ext^{m-n}(\mathcal{F}, \mathcal{E}). \tag{37}$$

Consider two holomorphic vector bundles E and F on X, with corresponding locally-free sheaves $\mathcal{E}$ and $\mathcal{F}$. The space of vector bundle morphisms $\mathrm{Hom}(E, F)$ is actually itself a vector bundle, with fiber defined by $\mathrm{Hom}(E, F)(x) = \mathrm{Hom}(E(x), F(x))$, for all $x \in X$. Since $\mathrm{Hom}(E, F)$ is a holomorphic vector bundle, there exists a corresponding locally-free sheaf which we denote as $\mathcal{H}om(\mathcal{E}, \mathcal{F})$. It is very easy to get mixed up here with the notation, so we briefly summarize,

$$\begin{aligned} \mathrm{Hom}(E, F) &= \text{the holomorphic vector bundle of bundle morphisms} \\ \mathcal{H}om(\mathcal{E}, \mathcal{F}) &= \text{locally-free sheaf associated to } \mathrm{Hom}(E, F) \\ \mathrm{Hom}(\mathcal{E}, \mathcal{F}) &= \text{abelian group of sheaf morphisms from } \mathcal{E} \text{ to } \mathcal{F}. \end{aligned}$$

Moreover, $\mathrm{Hom}(\mathcal{E}, \mathcal{F})$ is actually the abelian group of global sections of $\mathcal{H}om(\mathcal{E}, \mathcal{F})$. By the Čech-Dolbeault-Sheaf isomorphism,

$$H^{0,q}_{\bar{\partial}}(X, \mathrm{Hom}(E, F)) \cong \check{H}^q(X, \mathscr{H}om(\mathscr{E}, \mathscr{F})) \cong H^q(X, \mathscr{H}om(\mathscr{E}, \mathscr{F})). \quad (38)$$

Since $\mathbf{R}^0\mathrm{Hom}(\mathscr{O}_X, -)(\mathscr{F}) = \mathrm{Hom}(\mathscr{O}_X, \mathscr{F})$ gives the global sections of $\mathscr{F}$, and $\mathrm{Ext}^q(\mathscr{O}_X, \mathscr{F}) \cong H^q(X, \mathscr{F})$, we can conclude analogously that,

$$\mathrm{Ext}^q(\mathscr{E}, \mathscr{F}) \cong H^q(X, \mathscr{H}om(\mathscr{E}, \mathscr{F})) \cong H^{0,q}_{\bar{\partial}}(X, \mathrm{Hom}(E, F)) \quad (39)$$

With this, we have finally converted all complex geometry of D-branes into algebraic and sheaf-theoretic language. For X a Calabi–Yau threefold, we conclude from (39) and (17),

Given two stacks of D6-branes in the B-model with associated locally-free sheaves $\mathscr{E}$ and $\mathscr{F}$, the open strings states stretching from $\mathscr{E}$ to $\mathscr{F}$ with ghost number q, are given by the abelian group $\mathrm{Ext}^q(\mathscr{E}, \mathscr{F})$.

4 The Derived Category and Complexes of D-branes

For the time being, I would like to restrict attention to D6-branes modeled as locally-free sheaves, as opposed to more general coherent sheaves. We saw above that given two stacks of D6-branes wrapping a Calabi–Yau threefold X with corresponding holomorphic vector bundles E and F, we can ask about the morphisms between them. These were shown to be given by the Dolbeault cohomology $H^{0,q}_{\bar{\partial}}(X, \mathrm{Hom}(E, F))$ or equivalently, $\mathrm{Ext}^q(\mathscr{E}, \mathscr{F})$. We identify each morphism with a string state, and the *ghost number* or R-charge of the string corresponds to q. Mathematically, we can think of this q as providing a natural $\mathbb{Z}$-grading. Given a D-brane with holomorphic vector bundle E (or locally-free sheaf $\mathscr{E}$), we can consider all strings attached to it as being graded by an integer. Thus, it seems natural to initially consider direct sums of locally-free sheaves on X,

$$\mathscr{E} = \bigoplus_{n \in \mathbb{Z}} \mathscr{E}^n. \quad (40)$$

The above direct sum can trivially regarded as a complex with all maps being zero

$$\mathscr{E}^\bullet = \left(\dots \xrightarrow{0} \mathscr{E}^{-1} \xrightarrow{0} \mathscr{E}^0 \xrightarrow{0} \mathscr{E}^1 \xrightarrow{0} \dots\right). \quad (41)$$

Simply put, we want to deform away from the trivial case of direct sums by turning on non-zero maps between the $\mathscr{E}^i$ in the above sequence. These non-zero maps will be called *tachyons* for reasons to be explained shortly. Once we do this, the D-branes will correspond to elements in the category of complexes $\mathbf{Kom}(\mathscr{C})$, where $\mathscr{C}$ is the category of locally-free sheaves on X. However, physically, the string states correspond to elements in Q-cohomology, so we need to identify all states differing by a Q-exact terms. Remarkably, this identification on the physics side, corresponds precisely to identifying complexes up to homotopy in $\mathbf{Kom}(\mathscr{C})$. This places the B-model D-branes in correspondence with the homotopy category $\mathbf{K}(\mathscr{C})$.

This correspondence is certainly elegant, but there is a fundamental problem here. Most importantly, the homological algebra described just above requires that the category be *abelian*. The category of locally-free sheaves is additive, but not abelian. The resolution here will be to extend our consideration to the category of coherent sheaves $\mathrm{Coh}(X)$, which is an abelian category containing the category of locally-free sheaves $\mathscr{C}$. This seemingly dangerous problem was actually hinting that we weren't considering all of the branes that we need to. As we saw in an earlier section, the locally-free sheaves cannot describe D4, D2, nor D0-branes; these require torsion sheaves. Thus, extending to coherent sheaves is well-motivated both mathematically and physically.

Given the homotopy category $\mathbf{K}\mathrm{Coh}(X)$ of coherent sheaves, it is tempting to identify quasi-isomorphisms and arrive at the (bounded) derived category $D^b\mathrm{Coh}(X)$. But is there any physical motivation for this? Indeed, we will argue that two complexes which are quasi-isomorphic lie in the same universality class of the renormalization group flow. In other words, one complex can be thought of as *condensing* to another [11]. This explains the use of the term tachyon: in string theory, a tachyon is a particle which signifies an instability. This instability corresponds to the branes in a complex annihilating each other.

4.1 Deformation of Complexes

Let us begin by considering a stack of D6-branes on a threefold X given by a holomorphic vector bundle E (with associated sheaf $\mathscr{E}$), which decomposes as the direct sum

$$\mathscr{E} = \bigoplus_{n\in\mathbb{Z}} \mathscr{E}^n, \tag{42}$$

where each $\mathscr{E}^i$ is a locally-free sheaf on X. The open string states from $\mathscr{E}$ to itself, correspond to linear combinations of elements in $\mathrm{Ext}^*(\mathscr{E}, \mathscr{E})$. For all n, k the string states with ghost number q correspond to elements of $\mathrm{Ext}^k(\mathscr{E}^n, \mathscr{E}^{n-k+q})$. For example, when $k = 1$, the string with ghost number $q = 1$ correspond to elements in $\mathrm{Ext}^1(\mathscr{E}^n, \mathscr{E}^n)$, which describe deformations of the locally-free sheaf $\mathscr{E}$ associated to the vector bundle E. The more pressing case to consider is $k = 0$. Here, the ghost number $q = 1$ strings are elements of $\mathrm{Ext}^0(\mathscr{E}^n, \mathscr{E}^{n+1}) \simeq \mathrm{Hom}(\mathscr{E}^n, \mathscr{E}^{n+1})$. Let us define $d = \sum d_n \in \mathrm{Hom}(\mathscr{E}, \mathscr{E})$, where

$$d_n \in \mathrm{Hom}(\mathscr{E}^n, \mathscr{E}^{n+1}). \tag{43}$$

Thus, d_n is a holomorphic map from $\mathscr{E}^n$ to $\mathscr{E}^{n+1}$. We can use d to deform the physical sigma model action

$$\delta S = \oint_{\partial\Sigma} (\psi_+^i + \psi_-^i)\partial_i d, \tag{44}$$

and then prove that deforming the action by δS, requires a deformation of the BRST operator as well

$$Q = Q_0 + d. \tag{45}$$

We need to retain the nilpotence $Q^2 = 0$ of the BRST operator, which leads to the constraint

$$\{Q_0, d\} + d^2 = 0. \tag{46}$$

The two terms above must individually vanish. The constraint $\{Q_0, d\} = 0$ is merely the statement that d is a holomorphic map, recalling that in the B-model, the undeformed BRST operator is $Q_0 = \bar{\partial}$. The condition $d^2 = 0$ can be expanded in terms of successive maps, and we see that

$$d_{n+1}d_n = 0. \tag{47}$$

Thus, the nilpotence of the deformed BRST operator Q is translated into the conditions that each d_n must be a holomorphic map, and the consecutive application of two successive maps must vanish. That is to say $\mathscr{E}$ is deformed into the complex

$$\mathscr{E}^\bullet = \left(\dots \xrightarrow{d_{n-1}} \mathscr{E}^n \xrightarrow{d_n} \mathscr{E}^{n+1} \xrightarrow{d_{n+1}} \mathscr{E}^{n+2} \xrightarrow{d_{n+2}} \dots\right) \tag{48}$$

Consider now the slightly more general case of open strings stretching from a stack of D6-branes $\mathscr{E}^\bullet$ to another stack of D6-branes $\mathscr{F}^\bullet$. Let both $\mathscr{E}^\bullet$ and $\mathscr{F}^\bullet$ consist of a collection of objects (graded by ghost number) constituting a trivial complex; namely, $\mathscr{E}^\bullet$ and $\mathscr{F}^\bullet$ decompose into direct sums. We deform the theory by turning on the differentials $d^{\mathscr{E}}$ and $d^{\mathscr{F}}$, yielding two non-trivial complexes. The deformed BRST operator can be shown to be

$$Q = Q_0 + d^{\mathscr{E}} - d^{\mathscr{F}}. \tag{49}$$

Let $f^n : \mathscr{E}^n \to \mathscr{F}^n$ be a collection of maps from the elements of the complex $\mathscr{E}^\bullet$, to the complex $\mathscr{F}^\bullet$. These should be thought of intuitively as strings stretching from $\mathscr{E}^n$ to $\mathscr{F}^n$. What are the conditions that the map of complexes f is BRST invariant? We require

$$Qf^n = Q_0 f^n + f^{n+1}d^{\mathscr{E}} - d^{\mathscr{F}} f^n = 0. \tag{50}$$

Like above, this factors into two independent constraints. First, we require $Q_0 f^n = 0$ for all n. That is to say f^n is a holomorphic map, $f^n \in \text{Hom}(\mathscr{E}^n, \mathscr{F}^n)$. The second condition is that $f^{n+1}d^{\mathscr{E}} = d^{\mathscr{F}} f^n$. This is precisely what it means for f to define a morphism of complexes. Moreover, if two such maps f and f' differ by a Q-exact term,

$$f' = f + Qh, \tag{51}$$

then we see that f and f' are homotopic morphisms of complexes. Therefore, quotienting by homotopy equivalence is the mathematical manifestation of passing to Q-cohomology, which is well-motivated physically. In order that a map $f : \mathscr{E}^\bullet \to \mathscr{F}^\bullet$ be a genuine morphism of complexes, we require that f be BRST invariant (Q-closed) which is precisely the physical notion of corresponding to an allowed open string state. Moreover, two such states f and f' are deemed physically equivalent if and only if they differ by a Q-exact term, and this coincides with the definition of f and f' being homotopic chain maps! We conclude that the homotopy category $\mathbf{K}(\mathscr{C})$ naturally models stacks of D6-branes in the B-model.

4.2 Renormalization Group (RG) Flow and Quasi-Isomorphisms

Given the homotopy category $\mathbf{K}(\mathscr{C})$, it is clearly tempting to ask if there is any physical motivation to identify quasi-isomorphisms, landing us once and for all in the derived category. I hope to outline the state-of-the-art conjecture that D-branes in the B-model related by a quasi-isomorphism correspond to physical configurations related by worldsheet renormalization group (RG) flow; in some loose sense, the branes and anti-branes at least partially annihilate.

Branes, Anti-branes, and Tachyons

First, we introduce some terminology. Given a D-brane represented as a complex,

$$\mathscr{E}^\bullet = \left(\dots \xrightarrow{d_{n-1}} \mathscr{E}^n \xrightarrow{d_n} \mathscr{E}^{n+1} \xrightarrow{d_{n+1}} \mathscr{E}^{n+2} \xrightarrow{d_{n+2}} \dots\right) \tag{52}$$

we consider the entries of the complex to be alternating branes and anti-branes, and we call the maps d_n *tachyons*. In string theory, a tachyon indicates an instability in a physical system. Indeed, here we mean that non-zero tachyons d_n lead to an instability of the configuration of D-branes. In the trivial case where all d_n vanish, the system appears to be in a stable state, but with non-zero tachyons, the configuration may flow via the renormalization group to a more stable system.

In physics, two theories related by renormalization group flow are said to lie in the same *universality class*. The conjecture here is that the physical universality classes correspond to the equivalence classes of quasi-isomorphisms. It's crucial to note that two theories in the same universality class, are not equivalent physical theories. RG flow represents a flow in the "space of theories" to a completely different physical theory.

We seem to have argued that the category of D-branes in the B-model topological string is the (bounded) derived category $D^b(\mathscr{C})$ of *locally-free sheaves* on X. However, as mentioned earlier, the locally-free sheaves are not an abelian category. The natural guess is to pass to the coherent sheaves which are essentially the abelianization of the locally-free sheaves. Indeed, we have seen that a D-brane naïvely corresponds to a coherent sheaf, making this extension reasonable. Therefore, the conjectural conclusion provided by [2, 3] is,

The category of D-branes in the B-model topological string is the (bounded) derived category $D^b\mathrm{Coh}(X)$ of coherent sheaves on a Calabi–Yau threefold X. Quotienting by homotopy corresponds to identifying states up to Q-exact terms. Quasi-isomorphism corresponds to worldsheet RG flow and brane/anti-brane annihilation.

This is fundamentally built on the original conjecture of Kontsevich [1] relating homological algebra and mirror symmetry.

5 Examples

Example 1: Elementary Brane/Anti-Brane Annihilation
The following example is as simple as it gets, but it illustrates well all of the features of the discussion above. Consider the following complex,

$$\ldots \longrightarrow 0 \longrightarrow E \xrightarrow{\;c\;} E \longrightarrow 0 \longrightarrow \ldots \tag{53}$$

where E can be any coherent sheaf supported on a subvariety Z of X. If the above map c is identically zero, then we essentially have two copies of E which do not couple in any sense, and the complex decomposes into a direct sum $E \oplus E$. This can be thought of as two stacks of D-branes wrapped on Z which do not interact. If, however, we turn on the map $c \neq 0$, physically, we have added a VEV for a tachyon field, which indicates an unstable coupling between the branes. The E on the left represents an anti-brane while the E on the right represents a brane. Intuitively, we physically expect the branes to annihilate due to this instability. In other words, this sequence should be in the same universality class as the zero complex. Indeed, in this case this sequence is quasi-isomorphic to its cohomology, which is simply zero in every entry, for $c \neq 0$. And so the intuition is verified: the unstable configuration is quasi-isomorphic to the zero complex, signifying brane/anti-brane annihilation.

Example 2: D4-branes
Let $Z \subseteq X$ be a codimension one complex subvariety of X, i.e. a divisor. This means Z is cut out by a section of a line bundle $\mathcal{O}(-Z)$, which is given locally by the vanishing of a holomorphic function f. Since $\mathcal{O}(-Z)$ consists simply of all holomorphic functions vanishing identically on Z, it naturally injects into $\mathcal{O}_X$. The cokernel of this map $\mathcal{O}(-Z) \to \mathcal{O}_X$ is simply the structure sheaf of Z, denoted $\mathcal{O}_Z$. Thus, we have the following short exact sequence of sheaves,

$$0 \longrightarrow \mathcal{O}(-Z) \xrightarrow{\;f\;} \mathcal{O}_X \longrightarrow \mathcal{O}_Z \longrightarrow 0. \tag{54}$$

We can reinterpret this exact sequence. Let us define the complex (*not* exact sequence) $\mathscr{E}^\bullet$,

$$\mathscr{E}^\bullet = (\dots \longrightarrow 0 \longrightarrow \mathscr{O}(-Z) \xrightarrow{f} \mathscr{O}_X \longrightarrow 0 \longrightarrow \dots), \quad (55)$$

where we take $\mathscr{O}_X$ to be in the zeroth degree slot. It's straightforward to compute the cohomology of this complex, where we get zero in all degrees except $\mathscr{H}^0(\mathscr{E}^\bullet) = \mathscr{O}_X/\mathscr{O}(-Z) = \mathscr{O}_Z$. Since we are working in codimension one, $\mathscr{O}(-Z)$ is a locally-free sheaf, so we see that we have recovered the coherent sheaf $\mathscr{O}_Z$ as the cohomology of a complex of locally-free sheaves. Thus, we can interpret the complex $\mathscr{E}^\bullet$ as consisting of a brane and an anti-brane annihilating to yield simply $\mathscr{O}_Z$ as the endpoint of renormalization group flow. It's natural to consider $\mathscr{O}_Z$ as associated to a D4-brane, since it is supported on a four real dimensional manifold Z.

Interpreting this example in another light, we regard $\mathscr{O}(-Z) \to \mathscr{O}_X$ as a locally-free resolution of the torsion coherent sheaf $\mathscr{O}_Z$. Of course, the cohomology of a resolution coincides with the original object, itself. In this way, we can imagine resolving any coherent sheaf supported on a complex subvariety by locally-free sheaves. This is what we meant above, when we mentioned that coherent sheaves arise naturally from complexes of locally-free sheaves, under RG flow.

Example 3: D0-branes

Let us take X to be a Calabi–Yau threefold which we can study locally as $\mathbb{C}^3$ with coordinates (x, y, z). We define a map

$$\mathscr{O}_X^{\oplus 3} \xrightarrow{(x\ y\ z)} \mathscr{O}_X, \quad (56)$$

defined by taking a triple of holomorphic functions (f_1, f_2, f_3) to the holomorphic function $xf_1 + yf_2 + zf_3$. Since the cokernel of this map should be $\mathscr{O}_X$ modulo the image of this map, we expect that a section of the cokernel must vanish away from the origin in $\mathbb{C}^3$, but can take any complex value at the origin. Letting p denote the origin in $\mathbb{C}^3$, we see that the cokernel of the above map is simply the skyscraper sheaf $\mathscr{O}_p$ at p. Moreover, we naturally have a surjective map $\mathscr{O}_X \to \mathscr{O}_p$ arising from evaluating a holomorphic function at p. Thus, the skyscraper sheaf $\mathscr{O}_p$ is a coherent sheaf. We have the exact sequence

$$\mathscr{O}_X^{\oplus 3} \xrightarrow{(x\ y\ z)} \mathscr{O}_X \longrightarrow \mathscr{O}_p \longrightarrow 0. \quad (57)$$

Indeed, this map has a kernel, corresponding to the sheaf of all functions vanishing at p; this is the ideal sheaf of p. Finally, this gives the short exact sequence

$$0 \longrightarrow \mathscr{I}_p \longrightarrow \mathscr{O}_X \longrightarrow \mathscr{O}_p \longrightarrow 0. \quad (58)$$

Just like in the previous example, we can define the complex $\mathscr{E}^\bullet$,

$$\mathscr{E}^{\bullet} = \big(\ldots \longrightarrow 0 \longrightarrow \mathscr{I}_p \longrightarrow \mathscr{O}_X \longrightarrow 0 \longrightarrow \ldots\big), \tag{59}$$

and trivially compute the cohomology of the complex to vanish in all degrees except $\mathscr{H}^0(\mathscr{E}^{\bullet}) = \mathscr{O}_X/\mathscr{I}_p = \mathscr{O}_p$. Once again, we recover a general coherent sheaf as the cohomology of a complex of coherent sheaves. The only difference here is that $\mathscr{I}_p$ is no longer a locally-free sheaf. Rather, it can be though of roughly as a trivial line bundle outside the origin, where there is no fiber.

Example 4: Branes Wrapping Curves and Points

The setting most closely aligned with modern enumerative geometry and string theory consists of studying one-dimensional sheaves on a Calabi–Yau threefold X. These are sheaves which have a complex one-dimensional support. The Gromov-Witten, Donaldson–Thomas, and Gopakumar–Vafa invariants often package themselves into partition functions exhibiting remarkable properties, and uncovering surprising connections to subjects like modular forms, and representation theory, to name just a few. As a final example, I would like to briefly outline the connection two of these invariants have with the content of this article.

The *Donaldson–Thomas invariants* are a (virtual) count of subschemes $Z \subseteq X$ supported on a fixed homology class $\beta \in H_2(X, \mathbb{Z})$. and whose structure sheaf $\mathscr{O}_Z$ has a fixed holomorphic Euler characteristic. Such subschemes can be supported on both curves and points. We therefore think of $\mathscr{O}_Z$ as a *bound state of D2-D0 branes*. However, there is necessarily a surjective map $\mathscr{O}_X \to \mathscr{O}_Z$, with kernel $\mathscr{I}_Z$. Therefore, one often hears the Donaldson–Thomas invariants described as enumerating *bound states of D2-D0 (anti) branes within a single D6-brane*.

The *Gopakumar–Vafa invariants* are integers which count BPS states of D2-branes wrapped on curves in X. Given a class $\beta \in H_2(X, \mathbb{Z})$ we can consider the moduli space $\mathscr{M}(0, 0, \beta, 1)$ of pure one-dimensional sheaves $\mathscr{F}$ supported on class β with D-brane charge $\mathscr{Q}(\mathscr{F}) = (0, 0, \beta, 1)$. Recently, a proposal emerged [12] for extracting the Gopakumar–Vafa invariants from $\mathscr{M}(0, 0, \beta, 1)$ consistent with their known relation to Gromov-Witten theory.

Due to the combined shortness of my talk, and immense breadth of this subject, I necessarily had to omit certain important topics and examples. Particularly, some explicit computations of open string states as Ext groups. These are covered excellently in [2, 3] *to which I refer the reader. In particular,* [2] *contains quite a few very enlightening examples. Another topic I had to omit is spectral sequences; a great discussion can be found in* [3].

Acknowledgements I am extremely grateful for the opportunity to attend the PIMS 2016 Superschool on Derived Categories and D-branes. I'd like to thank the organizers Matthew Ballard, Charles Doran, and David Favero. First and foremost, I am especially indebted to Eric Sharpe. Without Eric's immense patience and lengthy responses to my questions, I would not have had the opportunity to understand the physical aspects of this subject. I'd also like to thank all the attendees for an excellent week, particularly Jake Bian who provided valuable input and discussion. Finally, I am thankful to my advisor Jim Bryan for reading a draft and offering extremely helpful suggestions.

References

1. Maxim Kontsevich. Homological Algebra of Mirror Symmetry. In *in Proceedings of the International Congress of Mathematicians*, pages 120–139, 1994.
2. E. Sharpe. Lectures on D-branes and Sheaves, July 2003. arXiv:hep-th/0307245.
3. Paul S. Aspinwall. D-Branes on Calabi-Yau Manifolds, March 2004. arXiv:hep-th/0403166.
4. Edward Witten. Mirror Manifolds And Topological Field Theory, December 1991. arXiv:hep-th/9112056.
5. Jin Dai, R.G. Leigh, and Joseph Polchinski. New connections between string theories. *Modern Physics Letters A*, 04(21):2073–2083, October 1989.
6. Robin Hartshorne. *Algebraic Geometry*. Springer, New York, NY, 1st ed. 1977. corr. 8th printing 1997 edition edition, April 1997.
7. Phillip Griffiths. *Principles of Algebraic Geometry*. Wiley India, 1 edition edition, 2014.
8. Jeffrey A. Harvey and Gregory Moore. On the algebras of BPS states. *Communications in Mathematical Physics*, 197(3):489–519, October 1998. arXiv:hep-th/9609017.
9. James Halverson, Hans Jockers, Joshua M. Lapan, and David R. Morrison. Perturbative Corrections to Kahler Moduli Spaces. *Communications in Mathematical Physics*, 333(3):1563–1584, February 2015. arXiv:1308.2157.
10. William Fulton. *Intersection Theory*. Springer, New York, 2nd ed. 1998 edition, June 1998.
11. Ashoke Sen. Tachyon Condensation on the Brane Antibrane System. *Journal of High Energy Physics*, 1998(08):012–012, August 1998. arXiv:hep-th/9805170.
12. Davesh Maulik and Yukinobu Toda. Gopakumar-Vafa invariants via vanishing cycles. *[hep-th]*, October 2016. arXiv:1610.07303

Introduction to Topological String Theories

Kento Osuga

1 Mathematical Background

In this section, we give mathematical definitions and some results that physicists might not be familiar with, but these are necessary to understand the topological A-model and B-model and mirror symmetry. See [1, 2] for more details.

1.1 Complex Manifolds

An m-dimensional complex manifold is defined to be a $2m$-dimensional real manifold which locally looks $\mathbb{C}^m$ with holomorphic transition functions, hence any complex manifold is a real manifold. The converse is however not always true and we need to introduce the concept of *complex structure*. Let M be an $2m$-dimensional real manifold with tangent bundle TM and cotangent bundle T^*M, and we denote by $\Gamma(\otimes^k TM \otimes^l T^*M)$ the space of tensor fields of rank (k, l). We define an *almost complex structure* $J \in \Gamma(TM \otimes T^*M)$ to be a smooth tensor field of rank $(1,1)$ on M satisfying $J^a_c J^c_b = -\delta^a_b$. Then we define the *Nijenhuis tensor* N which locally takes the form

$$N^a_{bc} = J^d_b(\partial_d J^a_c - \partial_c J^a_d) - J^d_c(\partial_d J^a_b - \partial_b J^a_d). \tag{1}$$

If $N = 0$ everywhere, J is called a *complex structure*. It is proven that an $2m$-dimensional real manifold can be considered to be an m-dimensional complex manifold only if it admits a complex structure J. Roughly speaking, a complex structure tells how to mix local coordinates $(z_i, \bar{z}_i)$.

K. Osuga (✉)
Department of Physics, University of Alberta, Edmonton, Alberta T6G 2E1, Canada
e-mail: osuga@ualberta.ca

M. Ballard et al. (eds.), *Superschool on Derived Categories and D-branes*,
Springer Proceedings in Mathematics & Statistics 240,
https://doi.org/10.1007/978-3-319-91626-2_15

1.2 Calabi–Yau Manifold

A *Hermitian metric* h on a complex vector bundle E over a complex manifold M is a smooth section $h \in \Gamma(E \otimes \overline{E})$ satisfying

$$h(u, \bar{v}) = h(\bar{u}, v), \qquad h(u, \bar{u}) \geq 0, \qquad u, v \in E. \tag{2}$$

Locally h can be written as

$$h = h_{a\bar{b}} dz^a \otimes dz^{\bar{b}}. \tag{3}$$

The Riemannian metric g on complexified cotangent bundle $T^*_C M$ is defined to be the real part of the Hermitian metric

$$g = \frac{1}{2}(h + \bar{h}). \tag{4}$$

On the other hand, the imaginary part

$$\omega = \frac{i}{2}(h - \bar{h}), \tag{5}$$

is called the *Hermitian form*. All h, g, ω are compatible with a complex structure J, i.e., $h(u, \bar{v}) = h(Ju, J\bar{v})$. Also note that any of these three uniquely determines the other two. A *Kähler manifold* is defined to be a complex manifold with the non-degenerate closed Hermitian form $d\omega = 0$ and ω in this case is called a *Kähler form*. It is known that locally a Kähler form is given by a so-called Kähler potential K as

$$\omega_{a\bar{b}} = i g_{a\bar{b}} = i \partial_a \bar{\partial}_{\bar{b}} K. \tag{6}$$

One can calculate the Riemann tensors by the Riemannian metric g and it turns out that all $R_{ab} = R_{\bar{a}\bar{b}} = 0$ on a Kähler manifold. A *Calabi–Yau manifold* is defined to be a compact Ricci flat Kähler manifold, which is our interest in topological string theories.

1.3 Cohomologies

On a complex manifold M, we define a (p,q)-form[1] which locally is given as

$$A = A_{a_1 \cdots a_p b_1 \cdots b_q}(z, \bar{z}) dz^{a_1} \wedge \cdots \wedge dz^{a_p} \wedge d\bar{z}^{b_1} \wedge \cdots \wedge d\bar{z}^{b_q}. \tag{7}$$

[1]Do not be confused with type (k.l) tensors.

The space of p-forms on M is the direct sum of the space of $(p-q, q)$ forms over q. Accordingly there exists three different exterior derivatives, namely d which maps p-forms to $(p+1)$-forms, ∂ which maps (p, q)-forms to $(p+1, q)$-forms and $\bar{\partial}$ which maps (p, q) to $(p, q+1)$-froms. They are all nilpotent.

Let us denote d-cohomology, ∂-cohomology and $\bar{\partial}$-cohomology by $H^p(M)$, $H^{p,q}_{\partial}(M)$, and $H^{p,q}_{\bar{\partial}}(M)$ respectively. Then a Kähler form ω is, for example, in $H^2(M)$ and also in $H^{1,1}(M)$. There is no relation among these cohomologies in general but if M is a Kähler manifold, it is known that $H^{p,q}_{\bar{\partial}}(M) = H^{p,q}_{\bar{\partial}}(M)$ and

$$H^p(M) = \bigoplus H^{p-q,q}(M). \tag{8}$$

We call $h^{p,q} = \dim H^{p,q}(M)$ the *Hodge numbers* of M and for a Kähler manifold they satisfy the following relations

$$h^{(p,q)} = h^{(q,p)}, \qquad h^{(p,q)} = h^{(m-p,m-q)}. \tag{9}$$

1.4 Chern Class

Let us consider a connection form[2] $\tilde{\omega}$ on M and define the curvature form Ω as

$$\Omega = d\tilde{\omega} + \tilde{\omega} \wedge \tilde{\omega}. \tag{10}$$

Then the *Chern class* is defined as

$$c(M) = \det\left(I + \frac{i\Omega}{2\pi}\right) = c_0(M) + c_1(M) + c_2(M) + \cdots, \tag{11}$$

where n^{th} Chern class $c_n(M)$ is given by the term with n powers of Ω.

In particular, we find

$$c_0(M) = 1, \qquad c_1(M) = \frac{i\mathrm{Tr}(\Omega)}{2\pi}. \tag{12}$$

For a Calabi–Yau manifold, it is known that $c_1(M) = 0$, which becomes a key to define the B-model in Sect. 6. Note that the requirement of the Ricci flatness is indeed equivalent to the requirement $c_1(M) = 0$.

One may be concerned that this definition relies on a connection $\tilde{\omega}$ so different choices of a connection gives different Chern classes. However, this turns out to be an overthinking. The Chern classes are independent of the choice of connection.

[2]This is called a spin connection in supergravity.

1.5 Moduli Spaces of Calabi–Yau manifolds

We denote a Calabi–Yau manifold by M_C within this subsection. There is a theorem by Calabi and Yau that given a complex manifold with vanishing first Chern class, there is precisely one Calabi–Yau manifold in each Kähler class. We thus define the moduli space $\mathscr{M}_C$ of Calabi–Yau manifolds to be a space of all possible Kähler classes and complex structures on M_C. It is shown that $h^{2,0} = h^{0,2}$ are fixed by $\dim(M_C)$, and especially they are zero if $\dim(M_C) \geq 3$.

It is suggested that the Kähler part of $\mathscr{M}$ is related to $H^{1,1}(M_C)$ since $\omega \in H^{1,1}(M_C)$. More precisely, the tangent space of it is isomorphic to $H^{1,1}(M_C)$. On the other hand, the complex structure part of $\mathscr{M}$ is more complicated in general so let us stick on M of dimensions three. In this case, the tangent space of infinitesimal deformation of complex structure of $\mathscr{M}$ is shown to be isomorphic to $H^{2,1}(M)$. In fact one can explicitly calculate some of the Hodge numbers for the Calabi–Yau 3-fold and the hodge diamond becomes

$$\begin{array}{ccccccc} & & & 1 & & & \\ & & 0 & & 0 & & \\ & 0 & & h^{1,1} & & 0 & \\ 1 & & h^{2,1} & & h^{2,1} & & 1 \\ & 0 & & h^{1,1} & & 0 & \\ & & 0 & & 0 & & \\ & & & 1 & & & \end{array} \tag{13}$$

The *mirror symmetry* between two Calabi–Yau threefolds is a duality under reflection along the diagonal line, i.e. by interchanging $h^{1,1}$ and $h^{2,1}$, or in other words $H^{1,1}(M)$ (Kähler classes) and $H^{2,1}(M)$ (complex structures) of two mirror pair theories.

2 Topological and Cohomological Field Theory

Discussions in this section and here after are based on [2, 3]. Our interests in quantum field theories are correlation functions, or observables, of physical operators in some certain background. Here a background includes a choice of a manifold, metric and coupling constants. A *topological field theory*, TFT, is defined to be a theory if all physical observables are independent of the choice of the metric. So obviously any observable has no explicit dependence of the metric, though it implicitly can.

This is a quite powerful requirement. Since one can freely change the metric and coordinates in TFT, and these two end up with changing insertion points of local operators without varying observables. That is, order of operators does not matter and observables are independent of insertion points in contrast to standard quantum field theories.

A *cohomological field theory*, CohFT or sometimes called a Witten's type TFT, is a TFT with the following properties.

- There is a global Grassmann scalar symmetry operator Q such that $Q^2 = 0$.
- All physical operators $\mathcal{O}_i$ are Q-closed, $\{Q, \mathcal{O}_i\} = 0$.
- The vacuum state is Q-symmetric, $Q|0 = 0$.
- The EM tensor $T_{\mu\nu}$ is Q-exact, $T_{\mu\nu} = \{Q, G_{\mu\nu}\}$, where $G_{\mu\nu}$ is some operator.

The first three are properties of any quantum field theory with a BRST charge. These three ensure that observables of physical operators $\mathcal{O}_i$ are invariant under a Q-exact shift with some operator Λ

$$\mathcal{O}_i \sim \mathcal{O}_i + \{Q, \Lambda_i\}. \tag{14}$$

The fourth one is crucial to make a theory topological. Let us consider a functional derivative of an observable with respect to the metric g. Since physical operators are required to be independent of the metric, we have

$$\begin{aligned} \frac{\delta}{\delta g^{\mu\nu}} \langle \mathcal{O}_{i_1} \cdots \mathcal{O}_{i_n} \rangle &= i \int D\phi \mathcal{O}_{i_1} \cdots \mathcal{O}_{i_n} \frac{\delta S}{\delta g^{\mu\nu}} e^{iS[\phi]}, \\ &= i \langle \mathcal{O}_{i_1} \cdots \mathcal{O}_{i_n} \{Q, G_{\mu\nu}\} \rangle, \\ &= 0, \end{aligned} \tag{15}$$

where ϕ is a field in the theory.

As a simple example, if the action is Q-exact, the theory is a CohFT since the functional derivative with respect to the metric should be also Q-exact. Further since the action is given by

$$\exp \frac{i}{h} \left\{ Q, \int_M V \right\}, \tag{16}$$

we have

$$\frac{d}{dh} \langle \mathcal{O}_{i_1} \cdots \mathcal{O}_{i_n} \rangle = 0. \tag{17}$$

Therefore in this case, all correlation functions are given in the classical limit $(h \to 0)$.

2.1 Nonlocal Operators

Consider an observable of physical local operators $\{\mathcal{O}_i(x_i)\}$. Since topological invariance of the theory implies that it is independent of insertion points x_i, the derivative with respect to, for example, x_1 vanishes

$$d_{x1}\,\langle \mathcal{O}_1(x_1)\cdots \mathcal{O}_k(x_k)\rangle = \langle d\mathcal{O}_1(x_1)\cdots \mathcal{O}_k(x_k)\rangle = 0. \tag{18}$$

This means that $d\mathcal{O}$ must be Q-exact

$$d\mathcal{O}^{(0)}(x_1) = \{Q, \mathcal{O}^{(1)}(x_1)\}, \tag{19}$$

where we denote the original physical operator by $\mathcal{O}^{(0)}$ and the associated local operator by $\mathcal{O}^{(1)}$. If C is a closed circle in M, then

$$U(C) = \oint_C \mathcal{O}^{(1)} \tag{20}$$

is Q-closed. In fact,

$$\{Q, U(C)\} = \oint_C d\mathcal{O}^{(0)} = 0. \tag{21}$$

Topological invariance implies that $\delta U(C)$ under small displacements of C should be Q-exact. Since Stoke's theorem gives for a small area A with two boundaries C_1, C_2,

$$\oint_{C_1} \mathcal{O}^{(1)} - \oint_{C_2} \mathcal{O}^{(1)} = \int_A d\mathcal{O}^{(1)}, \tag{22}$$

it is again shown that $d\mathcal{O}^{(1)}$ must be Q-exact $d\mathcal{O}^{(1)} = \{Q, \mathcal{O}^{(2)}\}$. Thus we have another nonlocal operator for a closed two-dimensional surface S.

$$U(S) = \int_S \mathcal{O}^{(2)}. \tag{23}$$

One can of course repeat this procedure and eventually obtain a Q-closed nonlocal operator

$$U(M) = \int_M \mathcal{O}^{(m)}, \tag{24}$$

where m is dimensions of M. Since this is independent of the metric by construction and Q-closed, one can freely add $\mathcal{O}^{(m)}$ into the action S with some coupling constants without spoiling the cohomological property.

3 Twisted $N = 2$ Supersymmetry

Let us consider a two-dimensional $N = 2$ supersymmetric theory in the superfield formulation. There are two bosonic local variables $z, \bar{z}$ and two Grassmann variables $\theta_\pm$ and their complex conjugate $\bar{\theta}_\pm$. In our conventions, we define $\bar{\theta}_-$ to be a complex conjugate of θ_+. This is because under the Lorentz transformation $z \mapsto e^{2i\alpha} z$, Grassmann variables are changed as

$$\theta_\pm \mapsto e^{\pm i\alpha}\theta_\pm, \qquad \bar{\theta}_\mp \mapsto e^{\pm i\alpha}\bar{\theta}_\pm. \tag{25}$$

In the superfield formulation, the set of supersymmetry generators is represented by

$$\begin{aligned}
H &= -\frac{d}{i dx^0} = -i(\partial_z - \partial_{\bar{z}}), \\
P &= -\frac{d}{dx^1} = -i(\partial_z + \partial_{\bar{z}}), \\
M &= 2z\partial_z - 2\bar{z}\partial_{\bar{z}} + \theta_+ \frac{\partial}{\partial\theta_+} + \bar{\theta}_+ \frac{\partial}{\partial\bar{\theta}_+} - \theta_- \frac{\partial}{\partial\theta_-} - \bar{\theta}_- \frac{\partial}{\partial\bar{\theta}_-}, \\
Q_\pm &= \frac{\partial}{\partial\theta_\pm} + i\bar{\theta}_\pm\partial_\pm, \\
\overline{Q}_\pm &= -\frac{\partial}{\partial\bar{\theta}_\pm} - i\theta_\pm\partial_\pm,
\end{aligned} \tag{26}$$

where $z = x^1 + ix^0$. Note that the complex conjugate of $\partial_{\bar{\theta}_+}$ is $-\partial_{\theta_-}$. Their commutators are

$$\begin{aligned}
[M, H] &= -2P, \qquad [M, P] = -2H, \\
[M, Q_\pm] &= \mp Q_\pm, \qquad [M, \overline{Q}_\pm] = \mp\overline{Q}_\pm, \\
\{Q_\pm, \overline{Q}_\pm\} &= P \pm H,
\end{aligned} \tag{27}$$

and others are zero. Note that transformations of supercharges under the Lorentz operator M are a half of those of H, P which represents that they are spinorial quantities.

Let Φ be a superfield, then a super transformation is given by a Grassmann parameter $\varepsilon_\pm$ as $\delta\Phi = (\varepsilon_- Q_+ + \varepsilon_+ Q_-)\Phi$. However since there are no constant covariant spinors $\varepsilon_\pm$ for arbitrary manifolds, these supersymmetries $Q_\pm$ are not global in general.

Fortunately since we have two supersymmetries, there is an additional $U(1)$ symmetry, called an *R-symmetry*, between them and it plays an important role in constructing a CohFT. In particular, consider the following two independent R_V, R_A transformations

$$\begin{aligned}
R_V &: (\theta_+, \bar{\theta}_+) \mapsto (e^{-i\beta}\theta_+, e^{i\beta}\bar{\theta}_+), \qquad (\theta_-, \bar{\theta}_-) \mapsto (e^{-i\beta}\theta_-, e^{i\beta}\bar{\theta}_-), \\
R_A &: (\theta_+, \bar{\theta}_+) \mapsto (e^{-i\beta}\theta_+, e^{i\beta}\bar{\theta}_+), \qquad (\theta_-, \bar{\theta}_-) \mapsto (e^{i\beta}\theta_-, e^{-i\beta}\bar{\theta}_-),
\end{aligned} \tag{28}$$

and they leave $z, \bar{z}$ invariant. That is, these are rotation among Grassmann variables. Their generators in terms of θ and nonzero commutators are given by

$$F_V = -\theta_+ \frac{\partial}{\partial\theta_+} + \bar{\theta}_+ \frac{\partial}{\partial\bar{\theta}_+} - \theta_- \frac{\partial}{\partial\theta_-} - \bar{\theta}_+ \frac{\partial}{\partial\bar{\theta}_-},$$
$$F_A = -\theta_+ \frac{\partial}{\partial\theta_+} + \bar{\theta}_+ \frac{\partial}{\partial\bar{\theta}_+} + \theta_- \frac{\partial}{\partial\theta_-} - \bar{\theta}_- \frac{\partial}{\partial\bar{\theta}_-}, \tag{29}$$

$$[F_V, Q_\pm] = +Q_\pm, \qquad [F_V, \overline{Q}_\pm] = -\overline{Q}_\pm,$$
$$[F_A, Q_\pm] = \pm Q_\pm, \qquad [F_A, \overline{Q}_\pm] = \mp\overline{Q}_\pm. \tag{30}$$

Note that these commutators are different from those in (27), though they are similar as F_V, F_A are basically generators of $U(1)$.

Now we get to a crucial argument. Let us define a new Lorentz operator as

$$M_A = M - F_V, \quad \text{or} \quad M_B = M - F_A, \tag{31}$$

then their commutators with H, P remains unchanged while those with supercharges are

$$[M_A, Q_+] = -2Q_+, \; [M_B, Q_+] = -2Q_+,$$
$$[M_A, Q_-] = \;\; 0, \qquad [M_B, Q_+] = +2Q_+,$$
$$[M_A, \overline{Q}_+] = \;\; 0, \qquad [M_B, \overline{Q}_+] = \;\; 0,$$
$$[M_A, \overline{Q}_-] = +2\overline{Q}_-, \; [M_B, \overline{Q}_-] = \;\; 0. \tag{32}$$

A theory with the Lorentz generator M_A is called *A-twisted* and one with M_B is *B-twisted*. In an A-twisted theory, if one defines $Q_A = \overline{Q}_+ + Q_-$ then we have

$$[M_A, Q_A] = 0, \qquad \{Q_A, Q_A\} = 0, \tag{33}$$

The first commutator shows that Q_A transforms as a scalar under the Lorentz transformation M_A so does its associated parameter ε_A. That is, Q_A-symmetry is globally defined. The second equation suggests that one can construct a CohFT with Q_A.

Similarly in a B-twisted model, commutators with $Q_B = \overline{Q}_+ + \overline{Q}_-$ are

$$[M_B, Q_B] = 0, \qquad \{Q_B, Q_B\} = 0, \tag{34}$$

which suggests the existence of another CohFT. Note that these observations only guarantee the first condition for a CohFT. We will explicitly see in the next section that we can indeed construct two CohFTs from $N = 2$ supersymmetric theory.

4 Sigma Model and R-anomalies

We study a supersymmetric nonlinear sigma model in two dimensions, which gives the topological A-model and B-model after twisting. Let Σ be a Riemann surface of two dimensions and M be a target space of complex dimensions m with the metric g then the sigma model governs maps $\Phi : \Sigma \to M$. The on-shell action of this model is given as

$$S = 2t \int_\Sigma d^2z \Big(\frac{1}{2} g_{IJ} \partial_z \phi^I \partial_{\bar{z}} \phi^J + \frac{i}{2} g_{IJ} \psi_-^I \Delta_z \psi_-^J \\ + \frac{i}{2} g_{IJ} \psi_+^I \Delta_{\bar{z}} \psi_+^J + \frac{1}{4} R_{IJKL} \psi_+^I \psi_+^J \psi_-^K \psi_-^L \Big), \tag{35}$$

where t is a coupling constant and Δ is the covariant derivative with respect to the metric on both Σ and M.

Let $K, \overline{K}$ be the canonical, and anti-canonical line bundles of Σ and $T^{1,0}M$, $T^{0,1}M$ be the complexified tangent bundles of M respectively. Then each field lives in

$$\phi^a \in \Phi^*(T^{1,0}M), \tag{36}$$
$$\phi^{\bar{a}} \in \Phi^*(T^{0,1}M), \tag{37}$$
$$\psi_+^a \in K^{1/2} \otimes \Phi^*(T^{1,0}M), \tag{38}$$
$$\psi_+^{\bar{a}} \in K^{1/2} \otimes \Phi^*(T^{0,1}M), \tag{39}$$
$$\psi_-^a \in \overline{K}^{1/2} \otimes \Phi^*(T^{1,0}M), \tag{40}$$
$$\psi_-^{\bar{a}} \in \overline{K}^{1/2} \otimes \Phi^*(T^{0,1}M). \tag{41}$$

If M is Kähler, then it has the supersymmetry transformations listed below. (If M is not Kähler, it is still supersymmetric, just not with (2,2) supersymmetry, only (1,1).) Let $\varepsilon_-, \bar{\varepsilon}_- \in K^{-1/2}$ and $\varepsilon_-, \bar{\varepsilon}_- \in \overline{K}^{-1/2}$. Then the super transformation laws with these parameters are respectively

$$\begin{aligned}
\delta\phi^a &= i\varepsilon_- \psi_+^a + i\varepsilon_+ \psi_-^a, \\
\delta\phi^{\bar{a}} &= i\bar{\varepsilon}_- \psi_+^{\bar{a}} + i\bar{\varepsilon}_+ \psi_-^{\bar{a}}, \\
\delta\psi_+^a &= -\bar{\varepsilon}_- \partial_z \phi^a - i\varepsilon_+ \psi_-^b \Gamma^a_{bc} \psi_+^c, \\
\delta\psi_+^{\bar{a}} &= -\varepsilon_- \partial_z \phi^{\bar{a}} - i\bar{\varepsilon}_+ \psi_-^{\bar{b}} \Gamma^{\bar{a}}_{\bar{b}\bar{c}} \psi_+^{\bar{c}}, \\
\delta\psi_-^a &= -\bar{\varepsilon}_+ \partial_{\bar{z}} \phi^a - i\varepsilon_- \psi_+^b \Gamma^a_{bc} \psi_-^c, \\
\delta\psi_-^{\bar{a}} &= -\varepsilon_+ \partial_{\bar{z}} \phi^{\bar{a}} - i\bar{\varepsilon}_- \psi_+^{\bar{b}} \Gamma^{\bar{a}}_{\bar{b}\bar{c}} \psi_+^{\bar{c}}.
\end{aligned} \tag{42}$$

Now let us discuss the R-anomalies. For simplicity, we drop the a-indices and $\pm$-indices. In quantum field theories with fermions, one needs to be careful about their zero modes. In our model (35), the following part is problematic

$$\int D\psi D\overline{\psi} \exp(\overline{\psi} \Delta \psi). \tag{43}$$

If ψ is expanded as $\psi = \sum \psi^{(k)}$, it is given by

$$\prod_{k,l} \int d\psi^{(k)} d\overline{\psi}^{(l)} \exp\left(\overline{\psi}^{(l)} \Delta \psi^{(k)}\right). \tag{44}$$

Thus if ψ has some zero modes, the path integral vanishes since $\int d\theta = 0$ for any Grassmann variable θ. We give some facts below which we omit proofs since they are too technical:

- Except in some special cases, one can show that the number of zero modes of $\psi_\pm$ is $|k_\pm|$ and that of $\overline{\psi}_\pm$ is zero if $k_\pm$ is positive, while the number of zero modes of $\psi_\pm$ is zero and that of $\overline{\psi}_\pm$ is $|k_\pm|$ if $k_\pm$ is negative.
- $k_\pm$ satisfy $k_+ = -k_-$. Thus one can choose $k = k_-$ then there are k zero modes of ψ_-, $\overline{\psi}_+$, and no zero modes of ψ_+, $\overline{\psi}_-$ if $k \geq 0$.
- k is given by the first Chern class of the target space as

$$k = \int_{\phi(\Sigma)} c_1(M). \tag{45}$$

- The last term is small perturbation in string scale. However since small perturbative effect is not expected to give some change of the integer number $k_\pm$ in topological theories, we ignore the contribution from this term to $k_\pm$.

Therefore in this model, we need to insert local operators to have nonzero observables and those are given in the following form

$$\int D\psi_+ D\overline{\psi}_+ D\psi_- D\overline{\psi}_- W_{a_1 \cdots a_k \bar{b}_1 \cdots \bar{b}_k} \left(\prod_{i=1}^{k} \psi_-^{a_i} \overline{\psi}_+^{\bar{b}_i}\right) e^{iS}, \tag{46}$$

where we have assumed $k \geq 0$. Equation (28) shows that the product of ψ_- and $\overline{\psi}_+$ is invariant under the R_V-symmetry, while it is not under R_A. Thus the R_A-symmetry is broken unless $k = 0$. We arrive at the same conclusion if $k \leq 0$. This implies that the A-twisting is defined for any Kähler target space, but the B-twisting can only be defined for Calabi–Yau target spaces since otherwise the R_A symmetry is not well-defined. In the next section, we twist this sigma model into the A-model and B-model to investigate more details.

5 The A-model

In the A-model, the spinors live in the following bundles

$$\begin{aligned}\psi_z^a &:= \psi_+^a \in K \otimes \Phi^*(T^{1,0}(M)),\\ \chi^a &:= \psi_-^a \in \Phi^*(T^{1,0}(M)),\\ \chi^{\bar{a}} &:= \psi_+^{\bar{a}} \in \Phi^*(T^{0,1}(M)),\\ \psi_{\bar{z}}^{\bar{a}} &:= \psi_-^{\bar{a}} \in \overline{K} \otimes \Phi^*(T^{0,1}(M)).\end{aligned} \tag{47}$$

By (42), by setting $\varepsilon_- = \overline{\varepsilon}_+ = 0$ and $\varepsilon_+ = \varepsilon, \overline{\varepsilon}_- = \overline{\varepsilon}$ to be constants, we have

$$\begin{aligned}\delta\phi^a &= i\varepsilon\chi^a\\ \delta\phi^{\bar{a}} &= i\bar{\varepsilon}\chi^{\bar{a}}\\ \delta\chi^a &= \delta\chi^{\bar{a}} = 0\\ \delta\psi_z^a &= -\bar{\varepsilon}\partial_z\phi^a - i\varepsilon\chi_-^b\Gamma_{bc}^a\psi_z^c\\ \delta\psi_{\bar{z}}^{\bar{a}} &= -\varepsilon\partial_{\bar{z}}\phi^{\bar{a}} - i\bar{\varepsilon}\chi_-^{\bar{b}}\Gamma_{\bar{b}\bar{c}}^{\bar{a}}\psi_{\bar{z}}^{\bar{c}}\end{aligned} \tag{48}$$

Note that δ^2, or Q_A^2 vanishes up to the equations of motion. We can of course consider the off-shell formalism and then $Q^2 = 0$ without using the equations of motion.

The on-shell action becomes

$$\begin{aligned}S = 2t\int_\Sigma d^z\Big(&\frac{1}{2}g_{IJ}\partial_z\phi^I\partial_{\bar{z}}\phi^J + ig_{\bar{a}b}\psi_{\bar{z}}^{\bar{a}}\Delta_z\chi^b\\ &+ig_{a\bar{b}}\psi_z^a\Delta_{\bar{z}}\chi^{\bar{b}} + \frac{1}{2}R_{a\bar{b}c\bar{d}}\psi_z^a\psi_{\bar{z}}^{\bar{b}}\chi^c\chi^{\bar{d}}\Big),\\ \sim it\int_\Sigma &d^z\{Q_A, V\} + t\int_\Sigma \Phi^*(\omega),\end{aligned} \tag{49}$$

where the second term is the pull back of the target space Kähler form and the last line is true up to terms vanishing by the ψ-equations of motion. This does not make any difference in the A-model as shown shortly. V is given by

$$V = g_{a\bar{b}}\left(\psi_z^a\partial_z\phi^{\bar{b}} + \psi_{\bar{z}}^{\bar{b}}\partial_{\bar{z}}\phi^a\right). \tag{50}$$

The second term of (49) depends only on the cohomology class of ω and the homotopy class of the map Φ. Let us denote it by $it\beta\cdot\omega$ then the action is given by

$$S = -t\beta\cdot\omega + \int_\Sigma\{Q_A, V\}. \tag{51}$$

We can put $it\beta \cdot \omega$ out from the path integral so a physical observable takes the form

$$\langle \prod_a \mathcal{O}_a \rangle = e^{-t\beta\cdot\omega} \int D\phi D\chi D\psi \prod_a \mathcal{O}_a e^{it\{Q_A, \int V\}}. \tag{52}$$

As discussed in Sect. 2, the path integral part is independent of t hence we can calculate it in the classical limit as long as $\Re(t\beta \cdot \omega) > 0$. The t-dependence factor is called an instanton number.

The remaining terms in the Lagrangian can be written as

$$\{Q_A, V\} = L - 2tg_{a\bar{b}} \left(\partial_z \phi^a \partial_{\bar{z}} \phi^{\bar{b}} - \partial_z \phi^{\bar{b}} \partial_{\bar{z}} \phi^a \right), \tag{53}$$

In particular, it includes only $\partial_{\bar{z}} \phi^a$ and $\partial_z \phi^{\bar{a}}$. Then one realizes that L is minimized (classical limit) when ϕ is holomorphic

$$\partial_{\bar{z}} \phi^a = \partial_z \phi^{\bar{a}} = 0. \tag{54}$$

Thus the A-model sums over holomorphic maps from $\Sigma \to M$. In general, the space of such maps is finite hence the path integral reduces to a finite dimensional integral and it is known to be $m(1-g)$ for a Calabi–Yau manifold where g is the number of genus of Σ.

Note that the instanton factor obviously depends on the choice of Kähler classes while it is independent of complex structures of M hence all information about complex structures of M is embedded in the definition of V. If one modifies a complex structure of M, the variation of the action gives

$$\delta S = \{Q_A, \int_M \delta V\}, \tag{55}$$

which is irrelevant in CohFTs. Therefore, the A-model depends on the Kähler classes on M but not their complex structures.

5.1 Local Operators

In order to construct local operators independent of both the worldsheet metric and diffeomorphism, one can use only ϕ, χ but not ψ because ψ behaves as a vector and the z-indices should be either contracted by the metric or integrated out into a nonlocal operator. χ is on the other hand a fermion even after twisting hence a well-defined local operator is in the form

$$\mathcal{O}_A = A_{a_1 \cdots a_p \bar{b}_1 \cdots \bar{b}_q}(\phi) \chi^{a_1} \cdots \chi^{a_p} \chi^{\bar{b}_1} \cdots \chi^{\bar{b}_q}. \tag{56}$$

By using (48), a simple calculation shows

$$\{Q_A, \mathcal{O}_A\} = -\mathcal{O}_{dA} \tag{57}$$

with d the exterior derivative acting on A. Indeed the de Rham cohomology on M turns out to be isomorphic to the Q_A-cohomology of the A-model as long as we consider only local operators.

Let us come back to the question why there is no need of terms vanishing by the ψ-equations of motion in (49). One can define a new operator $\tilde{Q}_A$ such that the second line of (49) is given by equality but instead the transformation law for ψ in (48) is modified. Then one can of course consider another topological theory with $\tilde{Q}_A$ and what potentially changes is only the $\tilde{Q}_A$-cohomology, i.e. the form of local operators. However since $\tilde{Q}_A$-operations on ϕ, χ are precisely the same as Q_A-operations, local operators given in (56) is not modified at all. Therefore there is no topological difference between the A-model with Q_A and $\tilde{Q}_A$.

After twisting, spinors are not in the same bundle as before so that the number of zero modes is also different. For example, it is known that the number of zero modes of χ is

$$k = m(1 - g), \tag{58}$$

if the target space is a Calabi–Yau manifold, which is the same as the dimensions of the space of the map ϕ[3] and no zero modes of ψ for $k \geq 0$. If k is negative, one can regard that there is no zero mode of χ but $|k|$ zero modes of ψ. However in this case, one cannot construct local topological theories because ψ should be inserted which is either nonlocal or is contracted by the worldsheet metric. Thus nonzero observables in the local A-model are only the partition function if $g = 1$ and (m, m)-point functions if $g = 0$.

6 The *B*-model

In the B-model, the spinors live in the following bundles

$$\begin{aligned}
\psi_+^a &\in K \otimes \Phi^*(T^{1,0}(M)),\\
\psi_-^a &\in \overline{K} \otimes \Phi^*(T^{1,0}(M)),\\
\psi_+^{\bar{a}} &\in \Phi^*(T^{0,1}(M)),\\
\psi_-^{\bar{a}} &\in \Phi^*(T^{0,1}(M)).
\end{aligned} \tag{59}$$

[3]This is not a coincidence. One can intuitively see from the isomorphism between d-cohomology and Q_A-cohomology by using the transformation laws of ϕ, χ.

It is convenient to define spinors as

$$\begin{aligned} \eta^{\bar{a}} &= \psi_+^{\bar{a}} + \psi_-^{\bar{a}}, \\ \theta_a &= g_{a\bar{b}}(\psi_+^{\bar{a}} - \psi_-^{\bar{a}}), \\ \rho_z^a &= \psi_+^a, \qquad \rho_{\bar{z}}^a = \psi_-^a, \end{aligned} \tag{60}$$

then by setting $\varepsilon_- = \varepsilon_+ = 0$ and $\bar{\varepsilon}_- = \bar{\varepsilon}_+ = \varepsilon$ to be constants, the super transformations become much simpler than those of the A-model

$$\begin{aligned} \delta\phi^{\bar{a}} &= i\varepsilon\eta^{\bar{a}}, \\ \delta\rho &= -\varepsilon d\phi^a, \\ \delta\phi^a &= \delta\eta^{\bar{a}} = \delta\theta_a = 0. \end{aligned} \tag{61}$$

The B-model Lagrangian is

$$\begin{aligned} L = t\int_\Sigma d^2z\Big(& g_{IJ}\partial_z\phi^I\partial_{\bar{z}}\phi^J + ig_{a\bar{b}}\eta^{\bar{a}}(\Delta_z\rho_{\bar{z}}^b + \Delta_{\bar{z}}\rho_z^b) \\ & + i\theta_a(\Delta_{\bar{z}}\rho_z^a - \Delta_z\rho_{\bar{z}}^a) + R_{a\bar{b}c\bar{d}}\rho_z^a\rho_{\bar{z}}^c\eta^{\bar{b}}g^{\bar{d}e}\theta_e\Big), \\ = it\int_\Sigma & d^2z\{Q_B, V\} + tW, \end{aligned} \tag{62}$$

where

$$V = g_{a\bar{b}}\left(\rho_z^a\partial_{\bar{z}}\phi^{\bar{b}} + \rho_{\bar{z}}^a\partial_z\phi^{\bar{b}}\right), \tag{63}$$

$$W = -\int_\Sigma\left(\theta_a D\rho^a + \frac{i}{2}R_{a\bar{b}c\bar{d}}\rho^a\wedge\rho^c\eta^{\bar{b}}g^{\bar{d}e}\theta_e\right), \tag{64}$$

where Δ is the extended exterior derivative on Σ. Note that this is an equality and we did not use any equations of motion, unlike the A-model. Note that since W is an integral of a (1,1)-form over Σ, it is independent of the worldsheet metric. Therefore this model satisfies the requirements to be a CohFT.

Just like the A-model, any local operator should be consisted of ϕ, θ and η thus it takes the form

$$\mathscr{O}_B = B_{\bar{b}_1\cdots\bar{b}_q}^{a_1\cdots a_p}(\phi)\theta_{a_1}\cdots\theta_{a_p}\eta^{\bar{b}_1}\cdots\eta^{\bar{b}_q}, \tag{65}$$

and by (61), one simply gets

$$\{Q_B, \mathscr{O}_B\} = -\mathscr{O}_{\bar{\partial}B}. \tag{66}$$

In contrast to the A-model, local operators include θ so that we cannot use the θ equations of motion. Note that since $B^{a_1\cdots a_p}_{\bar{b}_1\cdots\bar{b}_q}$ in (65) has not only subscripts but also superscripts, the Q_B-cohomology is isomorphc to $\oplus_{p,q} H^q(M, \wedge^p T^{1,0}M)$.

Fortunately this θ-dependence of W makes the B-model rather simpler. Since θ is linear in W, and V is independent of it, one can redefine $\theta \mapsto \theta/t$ hence we can get rid of the t-dependence of the W term. The remaining V term is Q_B-exact hence it is independent of t. Accordingly any observable in B-model is proportional to some power of t coming from the path-integral measure and local operators.

The path integral part is calculated in the classical limit similar to the A-model. In the B-model, the V term has both $(\partial_{\bar{z}}\phi^a, \partial_z\phi^{\bar{a}})$ and $(\partial_z\phi^a, \partial_{\bar{z}}\phi^{\bar{a}})$ so that the Lagrangian is minimized when

$$\partial_{\bar{z}}\phi^a = \partial_z\phi^{\bar{a}} = \partial_z\phi^a = \partial_{\bar{z}}\phi^{\bar{a}} = 0. \tag{67}$$

This is just a set of constant maps $\Phi : \Sigma \to M$. The space of such maps is a copy of M hence the path integral simply reduces to an integral over M.

The number of fermion zero modes again changes after twisting. If the target space is a Calabi–Yau, it is known that the difference of the number of η, θ zero modes and that of ρ zero modes is $k = m(1-g)$. Note that the objects integrated over M is not a $(0, m)$-form but (m, m)-form thus it is natural to contract with a holomorphic $(m, 0)$-form Ω

$$B^{a_1\cdots a_m}_{\bar{b}_1\cdots\bar{b}_m} \mapsto B^{a_1\cdots a_p}_{\bar{b}_1\cdots\bar{b}_q}\Omega_{a_1\cdots a_p}\Omega_{a'_1\cdots a'_p}. \tag{68}$$

Therefore, an observable of the B-model is an integral of wedge products of forms B and Ω over M, which one can classically calculate. It is shown for Calabi–Yau manifolds that the space of holomorphic (m,0)-forms is isomorphic to that of $\wedge^d T^{1,0}M$.

All properties of the B-model discussed so far are much simpler than those of the A-model. The only thing which is not so clear yet to see is that it is independent of Kähler classes. In fact a tedious calculation shows that a modification of the Kähler metric on M changes W in (62) by $\{Q, \cdots\}$. On the other hand, it is easy to see by the above argument that it depends on complex structures since observables are determined by the choice of the holomorphic $(m, 0)$-form Ω. As a conclusion the A-model and B-model are a mirror pair under interchange of their Kähler class and complex structure.

7 The Fixed Point Theorem

We explain why the maps ϕ reduce to holomorphic maps in the A-model and constant maps in the B-model here in an alternative way.

Consider an arbitrary quantum field theory with a group of symmetry G. Let F be the configuration space of all fields in the theory then the path integral of some operator $\mathcal{O}$ is

$$\int_F \mathcal{O}e^{-S} = \mathrm{Vol}(G) \cdot \int_{F/G} \mathcal{O}e^{-S} + \int_{F_0} \mathcal{O}e^{-S}, \tag{69}$$

where F_0 is a subspace invariant under the G-action. Notice that if G is Grassmann, $\mathrm{Vol}(F)$ should be also Grassmann and which implies that it vanishes because $\int d\theta = 0$. Therefore the first term is zero.

Now let us consider F to be a nilpotent group then F_0 is defined by fields such that $\delta\Phi = 0$. In the A-model (48) gives

$$\partial_{\bar{z}}\phi^a = \partial_z\phi^{\bar{a}} = 0, \tag{70}$$

while in the B-model we have by (61)

$$d\phi^a = 0. \tag{71}$$

Thus F_0 is the space of holomorphic maps in the A-model and the space of constants maps in the B-model respectively.

8 Topological String Theories

We only focus on closed string theories so that there is no need to worry about boundary conditions. The main difference between topological field theories and topological string theories is whether or not we path-integrate over the worldsheet metric $h_{\mu\nu}$. This makes theories more interesting.

8.1 R-anomalies

Note that the sigma model given in Sect. 4 becomes a super-conformal field theory once we couple the worldsheet metric in the action. There are three (bosonic) local symmetries, namely two diffeomorphisms and the Weyl symmetry, in two-dimensional CFT and the number of independent components of the metric is also three. Thus one can always locally gauge-fix the metric in the flat form

$$h_{\mu\nu} = \eta_{\mu\nu}. \tag{72}$$

However this is globally impossible since there are parameters that cannot be gauged away. In general, there is no parameter for a sphere, one parameter for a torus, the famous modular parameter τ, and for higher genus there are $m_g = 3(g-1)$ parameters left.

Let us first consider the $g > 1$ case, for which the number of parameters is $3(g-1)$. Conformal transformations in two dimensions are equivalent to holomorphic transformations hence these modular parameters in fact describe change of complex structure on Σ, which can be parametrized by $\mu_{\bar{z}}^{z}$, $\bar{\mu}_{z}^{\bar{z}}$ as

$$dz \mapsto dz + c\mu_{\bar{z}}^{z} d\bar{z}, \qquad d\bar{z} \mapsto d\bar{z} + \bar{c}\bar{\mu}_{z}^{\bar{z}} dz \tag{73}$$

where $c, \bar{c}$ are infinitesimal constants. After gauge-fixing the metric, one still need to integrate over this $3(g-1)$-dimensional moduli space $\tilde{M}_g$. The measure of $\tilde{M}_g$ should be invariant under coordinate transformations of Σ so $\mu_{\bar{z}}^{z}$, $\bar{\mu}_{z}^{\bar{z}}$ should be contracted by $G_{zz}, \overline{G}_{\bar{z}\bar{z}}$ and integrated over Σ, where G is the Q-partner of the EM tensor[4]. Thus, let $(m_i, \bar{m}_i)$ be coordinates on $\tilde{M}_g$ then it is natural to guess the form of the measure as

$$\int_{\tilde{M}_g} \prod_{i}^{3(g-1)} dm_i d\bar{m}_i \int_{\Sigma} G_{zz}\mu_{\bar{z}(i)}^{z} \int_{\Sigma} \overline{G}_{\bar{z}\bar{z}}\bar{\mu}_{z(i)}^{\bar{z}}. \tag{74}$$

It is indeed proven that this is correct, that is, this is invariant under coordinate transformations on $\tilde{M}_g$. We have arrived at the first crucial point of topological string theories. Even though the metric itself is independent of R-transformation, its path integral measure is not invariant under R-transformation because of fermionic fields $G, \overline{G}$. One can then see that the product of these two $G, \overline{G}$ has no R_V-charge while the R_A-charge is 2, thus the total R_A-charge is $6(g-1)$. On the other hand as discussed before, the fermion zero mode requires $2m(1-g)$ R_A-charges after twisting to be nonzero. Therefore the partition function vanishes for any genus $g > 1$ unless $m = 3$, a Calabi–Yau threefold.

For a sphere, there is no modular parameter so one can copy all results from topological filed theories and observables of (3, 3)-forms are evaluated. The only difference is that one needs to fix three rotational symmetries of a sphere. In particular for a Calabi–Yau threefold, this can be done to consider three -point functions with three marked points. That is, these points are fixed as a gauge choice.

For a torus, there is one modular parameter so we need to insert one local (1,1)-form to have nonzero observables because then the R_A-charge is consistent. Similar to the case for a sphere, there is a axial symmetry on a tours hence the insertion point of the local operator should be fixed.

As a summary for Calabi–Yau threefolds, nonzero observables of local operators are a three-point function of (3,3)-forms on a sphere, a one-point function of (1,1)-forms on a torus and partition functions for any higher genus.

[4]This result should be rigorously achieved by the Fadeev-Popov method so that the contracting tensor is suggested to be fermionic and G_{zz}, $\bar{G}_{\bar{z}\bar{z}}$ are the most natural choice.

8.2 Weyl Anomaly

In CFT, there is another anomaly we have to consider carefully, which is the *Weyl anomaly* coming from the central charge of the Virasoro algebra

$$[L_m, L_n] = (m-n)L_{m+n} + \frac{c}{12}m(m^2-1)\delta_{m+n}, \tag{75}$$

and similarly for the right moving modes $\bar{L}_m$.

To twist a theory, we need to use the R-symmetry, which is a $U(1)$ symmetry. Thus by the Noether theorem, there exist associated conserved currents $J(z)$, $\bar{J}(\bar{z})$. For open strings, $\bar{J}(\bar{z})$ is the complex conjugate of $J(z)$, on the other hand for closed strings, they are independent. The modes of J satisfy

$$[L_m, J_n] = -nJ_{m+n}, \qquad [J_m, J_n] = \frac{c}{3}m\delta_{m+n}, \tag{76}$$

and similarly for $\overline{J}_n$. Thus by using these currents J, $\overline{J}$, we can define the new stress tensors as

$$\tilde{T}^{\pm} = T \pm \frac{1}{2}\partial J, \qquad \tilde{\overline{T}}^{\pm} = \overline{T} \pm \frac{1}{2}\bar{\partial}\overline{J}, \tag{77}$$

and we denote their modes as $\tilde{L}_m^{\pm}$, $\tilde{\overline{L}}{}_m^{\pm}$. This is another important point of topological string theories that the new modes simply obey the Witt algebra, i.e. no central charge.

$$[\tilde{L}_m^{\pm}, \tilde{L}_n^{\pm}] = (m-n)\tilde{L}_{m+n}^{\pm}, \qquad [\tilde{\overline{L}}{}_m^{\pm}, \tilde{\overline{L}}{}_n^{\pm}] = (m-n)\tilde{\overline{L}}{}_{m+n}^{\pm}. \tag{78}$$

As a result, there is no Weyl anomaly.

It turns out that this shift of the stress tensors is equivalent to the A-twisting or B-twisting. In this sense, twisted string theories are somewhat more fundamental to define anomaly-free consistent theories. For simplicity, let us choose $+$ for both definitions in (77) then the zero modes are

$$\tilde{L}_0 = L_0 - \frac{1}{2}J_0, \tag{79}$$

and similarly for the left-moving mode. The generators of the R-symmetry and the new Lorentz symmetry $\tilde{M}$ are defined as

$$F_L = 2\pi i J_0, \qquad F_R = 2\pi i \bar{J}_0, \tag{80}$$

$$\tilde{M} = 2\pi i(\tilde{L}_0 - \tilde{\overline{L}}_0) = M - \frac{1}{2}(F_L - F_R), \tag{81}$$

where M is the generator of the Lorentz symmetry before the shift. (29) implies that $F_V + F_A$ only acts on $+$-indices, i.e. left-moving indices and $F_V - F_A$ on left-moving indices. That is, they are the generators of left- and right-moving currents and it is natural to identify them with F_L and F_R. More accurately, they are identified as

$$F_V = \frac{1}{2}(F_L + F_R), \qquad F_A = \frac{1}{2}(F_L - F_R). \tag{82}$$

Thus (81) is none other than the Lorentz generator for the B-model and a similar argument works for the A-model if one choose − sigh in the $\tilde{\overline{T}}$ shift (77).

Further discussions about topological string theories, in particular nonlocal operators, are left to [2].

References

1. Bouchard V. Lectures on complex geometry, Calabi–Yau manifolds and toric geometry. arXiv preprint hep-th/0702063. 2007 Feb 8.
2. Vonk M. A mini-course on topological strings. arXiv preprint hep-th/0504147. 2005 Apr 18.
3. Witten E. Mirror manifolds and topological field theory. arXiv preprint hep-th/9112056. 1991 Dec 19.

An Overview of B-branes in Gauged Linear Sigma Models

Nafiz Ishtiaque

1 Introduction

Two dimensional superconformal field theories (SCFTs) with $\mathcal{N} = (2, 2)$ supersymmetry have been a topic of much investigation for a long time, partly due to their essential role in string theory. Some natural objects to study in any field theory are boundaries in the space-time manifold preserving various amount of symmetry. In case of supersymmetric theories, in the presence of boundaries, at most half of the supersymmetry can be preserved. The boundaries preserving this maximal amount of supersymmetry are known as BPS D-branes. The BPS D-branes in $\mathcal{N} = (2, 2)$ SCFTs are particularly interesting objects as they provide the best (most manageable) probes of "stringy" geometry that we have so far. In particular, they have played a central role in our understanding of string dualities.[1] There are two inequivalent subalgebras of $\mathcal{N} = (2, 2)$ superconformal algebras (SCAs) containing half of the supersymmetry, they are known as $\mathcal{N} = 2_A$ and $\mathcal{N} = 2_B$ SCAs, and the BPS D-branes that preserve these subalgebras are called the *A-branes* and the *B-branes* respectively. The principal objects of study in this review are the B-branes.

The $\mathcal{N} = (2, 2)$ SCFTs have a moduli space with a product structure: $\mathcal{M}_{\text{SCFT}}^{\mathcal{N}=(2,2)} = \mathcal{M}_K \times \mathcal{M}_C$, where $\mathcal{M}_K$ and $\mathcal{M}_C$ correspond to the Kähler and complex structure moduli of some Calabi–Yau (CY) manifold. The infinitesimal deformations (preserving superconformal symmetry) of these SCFTs are generated by exactly marginal operators which correspond to the tangents to these moduli spaces. It is usually difficult, beyond perturbation theory, to study the algebraic structures of operators in full generality of a quantum field theory (QFT). But, the exactly marginal operators

N. Ishtiaque (✉)
Perimeter Institute for Theoretical Physics, University of Waterloo, Waterloo, ON, Canada
e-mail: nishtiaque@perimeterinstitute.ca

[1]Of particular relevance to the topic of this review is the mirror symmetry, where D-branes underlie both the homological approach of Kontsevich [1] and the T-duality perspective of Strominger, Yau, and Zaslow [2].

M. Ballard et al. (eds.), *Superschool on Derived Categories and D-branes*, Springer Proceedings in Mathematics & Statistics 240,
https://doi.org/10.1007/978-3-319-91626-2_16

corresponding to $\mathscr{M}_K$ and $\mathscr{M}_C$ belong to two special subsectors of operators, known respectively as the twisted chiral ring ($\mathscr{R}_{\text{tc}}$) and the chiral ring ($\mathscr{R}_{\text{c}}$), that can be studied in isolation with exact precision. These rings, $\mathscr{R}_{\text{c}}$ and $\mathscr{R}_{\text{tc}}$, are independent of Kähler deformations and complex structure deformations respectively. Another defining feature of $\mathscr{R}_{\text{c}}$ and $\mathscr{R}_{\text{tc}}$ is that they correspond to the $\boldsymbol{Q}$-cohomology of operators for a supercharge $\boldsymbol{Q}$ in $\mathscr{N} = 2_B$ and $\mathscr{N} = 2_A$ algebras respectively. This leads us to the fact that the *boundary chiral ring* and the *boundary twisted chiral ring*[2] contain some supersymmetry invariant data about the B-branes and the A-branes respectively. These supersymmetric data are of particular interest due to the fact that they are protected under renormalization (RG), which facilitates many exact computations. Furthermore, the chiral (twisted-chiral) sector of the B-branes (A-branes) should be independent of the Kähler (complex structure) moduli and should vary with the complex structure (Kähler) moduli.

The independence of the boundary chiral ring from the Kähler moduli in presence of a B-brane does not mean that the B-branes are completely unaffected by a change in the Kähler moduli, what it means is that the chiral sectors of the B-branes over different points of the Kähler moduli space are *isomorphic* to each other,[3] often quite non-trivially. Finding a unifying description of the chiral sectors of the B-branes over the bulk of the Kähler moduli space was the principal motivation and result of the comprehensive work [3] and is the topic of this review.

A priori, it is a difficult task to relate the B-branes at different points of the Kähler moduli space $\mathscr{M}_K$, because at different points, the low energy theory can look drastically different with different looking descriptions for the B-branes which are not trivially identified. For example, in some region of $\mathscr{M}_K$, the theory may be described by a non-linear sigma model on a non-compact Calabi–Yau and in another region by an orbifold theory with a discrete gauge group. The B-branes in these two theories are described as objects in some derived categories associated to the Calabi–Yau and the orbifold. The fact that these two categories are equivalent is a non-trivial statement which is familiar in the math literature as categorical McKay correspondence [4, 5]. Another example of such equivalence is when the low energy theory at two different regions are described by a non-linear sigma model on a compact Calabi–Yau and a Landau-Ginzburg (LG) type theory. The relevant categorical equivalence in this case, between a derived category of CY hypersurfaces defined by some polynomial and the category of matrix factroization of the same polynomial, was established by Orlov [6]. Explaining these equivalences from a physical definition of the B-branes and their chiral sectors was part of the motivation for the paper [3].

The main tool in overcoming the obstacle of having different looking theories at different places in $\mathscr{M}_K$, is to use an ultraviolet (UV) description of the theories, called the gauged linear sigma model (GLSM) [7], which includes as parameters the coordinates of $\mathscr{M}_K$. Depending on the values of these parameters it flows to the various infrared (IR) descriptions under renormalization. Therefore, having a description of the B-branes in the GLSM allows to see the fate of these B-branes

[2]Elements of the chiral and the twisted chiral rings that can be placed at a boundary.

[3]In a categorical sense, which we will make a bit clearer over the course of the review.

under RG flow and makes the connection between the IR B-branes in different regions of $\mathcal{M}_K$ physically transparent.

In this spirit we begin, in Sect. 2 by reviewing some basic facts about GLSMs and their low energy description in the absence of boundary. All of this and much more was discussed at length by Witten in [7]. In Sect. 3 we review the characterization of $\mathcal{N} = 2_B$ preserving boundaries in GLSM and the boundary chiral rings. We will see that these branes can be thought of as objects in a homotopy category of graded modules for certain rings and the chiral ring elements correspond to the morphisms in the category. In the last section, Sect. 4, we review the identification of the GLSM B-branes and the IR B-branes, and the mechanism of relating the chiral sectors of these IR B-branes in different regions of $\mathcal{M}_K$. It turns out that the relation between the GLSM B-branes and the IR B-branes is many to one and the IR branes are better described as objects in some derived category. These last two sections comprise a much abridged summary of a portion of the extensive work [3]. In the appendix we collect some background information about the $\mathcal{N} = (2, 2)$ supersymmetry algebra and some general remarks about Chan–Paton spaces relevant for describing the branes.

We should mention as a disclaimer that the main focus of this review is to point out the emergence of some mathematical notions purely from physical considerations. We will not go into much details of the relevant mathematics and we will often be rather schematic. A good (physics friendly) reference for much of the mathematics involved is [8].

2 GLSM Without Boundary

On a $(1 + 1)$D space-time Σ (also called the *world-sheet*), $\mathcal{N} = (2, 2)$ GLSMs are supersymmetric QFTs of maps:

$$x : \Sigma \to \mathbb{C}^N , \tag{1}$$

with some gauge group T. Using coordinates for $\mathbb{C}^N$, the map x can be thought of as a collection of N maps $x = (x^1, \ldots, x^N)$ where for any $p \in \Sigma$, $x^i(p)$ is the ith coordinate of $x(p)$. We will only consider compact abelian gauge groups $T \cong U(1)_1 \times \cdots \times U(1)_k$. Let us denote the charge of x^i under $U(1)_a$ by Q_i^a, i.e., an arbitrary element $g := (e^{i\phi_a}, \ldots, e^{i\phi_k}) \in T$ acts on the fields (1) as:

$$g : x^i \mapsto e^{i \sum_{a=1}^{k} Q_i^a \phi_a} x^i . \tag{2}$$

Note that for $k \geq N$, a $(k - N)$-dimensional subalgebra of the gauge algebra $\bigoplus_{a=1}^{k} u(1)_a$ acts trivially on the fields, therefore without loss of generality we assume $k \leq N$. Supersymmetry introduces superpartners for the fields x^i and we end up with *chiral multiplets* $\Phi^i = (x^i, \psi^i, F^i)$ where ψ^i's are Dirac fermions and F^i's are complex scalars. Similarly the gauge fields for the gauge group sit in vector multiplets $V_a = (\sigma_a, v_{a,\mu}, \lambda_a, D_a)$ for $a \in \{1, \ldots, k\}$, where σ_a are complex scalars, v_a is the 1-form gauge field for $U(1)_a$, λ_a is a Dirac fermion, and D_a's are real scalars. If there exists a gauge invariant holomorphic polynomial $W : \mathbb{C}^n \to \mathbb{C}$ then this can be included as additional data defining the theory. With all these data given, and upon defining the *field strength superfields* $\Sigma_a := \overline{D}_+ D_- V_a$ (which are twisted chiral), the Lagrangian of the theory, *in the absence of space-time boundary*, can be written most conveniently as a sum of several superspace integrals (Berezin integrals):

$$\mathscr{L}_{\text{bulk}} = \int \mathrm{d}^2\theta \mathrm{d}^2\overline{\theta} \left(-\frac{1}{2} \sum_{a,b=1}^{k} (e^{-2})_{ab} \overline{\Sigma}_a \Sigma_b + \sum_{i=1}^{N} \overline{\Phi}^i e^{\sum_{a=1}^{k} Q_i^a V_a} \Phi^i \right)$$
$$+ \Re \int \mathrm{d}\theta^+ \mathrm{d}\overline{\theta}^- \left(- \sum_{a=1}^{k} t^a \Sigma_a \right) + \Re \int \mathrm{d}\theta^+ \mathrm{d}\theta^- \, W(\Phi) \,. \tag{3}$$

In order to write the above Lagrangian we have introduced the gauge couplings e^2_{ab} (which are real numbers) and (e^{-2}) refers to the inverse of the $k \times k$-matrix e^2_{ab}. We have also introduced the complexified *Fayet-Iliopoulos* (FI) parameters:

$$t^a := r^a - \frac{i\theta^a}{2\pi} \,, \tag{4}$$

where r^a's are called the *real* FI parameters and θ^a's the topological theta angles.

2.1 Symmetries

The Lagrangian (3) has manifest (2, 2) supersymmetry since it is composed of (2, 2) superspace integrals.[4] Other global symmetries include space-time isometry (the Lorentz symmetry) $U(1)_L$ and an *axial* $U(1)_A$ R-symmetry.[5] If the superpotential is quasi-homogeneous, i.e., if it satisfies:

$$W(\xi^{p_i} x^i) = \xi^2 W(x^i) \,, \tag{5}$$

for some numbers $p_i \in \mathbb{R}$, then there is a *vector* $U(1)_V$ R-symmetry as well. The action of these symmetry groups and the gauge group can be summerized by mentioning the respective charges of various objects:

---

[4]Which implies that the supersymmetry variation of this Lagrangian is a total derivative, and in the absence of boundaries the action is invariant.

[5]An *R-symmetry* is a symmetry that acts nontrivially on the fermionic coordinates of the superspace and is not part of the space-time isometry.

	Φ^i	Σ	$\theta^\pm$	$\overline{\theta}^\pm$
$U(1)_L$	0	0	$\pm$	$\pm$
$U(1)_a\,,\ a \in \{1,\ldots,k\}$	Q_i^a	0	0	0
$U(1)_V$	p_i	0	$+$	$-$
$U(1)_A$	0	2	$\pm$	$\mp$

(6)

Though the axial R-symmetry is always present classicaly, it is generically broken quantum mechanically due to anomaly of the fermion (of the chiral multiplets) measure:

$$\delta_{u(1)_A}\left(\prod_{i=1}^N \mathscr{D}\psi^i \mathscr{D}\overline{\psi}^i\right) \propto \sum_{a=1}^k \sum_{i=1}^N Q_i^a c_1(E_a) \prod_{i=1}^N \mathscr{D}\psi^i \mathscr{D}\overline{\psi}^i\,, \tag{7}$$

where E_a denotes the $U(1)_a$-bundle over Σ and $c_1(E_a) \in \mathbb{Z}$ refers to its first Chern number. Note that, the twisted chiral superspace integral from the Lagrangian (3), in terms of the component fields, become:

$$\Re \int d\theta^+ d\overline{\theta}^- \left(-\sum_{a=1}^k t^a \Sigma_a\right) = -\sum_{a=1}^k \left(r^a D_a + \frac{\theta^a}{2\pi} v_{a,01}\right)\,, \tag{8}$$

where $v_{a,01} := \partial_0 v_{a,1} - \partial_1 v_{a,0}$ is the gauge field strength. Since

$$\frac{1}{2\pi}\int_\Sigma d^2y\, v_{a,01} = c_1(E_a) \in \mathbb{Z}\,, \tag{9}$$

and in a path integral θ-angles will only appear in exponentials

$$\exp\left(\frac{i\theta_a}{2\pi}\int d^2y\, v_{a,01}\right)\,, \tag{10}$$

the θ-angles are truly angular variables, $\theta_a \sim \theta_a + 2\pi$. Furthermore, we see that a change of the θ-angles is equivalent to a field redefinition via an axial R-rotation when (7) is nonzero. Therefore, in such cases the θ-angles are not actual parameters of the theory. They become true parameters when the following condition is met:

$$\sum_{i=1}^N Q_i^a = 0 \quad \forall a \in \{1,\ldots,k\}\,. \tag{11}$$

This condition is called the *Calabi–Yau condition* because when this condition holds, for some range of the FI parameters, the linear sigma model flows in the infrared to a non-linear sigma model with a Calabi–Yau target space. Unlike the axial R-symmetry, the vector R-symmetry is anomaly free whenever it exists as a classical symmetry (i.e., when the superpotential is quasi-homogeneous).

The $\mathcal{N} = (2, 2)$ superconformal algebra necessarily includes both the axial and the vector R-symmetry [9]. Therefore, unless both of these R-symmetries exist as quantum symmetries in the UV GLSM, it will not flow to a non-trivial IR conformal fixed point. Since our primary goal is to describe B-branes in the IR SCFTs, we will only consider GLSMs with quasi-homogeneous superpotentials (5) and gauge charges satisfying the CY condition (11) to ensure that both of the R-symmetries are preserved quantum mechanically.

2.2 *Low Energy Phases*

Performing the superspace integrals we get the Lagrangian in terms of the component fields. Then we find that the component fields D_a and F^i do not have any kinetic terms and therefore in a path integral we can perform the integrations over these fields, the outcome of which is simply to replace them by solving their classical equations of motion, which are algebraic.[6] These fields are called auxiliary fields and after integrating them out the classical potential for the rest of the bosonic scalars become:

$$U(\sigma, x) := \sum_{i=1}^{N} \left| \sum_{a=1}^{k} Q_i^a \sigma_a x^i \right|^2 + \frac{e^2}{2} \sum_{a=1}^{k} \left(\sum_{i=1}^{N} Q_i^a |x^i|^2 - r^a \right)^2 + \sum_{i=1}^{N} \left| \frac{\partial W(x)}{\partial x^i} \right|^2 , \tag{12}$$

where $x = (x^1, \ldots, x^N) \in \mathbb{C}^N$ and we have assumed a simple form of the gauge couplings $e_{ab}^2 = e^2 \delta_{ab}$ which makes no qualitative difference in our discussion.

2.2.1 Classical Analysis

A classical analysis of the space of vacua sheds some light on the qualitative nature of the low energy description of the theory. The classical space of vacua is the space of solutions of the equation:

$$U(\sigma, x) = 0 . \tag{13}$$

The qualitative nature of this space of solutions depends on the real FI parameters. Since the three terms of (12) are positive semi-definite, they must all separately vanish for U to vanish. Let us pick some *generic* values of the real FI parameters $r := (r^1, \ldots, r^k)$. Now we set the second term in (12) to zero (the resulting equations are called the *D-term* equations):

[6] It may seem a little redundant to introduce them in the first place. They serve the purpose of closing the supersymmetry algebra (on the fields) off shell (without using equations of motion) which is of paramount importance in some cases, such as in discussing supersymmetry in curved backgrounds, but they play no significant role for us at the moment.

$$\sum_{i=1}^{N} Q_i^a |x^i|^2 = r^a \,, \quad \forall a \in \{1, \ldots, k\} \tag{14}$$

The space of solutions to this system of equations is (real) $(N-k)$ dimensional and therefore at most $(N-k)$ of the x^i's can be simultaneously zero (due to the generic nature of the r^a's). By setting the first term of (12) to zero we find the following equations:

$$\sum_{a=1}^{k} Q_i^a \sigma_a = 0 \quad \forall i \in \{1, \ldots, N\} \quad \text{such that} \quad x^i \neq 0 \,. \tag{15}$$

Thus we have a homegenous system of at least k linear equations in k variables, the only solution being:

$$\sigma_a = 0 \quad \forall a \in \{1, \ldots, k\} \,. \tag{16}$$

This shows that for generic r, the gauge group is either completely broken or reduces to some discrete subgroup at low energy (i.e. no continuous degree of freedom left in the gauge multiplets). Note that not all of the solutions of (14) are physically distinguishable since some of them can be related by the action of the gauge group. Therefore, we should consider the quotient space:

$$X_r := \left\{ (x^1, \ldots, x^N) \in \mathbb{C}^N \,\middle|\, \sum_{i=1}^{N} Q_i^a |x^i|^2 = r^a \right\} \Big/ T = \mathbb{C}^N /\!/ T \,. \tag{17}$$

This is a symplectic quotient of $\mathbb{C}^N$ by T, since the D-term equations (14) can be interpreted as setting the moment maps of the T-action on $\mathbb{C}^N$ to zero. Equivalently, X_r can be written as a *toric variety*, i.e., an ordinary quotient of $\mathbb{C}^N$ (minus some "bad" points) by the *complexified* gauge group (a complex algebraic torus):

$$X_r = \left(\mathbb{C}^N \backslash \Delta_r\right) / T_{\mathbb{C}} \,, \tag{18}$$

where the *deleted set* Δ_r is the set of points whose $T_{\mathbb{C}}$-orbits do not pass through the space of solutions to the D-term equations (14). In this form it is clear how the dependence on the real FI parameters enter into the determination of the classical space of vacua, the real FI parameters determine the deleted set. Finally by setting the last term of the potential (12) to zero we find that the classical space of vacua is given by the intersection[7]:

$$\mathrm{Vac}_r^{\mathrm{cl}} := X_r \cap \left\{ x = (x^1, \ldots, x^N) \in \mathbb{C}^N \,\middle|\, \mathrm{d}W(x) = 0 \right\} \tag{19}$$

[7]The equations $\mathrm{d}W = 0$ are called the *F-term* equations. Also note that, as defined, the classical space of vacua $\mathrm{Vac}_r^{\mathrm{cl}}$ has a metric induced from the Fubiny-Study type metric on the toric variety X_r. If we could add all quantum corrections corresponding to integrating out all the massive modes, we would find that this metric (which appears in the kinetic term for the x^i's) gets modified to a Ricci-Flat (Calabi–Yau) metric [10].

The vacuum configurations where fields from chiral multiplets[8] acquire non-zero values, such as in the above description, are referred to as the *Higgs branch* of vacua. These vacua are characterized by the possibility of breaking the gauge group completely, or to some discrete subgroup.[9]

The geometry[10] of $\mathrm{Vac}_r^{\mathrm{cl}}$ varies smoothly with varying r, except when r falls into some (real codimension one) hypersurfaces where solutions to the D-term equations (14) include more than $(N-k)$ of the x^i's vanishing. In such non-generic cases, some of the σ_a's become unconstrained. Such hypersurfaces are invariant under scaling,[11] so the space of real FI parameters, let it be denoted by $\mathbb{R}^k_{\mathrm{FI}}$, becomes divided into cones, the cones are called *phases* of the low energy theory. The boundaries of these cones, i.e. the hypersurfaces, are called *phase boundaries* or *singular loci*.[12] At least classically, we can expect the nature of the low energy theory to be qualitatively the same in a given phase, but the nature changes significantly across phases.

If $W=0$, then all excitations transverse to the classical space of vacua $\mathrm{Vac}_r^{\mathrm{cl}} = X_r$ acquire masses from the second term of the potential (12), so the low energy theory consists of excitations tangential to X_r. In the absence of any residual discrete gauge symmetry, such theories are known as non-linear sigma models with target X_r. If some discrete gauge symmetry Γ survives and the space of vacua can be realized as a global orbifold $X_r = X_r'/\Gamma$, the low energy theories are known as orbifold theories [11]: Γ-gauged sigma models on X_r'.

In the presence of a non-trivial superpotential, there are phases where some excitations tangential to X_r but transverse to $\mathrm{Vac}_r^{\mathrm{cl}}$ picks up masses from the superpotential. The extreme example of such cases is when all such excitations become massive, i.e., only excitations tangential to $\mathrm{Vac}_r^{\mathrm{cl}}$ remain massless, these are called *purely geometric* phases or non-linear sigma models on $\mathrm{Vac}_r^{\mathrm{cl}}$. The opposite extreme is the *Landau-Ginzburg* phase where $\mathrm{Vac}_r^{\mathrm{cl}}$ consists of just a single point and all excitations of the fields x^i become massless, possibly with some discrete gauge symmetry.

Thus, classically, the real FI parameters $\mathbb{R}^k_{\mathrm{FI}}$ and any parameters that my appear in the superpotential, parametrize the low energy theories. The FI parameters[13] become the Kähler moduli, denoted as $\mathcal{M}_K$, of the target space for the low energy theory and the superpotential paratemeters become the complex structure moduli, denoted as $\mathcal{M}_C$. As we have seen, the classical Kähler moduli space is divided into disconnected phases by the singular loci. This picutre changes significantly with quantum corrections.

---

[8]More generally, fields that transform under faithful representations of the gauge group.

[9]Unlike fields from vector multiplets which transform under the adjoint representation and therefore can break the gauge group only upto the maximal torus.

[10]More specifically, the Kähler geometry.

[11]Since, if x is a solution of (14) for some r then $\sqrt{\xi}x$ is a solution of (14) for ξr.

[12]The singular nature arises from integrating out the σ_a's (which become massless on these loci) and trying to keep only the chiral multiplet fields as dynamical, which only works well away from the singularity.

[13]Classically we can only see the effects of the real FI parameters but after including quantum corrections, the true parameters of the low energy theories will be the complexified FI parameters (4).

2.2.2 Quantum Corrections

Quantum mechanically the space of low energy theories depend in addtion to the real FI parameters, on the θ-angles. In our classical analysis the phase boundaries (i.e. the singular loci) were places in the Kähler moduli space $\mathscr{M}_K$ where a Coulomb branch could open up, i.e., a complex scalar σ_a from the vector multiplets became unconstrained (massless). When this happens we can go to arbitrary large[14] values for this σ_a and this gives the chiral fields x^i large masses by the first term in the potential (12). So we can look at the one loop effect of integrating out these massive fields.[15] To see how the picture of the moduli space changes in the quantum theory we can assume large values for σ_a and look at its one loop effective potential $U_{\text{eff}}(\sigma)$ after integrating out the x^i's. The zero locus of this potential in terms of the parameters is the quantum corrected singular loci in $\mathscr{M}_K$. We will illustrate this procedure with a concrete example and we will see that the effective potential U_{eff} depends on the complexified FI parameter $t^a = r^a - \frac{i\theta^a}{2\pi}$ and the singular regions comprise a *complex* hypersurface in the space of these complex FI parameters. Therefore, unlike the classical result where the singular regions divided the $\mathbb{R}^k_{\text{FI}}$ into disconnected phases, in the quantum corrected picture the phases are all continuously connected.

Example 1 The example we consider is a $U(1)$ gauge theory with the following chiral fields with gauge charges satisfying the Calabi–Yau condition (11):

Fields	x^1	x^2	x^3	x^4
Gauge charges, Q_i	1	1	1	−3

(20)

We consider the theory without a superpotential. There is one real FI parameter $r \in \mathbb{R}_{\text{FI}}$. In constructing the space of vacua X_r for $r > 0$ and $r < 0$, the respective deleted sets are:

$$\begin{aligned} \Delta_+ &= \{x^1 = x^2 = x^3 = 0\} \qquad \text{for } r > 0\,, \\ \Delta_- &= \{x^4 = 0\} \qquad \text{for } r < 0\,. \end{aligned} \tag{21}$$

At low energy the $r > 0$ phase is described by a non-linear sigma model on $X_+ := \mathscr{O}_{\mathbb{CP}^2}(-3)$ and the $r < 0$ phase is described by the orbifold theory on $X_- := \mathbb{C}^3/\mathbb{Z}_3$.[16] Classically these two phases are separated by the singularity at the origin.

For large σ, i.e. $|\sigma| \gg e$, the effect of integrating out the massive chiral fields is to introduce the following effective action for σ [7, 10]:

$$U_{\text{eff}}(\sigma) = \frac{e^2_{\text{eff}}}{2}\,|r_{\text{eff}} - i\theta_{\min}|^2\,. \tag{22}$$

[14] Compared to the scale of theory set by the gauge coupling e, which has mass dimension 1.

[15] The one loop contribution to the effective potential is *exact* due to a non-renormalization theorem for the twisted superpotential.

[16] For $r > 0$, the base $\mathbb{CP}^2$ is the quotient $\{\mathbb{C}^3\backslash\Delta_+\}/\mathbb{C}^\times$ by the complexified gauge group, where the $\mathbb{C}^3$ is spanned by (x^1, x^2, x^3), and x^4 becomes the coordinate on the fiber. For $r < 0$, $\{\mathbb{C}\backslash\Delta_-\}/\mathbb{C}^\times$ is a point which is invariant under a discrete gauge group $\mathbb{Z}_3 \subset U(1)$ and (x^1, x^2, x^3) spans the $\mathbb{C}^3$ carrying a nontrivial representation of $\mathbb{Z}_3$.

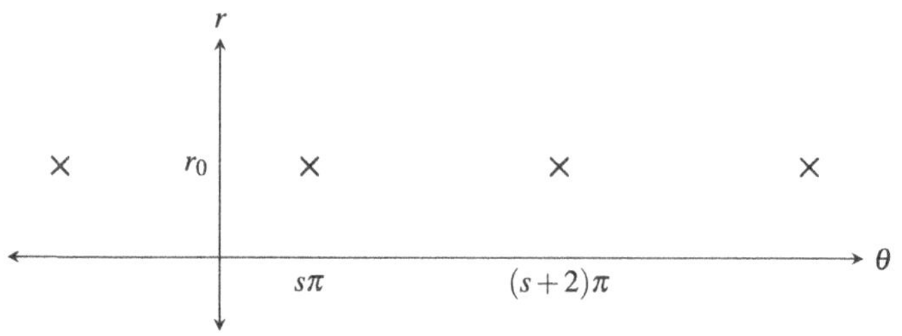

Fig. 1 A typical Kähler moduli space $\mathscr{M}_K$ with singularities, for some toric variety ($\dim_{\mathbb{C}} \mathscr{M}_K = 1$). The θ direction is 2π periodic

where, e_{eff} is the renormalized coupling, $r_{\text{eff}} = r + \sum_i Q_i \log|Q_i|$ and $|\theta_{\min}|^2 = \min_{n\in\mathbb{Z}} (\theta + s\pi + 2\pi n)^2$ with s being the sum of the positive charges, $s := \sum_{Q_i \geq 0} Q_i$. Thus, the solution to the equation $U_{\text{eff}}(\sigma) = 0$ is given by:

$$r = r_0 := -\sum_i Q_i \log|Q_i|\,, \qquad \theta = s\pi \ (\text{mod}\ 2\pi)\,. \tag{23}$$

So we see that the singular locus in the one complex dimensional Käler moduli space is a one complex codimensional hypersurface (Fig. 1).

3 Boundaries

Branes are subspaces of the target space where the world-sheet can end. Having boundaries of the world-sheet necessarily breaks some symmetries of the world-sheet theory. Partial symmetry can be preserved by introducing *boundary interactions*. We are interested in describing BPS D-branes preserving half ($\mathscr{N} = 2_B$ to be particular) of the world-sheet supersymmetry.

In the presence of boundaries the action $S_{\text{bulk}} = \int_\Sigma \mathrm{d}^2y\, \mathscr{L}_{\text{bulk}}$, where $\mathscr{L}_{\text{bulk}}$ is given by (3), is not supersymmetric since the supersymmetry variation of $\mathscr{L}_{\text{bulk}}$ is a total derivative and therefore the variation is a boundary integral. We add to it a boundary action:

$$S_{\text{bulk}} \ \to\ S = S_{\text{bulk}} + S_{\text{bdry}}\,, \qquad S_{\text{bdry}} = \int_{\partial\Sigma} \mathrm{d}y\, \mathscr{L}_{\text{bdry}}\,, \tag{24}$$

such that the variation of the boundary interaction cancels the variation of the bulk action for some supercharge $\mathscr{Q}$:

$$\delta_{\mathscr{Q}} S = 0\,. \tag{25}$$

If we can do this then the supercharge $\mathscr{Q}$ is preserved in presence of the boundary. It is only possible to preserve at most two of the four supercharges of the $\mathscr{N} = (2,2)$

algebra. In what follows we will be concerned with $\mathcal{N} = 2_B$ preserving boundaries, i.e., the B-branes. The explicit form of S_{bdry} that makes S 2_B-invariant can be found in [12, 13].

It turns out that the choice of boundary interaction, $\mathcal{L}_{\text{bdry}}$ in (24), is not unique. For two choices of boundary interactions, $\mathcal{L}^1_{\text{bdry}}$ and $\mathcal{L}^2_{\text{bdry}}$, the difference in the corresponding boundary actions, namely $\int_{\partial\Sigma} \mathrm{d}y \left(\mathcal{L}^1_{\text{bdry}} - \mathcal{L}^2_{\text{bdry}}\right)$, is necessarily 2_B-invariant on its own. Thus, different 2_B-invariant boundary actions give rise to different B-branes. In the following we discuss how to characterize these $\mathcal{N} = 2_B$ supersymmetric boundaries.

3.1 Chan–Paton Factors

Chan–Paton factors are essentially degrees of freedom associated to boundaries of our worldsheet. Various group actions can be defined on the boundary which leads to the introduction of these degrees of freedom and eventually associates rich structures to the boundaries which describe the branes. An introduction to these Chan–Paton spaces have been included in Sect. 1.B. Here we discuss the Chan–Paton spaces describing the GLSM B-branes.

3.1.1 Gauge Group, $U(1)^k$: Wilson Line Branes

In a gauge theory, a natural operator supported on a line defect or a boundary C is the Wilson line[17]:

$$\operatorname{Tr} W_\rho(C)\,, \qquad \text{with} \quad W_\rho(C) := \mathscr{P}\exp\left(i\int_C \rho_*(A)\right), \tag{26}$$

where A is the 1-form gauge field and ρ is a representation of the gauge group (so that the push forward ρ_* is a representation of the Lie algebra). For a line C with end points p_i and p_f, under a gauge transformation $A \to gAg^{-1} - igdg^{-1}$, the operator $W_\rho(C)$ transforms as:

$$W_\rho(C) \to \rho\left(g(p_f)\right) W_\rho(C)\rho\left(g^{-1}(t_i)\right)\,, \tag{27}$$

so for a closed loop C the operator $W_\rho(C)$ transforms under the adjoint representation of the gauge group and $\operatorname{Tr} W_\rho(C)$ is gauge invariant. In supersymmetric theories these operators are invariant under at most half of the supersymmetry or less, depending on the geometry of C and boundary conditions, and in order to preserve any amount

[17] The symbol $\mathscr{P}$ means *path ordered* which is a prescription to make sense of the exponential of the integral of a matrix valued connections in the cases where the matrices from different points are non-commuting.

of supersymmetry at all the exponential in (26) has to be properly modified, which can be done canonically.

In a $\mathcal{N}$-(2, 2) GLSM with gauge group $T = U(1)^k$ there are Wilson lines that are $\mathcal{N} = 2_B$ invariant. First let us consider irreducible representations of T. These representations are parametrized by k-tuples of weights, such as $q := (q^1, \ldots, q^k)$, where q^i is the weight for the ith $U(1)$ factor of the gauge group. Note that for a representation of $U(1)$, such as $U(1) \ni \lambda \mapsto \lambda^s$, to be truly a representation of $U(1)$ as opposed to that of a multiple cover of it, the weight s has to be an integer.[18] Therefore from now on we will assume that the weights q^i of the $U(1)$ factors of T are all integers, i.e., we assume $q \in \mathbb{Z}^k$. We refer to the irrep of $U(1)^k$ corresponding to the weight q as ρ^q:

$$\rho^q : U(1)^k \to U(\mathbb{C}) \cong U(1) , \qquad \rho^q : g \mapsto \prod_{a=1}^{k} g_a^{q^a} , \tag{28}$$

where $g := (g_1, \ldots, g_k)$ is an arbitrary element of $U(1)^k$. For concreteness of formulas, let us imagine a world-sheet with a boundary along the temporal direction. We are taking the y^0 and y^1 coordinates to correspond to time and space respectively and we are considering a boundary at $y^1 = 0$. Now, the Wilson line boundary interaction corresponding to this representation is (c.f. the log of (26)):

$$-\frac{i}{2} \int \mathrm{d}y^0 \int \mathrm{d}\theta \mathrm{d}\overline{\theta}\, \rho_*^q(V) = i \int \mathrm{d}y^0\, \rho_*^q(v_0 - \Re\sigma) , \tag{29}$$

where V is the vector superfield for $U(1)^k$ and the superspace integral is over the entire $\mathcal{N} = 2_B$ *boundary superspace* [13] making it manifestly 2_B-invariant. A brane characterized by such an elementary boundary interaction will be referred to as:

$$\mathscr{W}(q) . \tag{30}$$

More general representations of $U(1)^k$ are characterized by some number, say n, of k-tuple weights $\mathbf{q} := (q_1, \ldots, q_n)$ with $q_i \in \mathbb{Z}^k$, and a multiplicity for each of the weights. Let us denote the module for such a general representation by $\mathscr{V}_\mathbf{q}$ and the weight spaces for the weights q_i by $\mathscr{V}_{q_i}$. Then this representation can written as:

$$\rho^\mathbf{q} : U(1)^k \to U(\mathscr{V}_\mathbf{q}) , \qquad \rho^\mathbf{q} : g \mapsto \begin{pmatrix} \rho^{q_1}(g)\, \mathrm{id}_{\mathscr{V}_{q_1}} & & \\ & \ddots & \\ & & \rho^{q_n}(g)\, \mathrm{id}_{\mathscr{V}_{q_n}} \end{pmatrix} , \tag{31}$$

where ρ^{q_i}'s are the irreps from (28). We can interpret these representations by considering the irreducible representations as the basic boundary conditions and introduc-

[18] A representation $U(1) \ni \lambda \mapsto \lambda^{\frac{a}{b}}$ for some $a, b \in \mathbb{Z}$ with $\gcd(a, b) = 1$, is an $|a|$-fold representation of the $|b|$-fold cover of $U(1)$.

ing $\mathscr{V}_{\mathrm{q}}$ as a Chan–Paton space at the end points of the open strings (see Sect. 1.B).[19] The $\mathcal{N} = 2_B$ invariant Wilson line action for a general representation is a direct generalization of (29):

$$i \int \mathrm{d}y^0 \, \rho_*^{\mathrm{q}}(v_0 - \Re\sigma) \,. \tag{32}$$

Such a boundary action is a direct sum of elementary boundary actions, and the corresponding notation for a brane described by this boundary action is:

$$\bigoplus_{i=1}^{n} \mathscr{W}(q_i)^{\oplus \dim \mathscr{V}_{q_i}} \,. \tag{33}$$

This notation is limited in its expressibility of all the properties of the brane that we will describe in the rest of this section and for that reason we will not use this notation again until Sect. 4 where we will be primarily concerned with the gauge charges associated to a brane and this notation will be economic.

3.1.2 Fermion Number, $\mathbb{Z}_2$: Matrix Factorization

The Wilson line action (32) that we have discussed so far is $\mathcal{N} = 2_B$ invariant on its own and different choices of such terms correspond to different branes. What *all* B-branes have in common is the minimal boundary action that we must add to the bulk action to preserve $\mathcal{N} = 2_B$ supersymmetry. It turns out [3] that in the presence of nonzero superpotential, in order to preserve 2_B supersymmetry, we have to introduce a composite fermionic field, which we will call Q, living on the boundary of the world-sheet, and Q is a polynomial function of the chiral multiplet fields x^i. The Q-dependent part of the minimal boundary action is [3]:

$$\int \mathrm{d}y^0 \, \mathscr{I}_0 \,, \qquad \mathscr{I}_0 := -\frac{1}{2}\{Q, Q^\dagger\} + \frac{1}{2}\sum_{i=1}^{N}\left(\psi^i \frac{\partial}{\partial x^i} Q + \text{h.c.}\right) , \tag{34}$$

where $\psi^i := \psi_+^i + \psi_-^i$ and x^i are fields from the chiral multiplets restricted to the boundary. The fermionic field Q must satisfy some constraints in order to preserve 2_B supersymmetry which we will discuss shortly. But first note that, as a boundary operator Q acts on the Chan–Paton space $\mathscr{V}$ (just like the boundary operator $\rho_*^{\mathrm{q}}(v_0 - \Re\sigma)$ in the Wilson line boundary interaction (32)). A bit more precisely, since $Q \in \mathbb{C}[x^1, \dots, x^N] =: \mathbb{C}[x]$, Q acts as an operator on the $\mathbb{C}[x]$-module $\mathscr{V}_x := \mathbb{C}[x] \otimes_{\mathbb{C}} \mathscr{V}$. Since Q is a fermionic operator, it can be assigned a non-trivial charge under a $\mathbb{Z}_2$ group action under which Q^2 is bosonic. This makes the Chan–Paton space

[19] Note that when we mention string we simply mean that our space is one dimensional. In the honest string theory one needs to further couple the world-sheet SCFT to a ghost system to gauge world-sheet diffeomorphism.

$\mathbb{Z}_2$-graded with a graded action of Q:

$$\mathscr{V} = \mathscr{V}^{\text{od}} \oplus \mathscr{V}^{\text{ev}}, \qquad \mathscr{V}_x^{\text{od}} \underset{Q}{\overset{Q}{\rightleftarrows}} \mathscr{V}_x^{\text{ev}}, \tag{35}$$

where we have defined $\mathscr{V}_x^{\text{od/ev}} := \mathbb{C}[x] \otimes_{\mathbb{C}} \mathscr{V}^{\text{od/ev}}$. The $\mathbb{Z}_2$ charge measures the fermion number of a vector modulo 2, acting with the fermionic (odd) operator Q changes this charge by 1.

Now on to the constraints on Q. First of all, for the action (34) to be gauge invariant, we must impose on Q gauge equivariance:

$$\rho\left(g^{-1}\right) Q(g \cdot x) \rho\left(g\right) = Q(x)\,, \tag{36}$$

where ρ is the representation of $U(1)^k$ associated with the Chan–Paton space $\mathscr{V}$, $g \in U(1)^k$, and $x \in \mathbb{C}^N$. The whole purpose of the boundary action (34) is to cancel the variation of the bulk superpotential action (world-sheet integral of the last term in (3)).[20] To that end, let us note the variation of the superpotential action generated by a 2_B supersymmetry transformation parametrized by the Dirac spinor ε:

$$\delta\, \Re \int_{\Sigma} \mathrm{d}^2 y \int \mathrm{d}\theta^{+}\mathrm{d}\theta^{-}\; W(\Phi) = -\Re \int \mathrm{d}y^0 \sum_{i=1}^{N} \bar{\varepsilon}\psi^i \frac{\partial W(x)}{\partial x^i}\,. \tag{37}$$

Let us also write down the variation of the boundary interaction from (34) generated by ε (note that we are not assuming ε to be constant)[21]:

$$\delta \mathscr{I}_0 = \Re\left\{\sum_{i=1}^{N}\left(\bar{\varepsilon}\psi^i \frac{\partial Q^2(x)}{\partial x^i}\right) - \left[\bar{\varepsilon}Q^{\dagger}, Q^2\right]\right\} - i\mathscr{D}_0\left(\bar{\varepsilon}Q + \varepsilon Q^{\dagger}\right) + i\left(\dot{\bar{\varepsilon}}Q + \dot{\varepsilon}Q^{\dagger}\right), \tag{38}$$

where $\mathscr{D}_0$ is the gauge covariant derivative along the y^0 direction. The boundary integral of the first term in (38) cancels (37) if and only if Q satisfies:

$$Q^2(x) = W(x)\,\mathrm{id}_{\mathscr{V}}\,. \tag{39}$$

The above equation is referred to as a *matrix factorization* of W. The second term in (38) is a total derivative and does not contribute when integrated over the boundary.

[20] The rest of the minimal boundary action to cancel the variation of the world-sheet integral of the rest of the terms in (3) is not going to play any part in our discussion, so we are not going to talk about them. They can be found in [3, 12, 13].

[21] The actual 2_B supersymmetry is generated by constant ε, so in a variation generated by time dependent ε, any term that behaves as $\mathscr{O}(\dot{\varepsilon})$ can be ignored as far as symmetry preservation is concerned, but such terms help to recognize various contributions to the Noether charge associated to the symmetry.

The third term will vanish when we make the supersymmetry parameter ε constant, but it shows, via Noether's construction, that the supercharge generating $\mathcal{N} = 2_B$ supersymmetry gets modified at the boundary:

$$\mathcal{Q}_{\text{bulk}} \to \mathcal{Q}_B := \mathcal{Q}_{\text{bulk}} + \mathcal{Q}_{\text{bdry}}\,, \qquad \mathcal{Q}_{\text{bdry}} = iQ\big|_{y^1=0} - iQ\big|_{y^1=L}\,, \tag{40}$$

where we are considering two temporal boundaries at spatial positions $y^1 = 0$ and $y^1 = L$. Importantly, this boundary modifiation ensures the nilpotence of the total supercharge. The bulk supercharge, in the presence of a non-trivial superpotential and the boundary, is not nilpotent [3]:

$$\mathcal{Q}_{\text{bulk}}^2 = W\big|_{y^1=L} - W\big|_{y^1=0}\,, \tag{41}$$

and this is canceled by $\mathcal{Q}_{\text{bdry}}^2$ due to (39) so that we end up with the $\mathcal{N} = 2_B$ algebra (see (87)):

$$\mathcal{Q}_B^2 = 0\,. \tag{42}$$

Equations (36) and (39) are all the constraints that Q has to satisfy.

3.1.3 Vector R-Symmery, $U(1)_V$: Differential Grading

We are interested in GLSMs with vector $U(1)$ R-symmetry, denoted by $U(1)_V$, so that the theory flows to a non-trivial conformal fixed point in the extreme infrared. In that case, the bulk supercharge $\mathcal{Q}_{\text{bulk}}$ has $U(1)_V$-charge 1 (88). Since the boundary odd operator Q is now a part of the total supercharge along with $\mathcal{Q}_{\text{bulk}}$ (see (40)), we must assign the same $U(1)_V$-charge to Q. Consequently, the Chan–Paton space $\mathcal{V}$ must carry an action of $U(1)_V$. We pick a representation $R : U(1)_V \to U(\mathcal{V})$ with weights $(\nu_{\min}, \nu_{\min} + 1, \ldots, \nu_{\max})$ where $\nu_j \in \mathbb{Z}$. We require the representation to be such that the R-charge modulo 2 matches with the $\mathbb{Z}_2$ grading of $\mathcal{V}$, i.e., a $U(1)_V$-weight space decomposition of $\mathcal{V}$ satisfies:

$$\mathcal{V} = \bigoplus_{j=\nu_{\min}}^{\nu_{\max}} \mathcal{V}^j \qquad \text{with} \qquad \mathcal{V}^{\text{od}} = \bigoplus_{j\,:\,\text{odd}} \mathcal{V}^j\,, \quad \mathcal{V}^{\text{ev}} = \bigoplus_{j\,:\,\text{even}} \mathcal{V}^j\,, \tag{43}$$

where $\mathcal{V}^j$ is the $U(1)_V$-weight space with weight j so that the representation R can be explicitly written as:

$$R : \lambda \mapsto \begin{pmatrix} \lambda^{\nu_{\min}}\,\text{id}_{\mathcal{V}^{\nu_{\min}}} & & \\ & \ddots & \\ & & \lambda^{\nu_{\max}}\,\text{id}_{\mathcal{V}^{\nu_{\max}}} \end{pmatrix}, \qquad \lambda \in U(1)_V\,. \tag{44}$$

The R-symmetry commutes with the gauge symmetry, implying that each $\mathscr{V}^j$ is a direct sum of $U(1)^k$-weight spaces. Since Q increases the R-charge of a vector by 1, Q can be thought of as a collection of odd operators $Q = (\mathrm{d}_{\nu_{\min}}, \ldots, \mathrm{d}_{\nu_{\max}})$:

$$\cdots \xrightarrow{d_{j-1}} \mathscr{V}_x^j \xrightarrow{d_j} \mathscr{V}_x^{j+1} \xrightarrow{d_{j+1}} \cdots . \tag{45}$$

With a nonzero superpotential this is not a standard cochain complex since the differential Q is not nilpotent: $\mathrm{d}_{j+1} \circ \mathrm{d}_j = W\,\mathrm{id}_{\mathscr{V}_x^j}$ (due to (39)). This is sometimes referred to as a differential *twisted* by W. This can be remedied by defining the ring:

$$\mathscr{R}_W := \mathbb{C}[x^1, \ldots, x^n]/\langle W \rangle , \tag{46}$$

where $\langle W \rangle$ is the ideal generated by the polynomial W, and instead of working with the $\mathbb{C}[x]$-module $\mathscr{V}_x$, we now consider the following $\mathscr{R}_W$-module:

$$\mathscr{V}_W := \mathscr{R}_W \otimes_{\mathbb{C}} \mathscr{V} . \tag{47}$$

The good thing about this module is that W acts trivially on it, furthermore $\mathscr{V}_W$ inherits all the gradings of $\mathscr{V}$.[22] Therefore, we can build a genuine cochain complex using $\mathscr{V}_W$:

$$\cdots \xrightarrow{\mathrm{d}_{j-1}} \mathscr{V}_W^j \xrightarrow{\mathrm{d}_j} \mathscr{V}_W^{j+1} \xrightarrow{\mathrm{d}_{j+1}} \cdots . \tag{48}$$

Let us summerize our description of a GLSM B-brane:

A GLSM B-brane $\mathscr{B}$ is defined by the data $(\mathscr{V}, \rho, R, Q)$, where $\mathscr{V}$ is a Chan–Paton space carrying a representation ρ of the gauge group $T = U(1)^k$, a representation R of the $U(1)_V$ R-symmetry group that commutes with T, and a matrix factorization $Q^2 = W\,\mathrm{id}_{\mathscr{V}}$ of the superpotential W, where the T-equivariant operator Q has R-charge 1. All these data is encoded in the complex (48) of the *twisted* modules $\mathscr{V}_W^j = (\mathbb{C}[x]/\langle W \rangle) \otimes_{\mathbb{C}} \mathscr{V}^j$ where j refers to the $\mathbb{Z}$-grading by $U(1)_V$, i.e., $\mathscr{V}^j$ is the $U(1)_V$-weight space with weight j and the implicit $\mathbb{Z}^k$-gradings by T are compatible with this $\mathbb{Z}$-grading.

We will refer to the complex (48) associated to a B-brane $\mathscr{B}$ as $\mathscr{C}(\mathscr{B})$. In the next section we will relate the Q_B-invariant open string states to certain morphisms between these cochain complexes.

[22] We use the same notation to denote the gradings of $\mathscr{V}_W$ as we did for $\mathscr{V}$, just by replacing $\mathscr{V}$ with $\mathscr{V}_W$.

3.2 Open String States in the Chiral Sector

A satisfactory description of branes must provide two aspects: how to characterize the branes, and how to characterize the states in the Hilbert space that we get after quantizing a string[23] in the presence of these branes. By now we have some idea of how to do the former and in this section we discuss the latter in the limited context of BPS or chiral states.

3.2.1 Some Generalities and the Chiral Ring

The chiral sector of the open string Hilbert space consists of states that are invariant under $\boldsymbol{Q}_B$:

$$\boldsymbol{Q}_B|\Psi\rangle = 0\,. \tag{49}$$

This sector, owing to being $\boldsymbol{Q}_B$-invariant, is robust under various supersymmetry preserving deformations, in particular, it is robust under rescaling of the world-sheet metric,[24] i.e., renormalization. Therefore, the chiral states are particularly interesting if we care about the low energy dynamics of our theory. Analogous to (49), chiral operators are the $\boldsymbol{Q}_B$-invariant operators:

$$[\boldsymbol{Q}_B, \mathscr{O}] = 0\,. \tag{50}$$

The hermitian conjugates of (49) and the above equation define the *anti-chiral* states and operators respectively. There is in fact a one to one correspondence between chiral states and local chiral operators. This follows from the fact that the chiral states and operators are invariant under RG flow and therefore the state-operator correspondence that exists between them in the IR CFT works in the UV GLSM as well. In this review we are interested in chiral operators that can be inserted *on* the boundary. Once inserted, we can have two different $\mathcal{N} = 2_B$ boundary conditions on the two sides of this operator, and a path integral on a half-disc in a half-plane with such boundary conditions gives us the chiral state corresponding to this operator. This is essentially the state-operator correspondence between a chiral open string state between two boundaries of an infinite strip and a local chiral operator on the boundary of a half-plane with two boundary conditions on the two sides (see Fig. 2).

These chiral operators have non-singular operator product expansion among them [14] and this allows to define the following product of two chiral operators $\mathscr{O}_1$ and $\mathscr{O}_2$:

$$(\mathscr{O}_1\mathscr{O}_2)(y) := \lim_{y'\to y} \mathscr{O}_1(y)\mathscr{O}_2(y')\,. \tag{51}$$

[23] A time slice of the worldsheet.

[24] The chiral sector is invariant under any continuous world-sheet metric deformation.

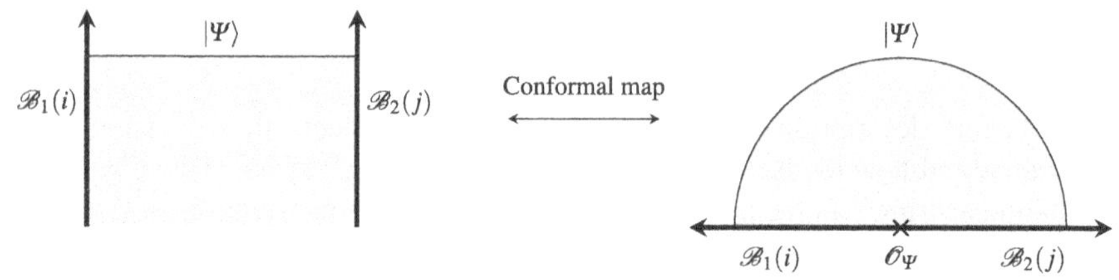

Fig. 2 State-operator correspondence: The operator $\mathscr{O}_\Psi$ on the right hand side is defined so that a path integral on the shaded region in the upper half-plane with the boundary conditions $\mathscr{B}_1(i)$ and $\mathscr{B}_2(j)$ produces the wave function for the state $|\Psi\rangle \in \mathscr{H}(\mathscr{B}_1(i), \mathscr{B}_2(j))$ (c.f. (89))

The product of two chiral operators is again a chiral operator and this turns the chiral sector into a ring called the *chiral ring*.[25] If we denote the chiral ring between two branes[26] $\mathscr{B}_i$ and $\mathscr{B}_j$ as $\mathscr{H}_{Q_B}(\mathscr{B}_i, \mathscr{B}_j)$ then the OPE defines a map as follows:

$$\mathscr{H}_{Q_B}(\mathscr{B}_1, \mathscr{B}_2) \times \mathscr{H}_{Q_B}(\mathscr{B}_2, \mathscr{B}_3) \to \mathscr{H}_{Q_B}(\mathscr{B}_1, \mathscr{B}_3)\,. \tag{52}$$

In constructing the chiral ring we are not only restricting to operators obeying (50), we also identify $\boldsymbol{Q}_B$-exact operators[27] with zero because all correlation functions involving a $\boldsymbol{Q}_B$-exact operator and other chiral operators are zero:

$$\langle [\boldsymbol{Q}_B, \mathscr{O}_1]\mathscr{O}_2\rangle = 0\,, \tag{53}$$

which follows from a Ward identity[28] for $\boldsymbol{Q}_B$ and the chirality of $\mathscr{O}_2$. Therefore, the chiral ring is really the $\boldsymbol{Q}_B$-cohomology of the local operators.

3.2.2 The Chiral Ring Between Two Branes

The local operators of the theory are arbitrary functions of the fields. In the presence of a boundary, in order to formulate a well posed initial value problem, the dynamical fields need to be constrained by boundary conditions (such as Dirichlet, Neuman or generalizations thereof). These boundary conditions can break some symmetries of the bulk theory, we of course want to preserve $\mathscr{N} = 2_B$ supersymmetry. Before mentioning the 2_B-invariant boundary conditions, let us take a simplifying limit.

[25] The chiral ring is also an aglebra over $\mathbb{C}$, but it is more commonly referred to as a ring.

[26] In the language of Sect. 1.B, each brane $\mathscr{B}_i$ comes with an index set $\mathscr{I}_i$ of the Chan–Paton indices, so that for each index $k \in \mathscr{I}_i$ we have a basic boundary condition $\mathscr{B}_i(k)$ associated to the brane $\mathscr{B}_i$, and the Hilbert space for the open string between $\mathscr{B}_1$ and $\mathscr{B}_2$ has the following decomposition: $\mathscr{H}(\mathscr{B}_1, \mathscr{B}_2) = \bigoplus_{\substack{i\in\mathscr{I}_1\\ j\in\mathscr{I}_2}} \mathscr{H}(\mathscr{B}_1(i), \mathscr{B}_2(j))$.

[27] Which are automatically chiral due to the nilpotence of $\boldsymbol{Q}_B$.

[28] Which simply says that the expectation value of the supersymmetry variation of an operator in a supersymmetric vacuum is zero.

We will construct the chiral boundary local operators in the far ultraviolet, which corresponds to the limit $e \to 0$. Even though the chiral ring is RG invariant, there are some subtleties involved with this limit, which we are going to ignore in this review, for details see [3]. In this limit, the coefficient in front of the vector multiplet kinetic terms (see (3)) diverges and those fields freeze to their classical values, becoming backgrounds. So we only have the chiral multiplet fields to construct the chiral ring with. In the limit $e \to 0$, the standard $\mathcal{N} = 2_B$ invariant boundary conditions are Neuman:

$$\partial_1 x^i = 0\,, \qquad \psi^i_+ = \psi^i_-\,, \qquad \partial_1(\psi^i_+ + \psi^i_-) = 0\,, \qquad F^i = 0\,. \tag{54}$$

We will make the further assumption that our strings are sufficiently small compared to the length scale of the target space. This allows us to treat the string as point particle and ignore stringy excitations. Since we are assuming the target space volume to be too large compared to the string length, this approximation is called the *large volume limit*. It is also known as *zero mode approximation* becuase, since we are ignoring the stringy excitations, we will only use the zero modes of the fields to construct our operators. It turns out though, that this approximation is exact for the chiral ring.[29] This can be argued by showing that the chiral ring is independent of any length scale on the world-sheet. With these approximations, the complex computing $\boldsymbol{Q}_B$-cohomology of the *bulk* chiral operators can be constructed just from the zero modes of x^i, $\overline{x}^i$, and $\overline{\psi}^i_\pm$ [15].[30] When we put these operators on the boundary, the boundary conditions (54) further reduce the set of fields by setting $\overline{\psi}^i_+ - \overline{\psi}^i_-$ to zero. The action of the bulk supercharge $\boldsymbol{Q}_{\text{bulk}}$ on the fields we have left is as follows[31]:

$$[i\boldsymbol{Q}_{\text{bulk}}, x^i] = 0\,, \qquad [i\boldsymbol{Q}_{\text{bulk}}, \overline{x}^i] = \overline{\psi}^i_+ + \overline{\psi}^i_-\,, \qquad \{i\boldsymbol{Q}_{\text{bulk}}, \overline{\psi}^i_+ + \overline{\psi}^i_-\} = 0\,. \tag{55}$$

If we identify $\overline{\psi}^i_+ + \overline{\psi}^i_-$ with $\mathrm{d}\overline{x}^i$, then the action of $\boldsymbol{Q}_{\text{bulk}}$ coincides with the action of the Dolbeault differential $\overline{\partial}$. The action of the full superchage $\boldsymbol{Q}_B$ will also include the action of the boundary supercharge $\boldsymbol{Q}_{\text{bdry}}$ which we will discuss later.

The operators we can construct from x^i, $\overline{x}^i$ and $\mathrm{d}\overline{x}^i$ correspond to anti-holomorphic forms in $\Omega^{0,\bullet}(\mathbb{C}^N)$ and to get the chiral ring we take the $\boldsymbol{Q}_B$-cohomology. Let us clarify all the data we have. We are interested in chiral operators interpolating between two branes, let us say $\mathscr{B}_1 = (\mathscr{V}_1, \rho_1, R_1, Q_1)$ and $\mathscr{B}_2 = (\mathscr{V}_2, \rho_2, R_2, Q_2)$. Then according to the general discussion of Sect. 1.B, the operators should be valued in $\mathrm{Hom}(\mathscr{V}_1, \mathscr{V}_2)$. The $\mathrm{Hom}(\mathscr{V}_1, \mathscr{V}_2)$-valued anti-holomorphic forms are charged

[29] This is a nontrivial result, for example, it is not true for the twisted chiral ring ($\boldsymbol{Q}_A$-cohomology) which receives world-sheet instanton corrections.

[30] Let us say $\psi^i_+\mathcal{O}$ is chiral for some operator $\mathcal{O}$. Then $0 = [\boldsymbol{Q}_{\text{bulk}}, \psi^i_+\mathcal{O}] = -2(\partial_+ x^i)\mathcal{O} - \psi^i_+[\boldsymbol{Q}_{\text{bulk}}, \mathcal{O}]$ where $2\partial_+ = \partial_0 + \partial_1$. If $\mathcal{O}$ is chiral then $2(\partial_+ x^i)\mathcal{O} = 0$ implies $\mathcal{O}$ is zero and so is $\psi^i_+\mathcal{O}$. If $\mathcal{O}$ is not chiral then $2(\partial_+ x^i)\mathcal{O} + \psi^i_+[\boldsymbol{Q}_{\text{bulk}}, \mathcal{O}] = 0$ implies $\mathcal{O}$ must be of the form $\psi^i_+\mathcal{O}'$ for some $\mathcal{O}'$ in which case $\psi^i_+\mathcal{O} = \psi^i_+\psi^i_+\mathcal{O}' = 0$ since ψ^i_+ is anti-commuting.

[31] The action of $\mathcal{N} = (2, 2)$ supersymmetry on the fields of GLSM can be found in [3, 7].

under both the gauge group and the $U(1)_V$ R-symmetry group. The R-charge is relevant for the differential grading since the supercharge $\boldsymbol{Q}_B$ has R-charge 1. So we will write $\mathrm{Hom}^\bullet(\mathscr{V}_1, \mathscr{V}_2)$ as a reminder that these maps carry $\mathbb{Z}$-grading by $U(1)_V$. Furthermore, only gauge invariant operators are physical. Now the chiral ring elements with R-charge p are given by the $\boldsymbol{Q}_B$-cohomology:

$$\mathscr{H}^p_{\boldsymbol{Q}_B}(\mathscr{B}_1, \mathscr{B}_2) = H^p_{\boldsymbol{Q}_B}\left(\Omega^{0,\bullet}\left(\mathbb{C}^N, \mathrm{Hom}^\bullet(\mathscr{V}_1, \mathscr{V}_2)\right)^T\right), \tag{56}$$

where the degree of the complex is a sum of the form degree and the degree of the morphisms in $\mathrm{Hom}^\bullet(\mathscr{V}_1, \mathscr{V}_2)$.

The action of $\boldsymbol{Q}_B$ on an operator $\mathscr{O}$ on the boundary between $\mathscr{B}_1$ and $\mathscr{B}_2$ is given by:

$$i\boldsymbol{Q}_B : \mathscr{O} \mapsto \overline{\partial}\mathscr{O} + Q_2\mathscr{O} - (-1)^{|\mathscr{O}|}\mathscr{O}Q_1, \tag{57}$$

where $|\mathscr{O}|$ is the R-charge[32] of $\mathscr{O}$ and the action of the boundary supercharges comes from Noether's construction of the supercharge in presence of the boundaries. Using the property of $\mathbb{C}^N$ that any $\overline{\partial}$-closed p-form on $\mathbb{C}^N$ is $\overline{\partial}$-exact for $p > 0$, it can be shown that the cohomologies (56) are concentrated on *holomorphic or polynomial functions* [3], i.e.:

$$\mathscr{H}^p_{\boldsymbol{Q}_B}(\mathscr{B}_1, \mathscr{B}_2) \cong H^p_{\boldsymbol{Q}_B^{\mathrm{hol}}}\left(\mathrm{Hom}^\bullet(\mathscr{V}_1, \mathscr{V}_2)^T\right). \tag{58}$$

Restricted to the subspace of holomorphic functions, the action of the supercharge (57) simplifies to

$$i\boldsymbol{Q}_B : \mathscr{O} \mapsto Q_2\mathscr{O} - (-1)^{|\mathscr{O}|}\mathscr{O}Q_1. \tag{59}$$

Now recall from the end of Sect. 3.1 that the data of a B-brane $\mathscr{B}$ can be encoded in a cochain complex that we called $\mathscr{C}(\mathscr{B})$ (48). These complexes were built out of $\mathscr{R}_W$-modules $\mathscr{V}_W$ (see (47) and (46)) which were $\mathbb{Z}^k$-graded (by the gauge group $T = U(1)^k$).[33] Gauge invariant chiral holmorphic functions valued in $\mathrm{Hom}(\mathscr{V}_1, \mathscr{V}_2)$ now precisely correspond to chain maps[34] between $\mathscr{C}(\mathscr{B}_1)$ and $\mathscr{C}(\mathscr{B}_2)$ as graded $\mathscr{R}_W$-modules.[35] Identifying a chain map of the form $[\boldsymbol{Q}_B, \mathscr{O}]$ with zero for some arbitrary local operator $\mathscr{O}$, as we do in the chiral ring, is then equivalent to identifying chain maps upto homotopy. Thus we reach the conclusion that the chiral ring elements between two branes $\mathscr{B}_1$ and $\mathscr{B}_2$ are in one to one correspondence with the cochain maps upto homotopy, between $\mathscr{C}(\mathscr{B}_1)$ and $\mathscr{C}(\mathscr{B}_2)$ where these are $\mathbb{Z}^k$-graded cochain

---

[32] Which is a sum of the form degree (as $\overline{\psi}^i_+ + \overline{\psi}^i_-$ has R-charge 1) and the degree of the $\mathrm{Hom}(\mathscr{V}_1, \mathscr{V}_1)$ part (which is due to the x^i's and $\overline{x}^i$'s).

[33] In addition to the (differential) $\mathbb{Z}$-grading by $U(1)_V$.

[34] The condition of chirality $\boldsymbol{Q}_B(\mathscr{O}) = [\boldsymbol{Q}_B, \mathscr{O}] = 0$ for $\mathrm{Hom}(\mathscr{V}_1, \mathscr{V}_2)$-valued holomorphic functions is equivalent to the condition for $\mathscr{O} : \mathscr{C}(\mathscr{B}_1) \to \mathscr{C}(\mathscr{B}_2)$ to be a cochain map (see (59)).

[35] A morphism between two $\mathbb{Z}^k$-graded $\mathscr{R}_W$-modules is a morphism that preserves the grading, i.e., the morphisms themselves have $\mathbb{Z}^k$-degree zero, in our context this means that the morphisms are gauge invariant.

complexes of $\mathscr{R}_W$-modules. One last thing to add is that, a cochain map will have differential degree, i.e. R-charge, zero, on the other hand the chiral ring contains elements of all possible degrees ($\mathbb{Z}$-graded). To allow this, given two branes $\mathscr{B}_1$ and $\mathscr{B}_2$ we must consider cochain maps between $\mathscr{C}(\mathscr{B}_1)$ and $\mathscr{C}(\mathscr{B}_2)[p]$ where $[p]$ denotes the same complex shifted p times to the left. Finally, our description of the chiral ring is:

The degree p chiral ring elements between two branes $\mathscr{B}_1$ and $\mathscr{B}_2$ are the morphisms between $\mathscr{C}(\mathscr{B}_1)$ and $\mathscr{C}(\mathscr{B}_2)[p]$ in the homotopy category of $\mathbb{Z}^k$-graded complexes of $\mathscr{R}_W$-modules:

$$\mathscr{H}^p_{Q_B}(\mathscr{B}_1, \mathscr{B}_1) \cong \mathrm{Hom}_{\mathbf{K}(\text{gr-}\mathscr{R}_W)}(\mathscr{C}(\mathscr{B}_1), \mathscr{C}(\mathscr{B}_2)[p]) \ . \tag{60}$$

From the discussion of this section and the last, it is clear that, as far as the chiral sector is concerned, the GLSM B-branes are well described by the homotopy category of some graded modules. The branes appear as objects of this category and the chiral ring between them corresponds to the morphisms.

4 Relations with the IR Branes

In this section we will define a projection from the GLSM B-branes to the IR B-branes and we will see that we get the *stable* IR branes if we identify the projections up to quasi-isomorphisms. This tells us that the low energy branes are objects in a derived category. We will only sketch out the key ideas here, and to avoid many technical complexities we will describe the relevant ideas for theories *without* a superpotential, so that at a generic point in the Kähler moduli space $\mathscr{M}_K$, the low energy theory is described by a non-linear sigma model on a non-compact toric variety X_r (17).

4.1 Tracking Branes to the Infrared

The gauge theory becomes strongly coupled as we go to the infrared, the deep infrared limit corresponds to taking $e \to \infty$. In this limit the kinetic terms for the vector multiplet fields vanish (see (3)) and therefore these fields acquire algebraic equations of motion. Let us illustrate this limit for a $U(1)$ gauge field with a Wilson line interaction on a boundary with charge q. Now, in the limit $e \to \infty$, the zeroth component v_0 of the gauge field appears in the action as:

$$\frac{1}{2\pi}\int_{\Sigma} d^2y \sum_{i=1}^{N} Q_i^2|x^i|^2 \left(v_0^2 - \frac{\sum_{j=1}^{N} iQ_j(\overline{x}^j\partial_0 x^j - \partial_0\overline{x}^j x^j)}{\sum_{i=1}^{N} Q_i^2|x^i|^2} v_0 + \cdots \right) - \int_{\partial\Sigma} \left(\frac{\theta}{2\pi} + q\right) v_0 \,, \tag{61}$$

where the $\cdots$ includes terms involving fermions and will vanish on a vacuum background. We can do the same computation for v_1 and after putting them together, the algebraic equation of motion for the one form v that we get by varying these expressions is that v becomes the pullback:

$$v = x^* A \,, \tag{62}$$

where A is the connection of the holomorphic line bundle $\mathscr{O}(1)$ on X_r. Then the Wilson line (29) with charge q corresponds to the holomorphic line bundle $\mathscr{O}(q)$, which is the low energy brane supported on X_r corresponding to the Wilson line brane in the GLSM. We thus get the following projection from GLSM B-branes to IR B-branes:

$$\pi : \mathscr{W}(q) \mapsto \mathscr{O}(q) \,. \tag{63}$$

Not all branes are stable in the infrared however. We have to find a zero of the potential $\{Q, Q^\dagger\}$ in the boundary action (34), just as we found the zero of the potential of the bulk action in Sect. 2.2. If $\{Q, Q^\dagger\}$ is positive everywhere on X_r then the entire brane is unstable and it will vanish by the mechanism of brane anti-brane annihilation [16]. Such a brane is called *trivial* in the infrared. For example, if we do not care about the $U(1)_V$ R-symmetry then we can have a constant superpotential $W = c$ and then the ring $\mathscr{R}_W$ (46) and consequently the modules $\mathscr{V}_W$ (47) will vanish, resulting in a null complex (brane) (48). More interesting examples of infrared trivial branes arise from the existence of the deleted sets Δ_r while constructing X_r as a quotient (18). If $\{Q, Q^\dagger\}$ is strictily positive everywhere in $\mathbb{C}^N \backslash \Delta_r$, then the corresponding brane will be infrared trivial. The fact that the deleted set varies from phase to phase now makes phase transition for branes rather interesting, since a UV brane that is trivial in one IR phase may be non-trivial in another. This makes the projection of a GLSM B-brane complex $\mathscr{C}(\mathscr{B})$ to an IR complex phase dependent, let us make it explicit in notation:

$$\pi_r : \mathscr{W}(q) \mapsto \mathscr{O}(q) \,, \qquad \pi_r : Q(x)\big|_{x\in\mathbb{C}^N} \mapsto Q(x)\big|_{x\in X_r} \,. \tag{64}$$

Similarly, for any cochain map $\phi : \mathscr{C}(\mathscr{B}_1) \to \mathscr{C}(\mathscr{B}_2)$ between two GLSM B-brane complexes, the projection $\pi_r(\phi) : \pi_r(\mathscr{C}(\mathscr{B}_1)) \to \pi_r(\mathscr{C}(\mathscr{B}_2))$ is defined by projecting $x \in \mathbb{C}^N$ to $x \in X_r$:

$$\pi_r : \phi(x)\big|_{x\in\mathbb{C}^N} \mapsto \phi(x)\big|_{x\in X_r} \,. \tag{65}$$

Given two UV B-branes we would like to know whether they will flow to the same IR branes or not. The non-trivial answer to this question is that[36]:

Two GLSM B-branes $\mathscr{B}_1$ and $\mathscr{B}_2$ flow to the same IR brane in a phase containing $r \in \mathscr{M}_K$, if there exists a *quasi-isomorphism* $\varphi : \pi_r(\mathscr{C}(\mathscr{B}_1)) \to \pi_r(\mathscr{C}(\mathscr{B}_2))$.

We will omit the proof of this here, which can be found in [3]. The key observations behind this proof are that, a brane can be deformed to another quasi-isomorphic brane by brane anti-brane annihilation and some deformations of the boundary actions that do not affect the low energy physics. We can observe that if two branes have quasi-isomorphic projections in some phase then we can turn one of them into an anti-brane and the combined brane anti-brane system is indeed trivial in the infrared (in that phase), signaling the fact that the two original branes are indistinguishable at low energy in that phase. Concretely, turning one of the branes, say $\mathscr{B}_1$, into anti-brane entails shifting the associated complex $\mathscr{C}(\mathscr{B}_1)$ by a step, i.e., we consider the complex $\mathscr{C}(\mathscr{B}_1)[1]$. Then the bound state of the two branes is the *mapping cone* of the respective complexes, which depends on the choice of a cochain map $\phi : \mathscr{C}(\mathscr{B}_1) \to \mathscr{C}(\mathscr{B}_2)$. We denote the cone as $C(\phi)$:

$$C(\phi) = \mathscr{C}(\mathscr{B}_1)[1] \oplus \mathscr{C}(\mathscr{B}_2) , \tag{66}$$

with the differential:

$$Q_{C(\phi)} := \begin{pmatrix} -Q_1 & 0 \\ \phi & Q_2 \end{pmatrix} . \tag{67}$$

It is not difficult to prove that the complex $\pi_r(C(\phi))$ is exact if and only if $\pi_r(\phi)$ is a quasi-isomorphism. An exact complex is quasi-isomorphic to the null complex by a trivial cochain map: $\pi_r(C(\phi)) \to 0$. Therefore we see that complete brane anti-brane annihilation takes place in the brane $\mathscr{C}(\mathscr{B}_1)[1] \oplus \mathscr{C}(\mathscr{B}_2)$ if and only if $\pi_r(\mathscr{B}_1)$ and $\pi_r(\mathscr{B}_2)$ are quasi-isomorphic.[37] We thus reach the conclusion that the low energy branes are better described as objects in a derived version of the homotopy category that describes the GLSM B-branes.[38]

Let us denote the homotopy category of GLSM B-branes (in a theory with N chiral multiplets and gauge group T) as $\mathfrak{D}(\mathbb{C}^N, T)$ and the derived category of IR

[36] A quasi-isomorphism is a cochain map that induces an isomorphism of cohomology. Recall that in physical terms, cochain maps are chiral ring elements.

[37] Perhaps we should say "if $\pi_r(\mathscr{C}(\mathscr{B}_1))$ and $\pi_r(\mathscr{C}(\mathscr{B}_2))$ are quasi-isomorphic," but we will not put much effort into distinguishing a brane from its complex.

[38] In passing from a homotopy category to the derived category one introduces formal inverses for the quasi-isomorphisms (and all compositions of morphisms involving these inverses) so that they can be treated as genuine isomorphisms. Also note that, this conclusion is consistent with the, by now standard, result that the low energy branes in a geometric phase are objects in the derived category of coherent sheaves supported on the CY target space of the IR sigma model [1, 17, 18].

B-branes on the toric variety X_r as $\mathrm{D}(X_r)$.[39] Then the projection (64) is a map:

$$\pi_r : \mathfrak{D}(\mathbb{C}^N, T) \to \mathrm{D}(X_r)\,. \tag{68}$$

Note that now we are thinking of this map as first taking the projection (64) and then *localizing* the quasi-isomorphisms. Localizing simply means adding formal inverses of the quasi-isomorphisms together with all the compositions involving these inverses. Then we can treat the quasi-isomorphisms as honest isomorphisms and the isomorphism classes of the objects are now in one to one correspondence with the stable IR B-branes.

Example 2 Let us look at a concrete example of quasi-isomorphism in the simple $U(1)$ gauge theory that we considered in Example 1. We consider the GLSM B-brane $\mathscr{B}$ in the phase $r > 0$[40]:

$$\mathscr{C}(\mathscr{B}): \quad \mathscr{W}^{-3}(-1) \xrightarrow{d_{-3}} \mathscr{W}^{-2}(0)^{\oplus 3} \xrightarrow{d_{-2}} \mathscr{W}^{-1}(1)^{\oplus 3} \xrightarrow{d_{-1}} \mathscr{W}^{0}(2)\,, \tag{69}$$

where the differentials are defined as:

$$d_{-3} := \begin{pmatrix} x^1 \\ x^2 \\ x^3 \end{pmatrix}, \qquad d_{-2} := \begin{pmatrix} 0 & x^3 & -x^2 \\ -x^3 & 0 & x^1 \\ x^2 & -x^1 & 0 \end{pmatrix}, \qquad d_{-1} := \begin{pmatrix} x^1 & x^2 & x^3 \end{pmatrix}. \tag{70}$$

The differential on the total complex $\mathscr{C}(\mathscr{B})$ is:

$$Q = \begin{pmatrix} 0 & & & \\ d_{-3} & & & \\ & d_{-2} & & \\ & & d_{-1} & 0 \end{pmatrix}, \quad \text{satisfying} \quad \{Q, Q^\dagger\} = \left(\sum_{i=1}^{3} |x^i|^2\right) \mathrm{id}_{\mathscr{C}(\mathscr{B})}\,. \tag{71}$$

Recalling that the deleted set in the geometric phase $r > 0$ was $\Delta_+ = \{x^1 = x^2 = x^3 = 0\}$, we see that the boundary potential $\{Q, Q^\dagger\}$ is strictly positive in this phase and therefore the brane $\mathscr{B}$ is trivial in the IR, i.e. $\pi_+(\mathscr{B}) \cong 0$. It can also be checked that the projection of the complex (69) in this phase is exact, reaching the same conclusion that it is quasi-isomorphic to the null complex. This implies that we can write this brane as a bound state of two branes $\mathscr{C}(\mathscr{B}_1)[1]$ and $\mathscr{C}(\mathscr{B}_2)$ so that $\pi_+(\mathscr{B}_1)$ and $\pi_+(\mathscr{B}_2)$ are quasi-isomorphic. This can be achieved by breaking the complex at an arrow and shifting one of the two resulting complexes, e.g.:

---

[39]If we have non-trivial superpotential we need to mention that as well in the notation for the GLSM B-brane category.

[40]The following complex represents the cochain complex (48), which in the absence of a superpotential is the same as (45). We are using the notation $\mathscr{W}$ introduced in the context of Wilson line branes (see "Wilson line branes" in Sect. 3.1) instead of using the Chan–Paton spaces to highlight the gauge charges, which plays a crucial role in the computations of the next section. The superscript on $\mathscr{W}$ denotes the R-charge.

$$
\begin{array}{llll}
\mathscr{C}(\mathscr{B}_1): & \mathscr{W}^{-2}(-1) \xrightarrow{d_{-3}} \mathscr{W}^{-1}(0)^{\oplus 3} \xrightarrow{d_{-2}} \mathscr{W}^{0}(1)^{\oplus 3} \\
 & \qquad\qquad\qquad\qquad\qquad\qquad\quad \downarrow d_{-1} \\
\mathscr{C}(\mathscr{B}_2): & \qquad\qquad\qquad\qquad\qquad\qquad\quad \mathscr{W}^{0}(2)
\end{array}
\tag{72}
$$

The cochain map $\pi_+(d_{-1}) : \pi_+(\mathscr{C}(\mathscr{B}_1)) \to \pi_+(\mathscr{C}(\mathscr{B}_2))$ being the quasi-isomorphism.

On the other hand, in the orbifold phase $r < 0$, the deleted set is $\Delta_- = \{x^4 = 0\}$ and the boundary potential is vanishing at the orbifold point $p := \{x^1 = x^2 = x^3 = 0\}$. Therefore, the brane $\mathscr{B}$ localizes at p at low energy and, in particular, is non-vanishing. This also implies that $\pi_-(\mathscr{B}_1)$ and $\pi_-(\mathscr{B}_2)$ are *not* quasi-isomorphic in this phase.

4.2 Grade Restriction Rule

The next step in the story of branes is to consider how to transport the IR B-branes from one phase to another. The GLSM approach to brane proves to be particularly suited to answer this question. The difficulties in transporting branes arise when we consider branes near a phase boundary. To precisely study the nature of the branes near such singular loci, we have to be careful about quantum corrections. We will proceed analogously as we did in computing quantum corrections for theories without boundary. We look at the effective action for the vector multiplet scalar after integrating out the chiral multiplet fields, this time with Wilson line boundary interaction. When the theory is formulated on a strip of width L, the result is [3]:

$$
U_{\text{eff}}(\sigma) = L\frac{e_{\text{eff}}^2}{2}|r_{\text{eff}} - i\theta_{\text{eff}}|^2 + 2\left[-\left(\frac{\theta}{2\pi} + q\right)\Re\sigma + \frac{s}{2}|\Re\sigma|\right], \tag{73}
$$

where $r_{\text{eff}} = r + \sum_i Q_i \log|Q_i|$ as before, $s = \sum_{Q_i \geq 0} Q_i$ as before, and $\theta_{\text{eff}} = \theta - \text{sgn}(\Re\sigma)s\pi$. The linear dependence on σ means that whenever there is a Coulomb branch, i.e. σ is unconstrained, the potential can become unbounded from below. Such potentials signal instability in a physical system since it appears that there is no stable vacuum. The potential becomes bounded from below and therefore the problem of instability is cured if the gauge charge of the Wilson line brane is constrained:

$$
-\frac{s}{2} < \frac{\theta}{2\pi} + q < \frac{s}{2}. \tag{74}
$$

Wilson line branes with charges satisfying this constraint can be safely transported through phase boundaries without becoming unstable. This criterion for stable brane transport in the Kähler moduli space was one of the key results of [3], where the constraint (74) was termed the *grade restriction rule*.

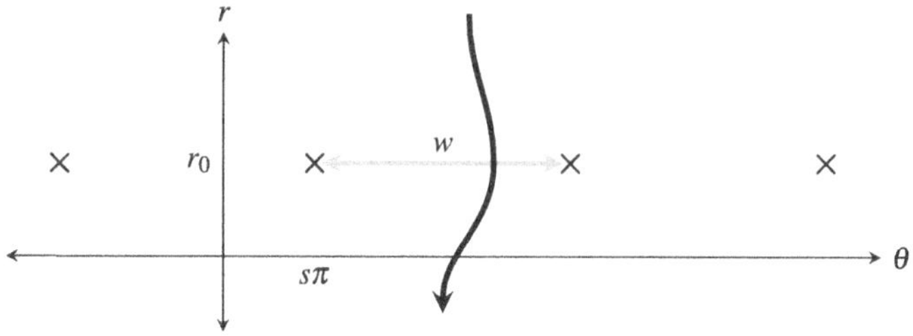

Fig. 3 Brane transport through a window in a Kähler moduli space. $r_0 = -\sum_i Q_i \log Q_i$, as in (23)

Since the theta angle θ is a periodic variable, we must specify a *window* in the complexified Kähler moduli space, through which we wish to transport a brane. Figure 3 shows such a window w and a schematic path to transport branes along. Given a window w, we define the set of gauge charges that satisfy the grade restriction rule (74):

$$N^w := \left\{ n \in \mathbb{Z} \mid \forall \theta \in w : -\frac{s}{2} < \frac{\theta}{2\pi} + q < \frac{s}{2} \right\}. \tag{75}$$

A GLSM B-brane with charge q will be called *grade restricted* with respect to a window w if $q \in N^w$. The set of all grade restricted brane with respect to a window w will be called $\mathscr{T}^w$.

The main goal is to find a one to one correspondence between IR branes in different phases. We find this correspondence by transporting branes across phase boundaries. The general prescription for doing so is the following: Suppose we are given a window w between two adjacent phases, one containing $r \in \mathscr{M}_K$ and the other $r' \in \mathscr{M}_K$. Then given an IR brane $\mathscr{B}_{\mathrm{IR}} \in \mathrm{D}(X_r)$ in the phase containing r, we first lift it to a GLSM brane $\mathscr{B} \in \mathfrak{D}(\mathbb{C}^N, T)$ satisfying $\pi_r(\mathscr{B}) = \mathscr{B}_{\mathrm{IR}}$. If $\mathscr{B} \notin \mathscr{T}^w$, then we find another brane $\mathscr{B}'$ such that $\mathscr{B}' \in \mathscr{T}^w$ and $\pi_r(\mathscr{B}')$ is quasi-isomorphic to $\pi_r(\mathscr{B})$. Now we can transport this brane $\mathscr{B}'$ across the window and then project down in the adjacent phase with the outcome $\pi_{r'}(\mathscr{B}')$.

Given any two adjacent phases containing $r, r' \in \mathscr{M}_K$ and a window w between them, there is a unique lift from $\mathrm{D}(X_r)$ and $\mathrm{D}(X_{r'})$ to $\mathscr{T}^w$ [3] which we will denote as $\omega^w_{r,r'}$ and $\omega^w_{r',r}$ respectively. We can draw all the relevant maps in the following diagram:

$$\begin{array}{ccccc} & & \mathfrak{D}(\mathbb{C}^N, T) & & \\ & \swarrow^{\pi_r} & \uparrow & \searrow^{\pi_{r'}} & \\ \mathrm{D}(X_r) & \xrightarrow[\omega^w_{r,r'}]{} & \mathscr{T}^w & \xleftarrow[\omega^w_{r',r}]{} & \mathrm{D}(X_{r'}) \end{array} \tag{76}$$

Brane transport from $\mathrm{D}(X_r)$ to $\mathrm{D}(X_{r'})$ is the composition:

$$\mathrm{D}(X_r) \xrightarrow{\omega^w_{r,r'}} \mathcal{T}^w \xrightarrow{\pi_{r'}} \mathrm{D}(X_{r'})\,. \tag{77}$$

Transport in the other direction is defined similarly.

Example 3 We illustrate this mechanism of brane transport in the context of our example $U(1)$ gauge theory (see Examples 1, 2). The singularities in the Kähler moduli space are at $r = 3\log 3$ and $\theta \in 3\pi + 2\pi\mathbb{Z}$. In the geometric phase $r > 0$, where branes are represented by holomorphic bundles (coherent sheaves to be more precise) we pick the brane[41] $\underline{\mathcal{O}(2)}$ and we wish to transport it from the phase $r > 0$ to the orbifold phase $r < 0$ through the window $w : -\pi < \theta < \pi$. The obvious lift of this brane to a GLSM brane is $\mathcal{W}(2)$. But note that in the window w the grade restricted gauge charges are $N^w = \{-1, 0, 1\}$ so the GLSM brane $\mathcal{W}(2)$ is not grade restricted. On the other hand in Example 2 we saw that the brane $\mathcal{C}(\mathcal{B}_1)$: $\mathcal{W}^{-2}(-1) \xrightarrow{d_{-3}} \mathcal{W}^{-1}(0)^{\oplus 3} \xrightarrow{d_{-2}} \mathcal{W}^0(1)^{\oplus 3}$ is quasi-isomorphic to $\mathcal{W}^0(2)$ at low energy in the phase $r > 0$, and $\mathcal{B}_1$ is grade restricted. Therefore, we can transport $\mathcal{B}_1$ over to the phase $r < 0$ and project down to $\pi_-(\mathcal{B}_1)$:

$$\pi_-(\mathcal{B}_1) : \mathcal{O}(-1) \xrightarrow{d_{-3}} \mathcal{O}(0)^{\oplus 3} \xrightarrow{d_{-2}} \underline{\mathcal{O}(1)}^{\oplus 3}\,. \tag{78}$$

The above brane is the image in $\mathrm{D}(X_-)$ of $\underline{\mathcal{O}(2)} \in \mathrm{D}(X_+)$ under brane transport.

More generally, consider a GLSM with abelian gauge group T that reduces, in a phase, to an orbifold theory $X_{\mathrm{orb}} = \mathbb{C}^N/\Gamma$ with a discrete gauge group $\Gamma \subset T$. The GLSM Chan–Paton space reduces to a representation of the discrete gauge group Γ and the low energy branes in the orbifold phase are complexes of Γ-equivariant vector bundles on $\mathbb{C}^N$, $\mathrm{D}(X_{\mathrm{orb}}) \cong \mathrm{D}_\Gamma(\mathbb{C}^N)$. Other phases of this GLSM are given by partial or complete *crepant resolutions* of the orbifold singularity. If we denote such a resolution as X_{res}, then the brane transport establishes a correspondence of D-branes between these phases:

$$\mathrm{D}_\Gamma(\mathbb{C}^N) \quad \xleftrightarrow{\text{brane transport}} \quad \mathrm{D}(X_{\mathrm{res}}) \tag{79}$$

Considering the mapping between the chiral sectors, this becomes an equivalence of derived categories which is known as *McKay correspondence* [4, 5].

In more general cases, with non-zero superpotential, complexes of sheaves are replaced by matrix factorizations of the superpotential and low energy computation becomes more involved as some of the fields can acquire masses from the superpotential and need to be integrated out. The main ideas remain unchanged. The low energy branes are still the GLSM branes upto phase dependent quasi-isomorphisms. The grade restriction rules are the same and brane transport works similarly. In the

41 We use an underline to point out the R-charge (differential degree) zero part of a complex.

math literature, the relevant equivalence of derived categories, that of CY hypersurface defined by a polynomial and of matrix factorization of the same polynomial, was established by Orlov [6]. Details including relation between the physical perspective of brane transport and Orlov's construction of categorical equivalence can be found in [3].

A $\mathcal{N} = (2, 2)$ Supersymmetry

Our world-sheet has, as time and space coordinates, y^0 and y^1 respectively, with a Lorentzian signature $\mathrm{d}s^2 = -(\mathrm{d}y^0)^2 + (\mathrm{d}y^1)^2$. It is convenient to introduce *light cone* coordinates and their derivatives:

$$z^{\pm} := y^0 \pm y^1 , \qquad \partial_{\pm} := \frac{\partial}{\partial z^{\pm}} = \frac{1}{2}(\partial_0 \pm \partial_1) . \tag{80}$$

The $\mathcal{N} = (2, 2)$ supersymmetry algebra contains four supercharges:

$$\boldsymbol{Q}_+ , \ \boldsymbol{Q}_- , \ \overline{\boldsymbol{Q}}_+ , \ \overline{\boldsymbol{Q}}_- , \tag{81}$$

and the following bosonic generators:

$$\text{Time translation: } H := \partial_0 , \tag{82a}$$

$$\text{Spatial translation: } P := \partial_1 , \tag{82b}$$

$$\text{Rotation: } M := x^0\partial_1 + x^1\partial_0 . \tag{82c}$$

Depending on the field theory, this algebra may be augmented by two $U(1)$ R-symmetry generators:

$$F_V , \ F_A , \tag{83}$$

corresponding to the vector R-symmetry ($U(1)_V$) and the axial R-symmetry ($U(1)_A$) respectively. The non-zero commutation relations are:

$$\{\boldsymbol{Q}_{\pm}, \overline{\boldsymbol{Q}}_{\pm}\} = H \pm P , \tag{84a}$$

$$[iM, \boldsymbol{Q}_{\pm}] = \mp \boldsymbol{Q}_{\pm} , \quad [iM, \overline{\boldsymbol{Q}}_{\pm}] = \mp \overline{\boldsymbol{Q}}_{\pm} , \tag{84b}$$

$$[F_V, \boldsymbol{Q}_{\pm}] = -\boldsymbol{Q}_{\pm} , \quad [F_V, \overline{\boldsymbol{Q}}_{\pm}] = \overline{\boldsymbol{Q}}_{\pm} , \tag{84c}$$

$$[F_A, \boldsymbol{Q}_{\pm}] = \mp \boldsymbol{Q}_{\pm} , \quad [F_A, \overline{\boldsymbol{Q}}_{\pm}] = \pm \overline{\boldsymbol{Q}}_{\pm} . \tag{84d}$$

The hermiticity property of the supercharges is:

$$\boldsymbol{Q}_{\pm}^{\dagger} = \overline{\boldsymbol{Q}}_{\pm} . \tag{85}$$

The bosonic operators are all hermitian.

Two distinguished subalgebras of the above algebra containing half of the supersymmetry are defined with the following supercharges:

$$\mathcal{N} = 2_A : \qquad Q_A, Q_A^\dagger, \qquad Q_A := \overline{Q}_+ + Q_-, \tag{86a}$$

$$\mathcal{N} = 2_B : \qquad Q_B, Q_B^\dagger, \qquad Q_B := \overline{Q}_+ + \overline{Q}_-. \tag{86b}$$

They satisfy the same anti-commutation relations, for $Q \in \{Q_A, Q_B\}$:

$$\{Q, Q^\dagger\} = 2H, \qquad Q^2 = Q^{\dagger 2} = 0. \tag{87}$$

Their charges under the R-symmetries play important role in our analysis:

$$[F_V, Q_A] = 0, \qquad [F_A, Q_A] = Q_A, \tag{88a}$$

$$[F_V, Q_B] = Q_B, \qquad [F_A, Q_B] = 0, \tag{88b}$$

which implies that a Q-complex (computing some cohomology of interest) will be $\mathbb{Z}$-graded by the $U(1)_V$ for $Q = Q_B$ and $U(1)_A$ for $Q = Q_A$, given that these R-symmetries are preserved by the quantum theory.

B Chan–Paton Space

Let us consider a general setup of having a QFT on a world-sheet (a surface) Σ with boundary. We may have a set of possible boundary conditions[42] that we can assign to the fields of the theory, let us call this set of *basic boundary conditions* B_0. Each connected component of the boundary $\partial\Sigma$ corresponds to a brane (to which the component is thought to be attached) or equivalently, each boundary condition corresponds to a brane. If we have a configuration of multiple branes with a boundary condition for each of them then we can label each brane with an index, let us call the index set $\mathcal{I}$, and collect all these boundary conditions in the map $\mathcal{B} : \mathcal{I} \to B_0$. The indices in the set $\mathcal{I}$ are called Chan–Paton indices. This way each end point of the strings will carry an index of its own. More generally it may be possible, and at times necessary, to consider linear combinations of indices attached to an end point, or to allow some group to act on them.

For example, if we have N coincident identical branes then there may be a symmetry rotating those branes (among each other). Once we introduce an index for each brane, $\mathcal{I} = \{1, \ldots, N\}$, we can define the map $\mathcal{B} : \mathcal{I} \to B_0$ to send each index to

[42] A boundary condition can involve literally boundary conditions for the fields at the boundary along with boundary actions supported on the boundary that help preserve some of the symmetries of the bulk theory.

the same boundary condition. Now some rotation subgroup[43] of $GL(\mathbb{C}^N)$ acting on the indices in $\mathscr{I}$ can become a symmetry of the theory. Another way to interpret this is to imagine, instead of a stack of N branes, just a single brane with a vector bundle like structure where the fiber over each point is $\mathbb{C}^N$. Then an open string that ends on this brane has not only a specific position on the brane attached to its end point but also a specific position along the fiber and some subgroup of $GL(\mathbb{C}^N)$ can act on this position along the fiber. In this picture it is said that the end points of the open strings always have a vector space attached to them, the vector spaces being the fibers over the points on the branes where the end points are attached. These vector spaces are called Chan–Paton spaces, we will use the letter $\mathscr{V}$ to refer to Chan–Paton spaces.

Let us schematically discuss some fairly general properties of open string Hilbert spaces in the presence of non-trivial Chan–Paton spaces. We will parametrize time and space with y^0 and y^1 respectively, then in the presence of two boundaries at $y^1 = 0$ and $y^1 = L$, computing correlation functions on the semi-infinite strip $[0, L] \times [-\infty, t]$ defines wave functions in the Hilbert space associated to the open string at time t (see Fig. 4). To do the path integral we have to use two boundary conditions for the two boundaries and specify a value for the fields at time t. These boundary conditions can include different boundary actions to preserve some symmetry. Corresponding to two boundary conditions $\mathscr{B}(i)$ and $\mathscr{B}(j)$ for $i, j \in \mathscr{I}$, let us choose two boundary actions $I_i(y^1 = 0) := \int_{-\infty}^{t} \mathrm{d}y^0\, \mathscr{L}_i$ and $I_j(y^1 = L) := \int_{-\infty}^{t} \mathrm{d}y^0\, \mathscr{L}_j$ supported on $y^1 = 0$ and $y^1 = L$ respectively. Now a state in the Hilbert space at time t is defined by the path integral with some operator insertion:

$$\Psi_i^{\,j}(\phi_0) := \int_{\phi(t)=\phi_0} \mathscr{D}\phi\; \mathscr{O}_\Psi(\phi) e^{i(S+I_i(y^1=0)+I_j(y^1=L))}\,, \tag{89}$$

where the path integral is over all field configurations that take the specific value ϕ_0 at time t and satisfy some boundary conditions (such as Dirichlet, Neuman etc.) at the boundaries. We have put the indices at different heights to keep track of the

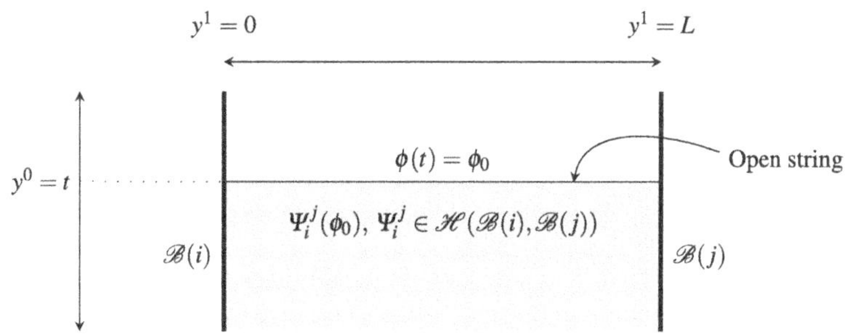

Fig. 4 Wave function on an open string with two boundaries

[43]We are using the term "rotation" loosely, the actual symmetry depends on the details of the theory, it can be $U(N)$, $SO(N)$, $Sp(N)$, etc.

orientation of the string. The Hilbert space consisting of all such wave functions for fixed i and j will be called $\mathscr{H}(\mathscr{B}(i), \mathscr{B}(j))$. The total open string Hilbert space is a direct sum over all possible boundary conditions:

$$\mathscr{H} = \bigoplus_{i,j\in\mathscr{I}} \mathscr{H}(\mathscr{B}(i), \mathscr{B}(j)). \tag{90}$$

Since the indices in $\mathscr{I}$ can be thought to represent the basis vectors of the Chan–Paton space $\mathscr{V}$, in an index free notation the wave function Ψ_i^j becomes valued in $\mathscr{V}^* \otimes \mathscr{V} = \mathrm{Hom}(\mathscr{V}, \mathscr{V})$.[44] More generally, we can consider two different configurations of branes for the two end points of an open string. Equivalently, we can have two different index sets and sets of basic boundary conditions, i.e., two different Chan–Paton spaces, $\mathscr{V}_1$ and $\mathscr{V}_2$, for the two ends. In such cases the open string states will be valued in $\mathrm{Hom}(\mathscr{V}_1, \mathscr{V}_2)$.

References

1. Maxim Kontsevich. Homological Algebra of Mirror Symmetry. 1994.
2. Andrew Strominger, Shing-Tung Yau, and Eric Zaslow. Mirror symmetry is T duality. Nucl. Phys., B479:243–259, 1996.
3. Manfred Herbst, Kentaro Hori, and David Page. Phases Of N = 2 Theories In 1+1 Dimensions With Boundary. 2008.
4. T. Bridgeland, A. King, and M. Reid. Mukai implies McKay: the McKay correspondence as an equivalence of derived categories. *ArXiv Mathematics e-prints*, August 1999.
5. Y. Kawamata. Log Crepant Birational Maps and Derived Categories. *ArXiv Mathematics e-prints*, November 2003.
6. D. Orlov. Derived Categories of Coherent Sheaves and Triangulated Categories of Singularities. *ArXiv Mathematics e-prints*, March 2005.
7. Edward Witten. Phases of N=2 theories in two-dimensions. *Nucl. Phys.*, B403:159–222, 1993. [AMS/IP Stud. Adv. Math.1,143(1996)].
8. Paul S. Aspinwall, Tom Bridgeland, Alastair Craw, Michael R. Douglas, Anton Kapustin, Gregory W. Moore, Mark Gross, Graeme Segal, Balázs Szendröi, and P. M. H. Wilson. Dirichlet branes and mirror symmetry, volume 4 of Clay Mathematics Monographs. AMS, Providence, RI, 2009.
9. A. Schwimmer and N. Seiberg. Comments on the N = 2, N = 3, N = 4 Superconformal Algebras in Two-Dimensions. Phys. Lett., B184:191–196, 1987.
10. Kentaro Hori, Sheldon Katz, Albrecht Klemm, Rahul Pandharipande, Richard Thomas, Cumrun Vafa, Ravi Vakil, and Eric Zaslow. *Mirror symmetry*, volume 1 of *Clay mathematics monographs*. AMS, Providence, RI, 2003. With a preface by Vafa.
11. Lance J. Dixon, Jeffrey A. Harvey, C. Vafa, and Edward Witten. Strings on Orbifolds. Nucl. Phys., B261:678–686, 1985.
12. Simeon Hellerman, Shamit Kachru, Albion E. Lawrence, and John McGreevy. Linear sigma models for open strings. JHEP, 07:002, 2002.

[44] The dual distinguishes the lower index from the upper in terms of representations of groups acting on $\mathscr{V}$. This is important for oriented strings, for unoriented strings the relevant representations must be self-dual.

13. Kentaro Hori. Linear models of supersymmetric D-branes. In *Symplectic geometry and mirror symmetry. Proceedings, 4th KIAS Annual International Conference, Seoul, South Korea, August 14-18, 2000*, pages 111–186, 2000.
14. Wolfgang Lerche, Cumrun Vafa, and Nicholas P. Warner. Chiral Rings in N = 2 Superconformal Theories. Nucl. Phys., B324:427–474, 1989.
15. Edward Witten. Mirror manifolds and topological field theory. 1991. [AMS/IP Stud. Adv. Math.9,121(1998)].
16. Ashoke Sen. Tachyon condensation on the brane anti-brane system. JHEP, 08:012, 1998.
17. Michael R. Douglas. D-branes, categories and N=1 supersymmetry. J. Math. Phys., 42:2818–2843, 2001.
18. Eric R. Sharpe. D-branes, derived categories, and Grothendieck groups. Nucl. Phys., B561:433–450, 1999.

The manufacturer's authorised representative in the EU is Springer Nature Customer Service Centre GmbH, Europaplatz 3, 69115 Heidelberg, Germany. If you have any concerns regarding our products, please contact ProductSafety@springernature.com

Printed and bound by CPI Group (UK) Ltd, Croydon, CR0 4YY
30/09/2026
02225659-0006